Ecological studies in tropical fish communities

D0930926

Ecological studies in

tropical fish

communities

R. H. LOWE-McCONNELL

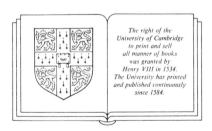

The right of the
University of Cambridge
to print and sell
all manner of books
was granted by
Henry VIII in 1534.
The University has printed
and published continuously
since 1584.

CAMBRIDGE UNIVERSITY PRESS
Cambridge
London   New York   New Rochelle
Melbourne   Sydney

Published by the Press Syndicate of the University of Cambridge
The Pitt Building, Trumpington Street, Cambridge CB2 1RP
32 East 57th Street, New York, NY 10022, USA
10 Stamford Road, Oakleigh, Melbourne 3166, Australia
© Cambridge University Press 1987

First published 1987

Printed in Great Britain at the University Press, Cambridge

British Library cataloguing in publication data

Lowe-McConnell, R. H.
Ecological studies in tropical fish
communities. – (Cambridge tropical
biology series)
1. Fishes – Tropics – Ecology
I. Title
597′.05′0913   QL637.5

Library of Congress cataloguing in publication data

Lowe-McConnell, R. H.
Ecological studies in tropical fish communities.

(Cambridge tropical biology series)
Bibliography
Includes index.
1. Fishes – Tropics – Ecology.   2. Animal communities – Tropics.
3. Fishes – Ecology.   I. Title.   II. Title: Fish communities.
III. Series.
QL637.5.L66   1987   597′.05′2623   86-21603

ISBN 0 521 23601 0 hard covers
ISBN 0 521 28064 8 paperback

Dedication

To the many without whom this book would
never have been written.

CONTENTS

PREFACE

This book, like its predecessor *Fish Communities in Tropical Freshwaters* (Lowe-McConnell, 1975), is based on field studies over a long series of years in many parts of the tropics, born of the delight of watching fishes in their natural environments. In such a book it is not possible to describe all the warmth and the colour, the sounds and the smells, the bird calls along the rivers, the glancing light – the many facets that make up the environmental whole. Yet often it is just the glancing light that gives a fish away, a ripple on the water surface, a shadow on the stream floor, leading predator to prey, woven into the complex pattern.

Why study tropical fish communities? Apart from the sheer pleasure of watching the behaviour of colourful fishes in clear warm waters of coral seas and freshwaters which provide natural aquaria, fishes are very important sources of protein in the diets of indigenous peoples throughout the tropics. Human populations are increasing at such an alarming rate that there is great pressure to develop fishery resources. It is vitally important that fish stocks should not be damaged by overexploitation and to avoid this we need to understand the biology of the fishes and factors governing fish production. Furthermore, many tropical fish communities, such as those of the African Great Lakes, are of the greatest scientific interest as they provide outstanding opportunities for studying evolutionary processes and mechanisms whereby new species arise, and are a mine of information for those interested in theoretical aspects of community ecology. Reef fishes also provide excellent material for assessing costs and benefits of different types of social and reproductive behaviour.

The tropical fish fauna is a very rich one and many of the names will be unfamiliar to temperate zone readers. The families of fishes important in tropical communities are listed in the Appendix in systematic order. Some changes of names are inevitable as more is learnt about the relationships of these fishes; generic names in quotes are used here for some groups where systematic revision is in progress.

This book compares information on the ecology of fishes from tropical freshwaters and from tropical seas to examine general principles and the main gaps in our knowledge. Each of the distinct fields of study: in rivers, natural and man-made lakes, in the sea on coral reefs, continental shelves, in upwelling areas and the open ocean, has its own vast literature. Inevitably a survey limited for space can only cite the most fecund and relevant papers and those with full bibliographies. Examples have been selected to complement one another, and are mainly from areas where the author has had field experience. Apologies are given here to the many scientists who will feel that their particular contribution has not been cited as it should have been. The bibliography is designed to help ecologists find a way through the jungle of references in the widely scattered literature; it collates information from a multitude of sources, many little known and difficult to consult, and is designed primarily for the specialist reader.

Much of the material has been considered in two earlier books: *Fish Communities in Tropical Freshwaters* (Lowe-McConnell, 1975 (now out of print)) and the very brief *Ecology of Fishes in Tropical Waters* (Lowe-McConnell, 1977) which included some marine fish studies. Since then an immense amount of new information has become available. This present book updates these earlier ones, reinterpreting information in the light of more recent work. Complementary books dealing with certain aspects in more detail include: *Fisheries Ecology of Floodplain Rivers* (Welcomme, 1979, 1985); *The Inland Waters of Tropical Africa* (Beadle, 1981); *The Niger and its Neighbours* (Grove, 1985); three books on Amazon fishes – *The Fishes and the Forest* (Goulding, 1980), *Man and Fisheries on an Amazon Frontier* (Goulding 1981) and *Man, Fishes, and the Amazon* (Smith, 1981); and monographs on *Lake Chad* (Carmouze, Durand & Levêque, 1983) and on *The Amazon* (Sioli, 1984). Symposium papers published as volumes include *Evolutionary Ecology of Neotropical Freshwater Fishes* (Zaret, 1984c) and *Evolution of Fish Species Flocks* (Echelle & Kornfield, 1984). Other relevant volumes include *The Ecology of River Systems* (ed Davies & Walker), *The Ecology of Tropical Lakes and Rivers* (A. I. Payne) and *Biology and Ecology of African Freshwater Fishes* (ed Levêque, Bruton & Ssentongo).

ACKNOWLEDGEMENTS

'No man is an island' and this book, the product of field experience mainly in Africa and South America over more than 30 years, owes its being to so many helpful people that it would be impossible to name them all. My especial thanks go to Dr E. B. Worthington who first made it possible for me to work in Africa; to Dr E. Trewavas who taught me what fishes I should find there and who has throughout the years been a constant source of helpful information; to Dr P. H. Greenwood and other colleagues on the staffs of the East African Fisheries Research Organization and Fish Section of the British Museum (Natural History) for stimulating discussions. Also to the many who made the five years in Guyana (South America) so fruitful. A very great deal has been learnt from the fishermen of the lakes and rivers, from those living on the shores of Lake Malawi, where I first worked with tropical fishes, to the Amerindians fishing the forest streams of South America.

I am also very grateful to the Trustees of the British Museum (Natural History) for working space and the use of their excellent library facilities, to Ethelwynn Trewavas, Tony Ribbink, Gordon Howes, Robin Bruce and Michael Goulding for helpful comments on drafts of various chapters, and to Gordon Howes for drawing many of the fishes. My old friend Dr M. E. Varley (Peggy Brown) and my husband Richard have encouraged the enterprise throughout a very long gestation period.

R. H. Lowe-McConnell
January 1986

PART I

Introduction

1

Introduction

Tropical communities of both plants and animals are characteristically diverse, with large numbers of species and very complex inter-relationships compared with those in temperate zone communities. Fish faunas obey this general ecological rule, within both families and environments.

Fishes are the most ancient and numerous of vertebrates. Over 20 000 species are known, the majority of which live in the warm waters of the world. About 8000 species (40%) live on the continental shelves of warm seas in water less than 200 m deep, compared with only about 1130 species (5%) in similar habitats in cold seas. Freshwaters carry a surprisingly high total, about 8500 species (over 40%), the majority in the vast river systems and lakes of the tropics (Cohen, 1970). The Amazon system has over 1300 species, the Zaire (Congo) in Africa nearly 700, compared with but 250 species in the Mississippi system in North America and 192 in the whole of Europe. Each of the Great Lakes Victoria, Tanganyika and Malawi in eastern Africa has over 200 species, most of them endemic, i.e. found only in the particular lake. Of the 445 families of fish, the seven largest (which include *ca* 30% of the total) – Cyprinidae, Gobiidae, Characidae, Cichlidae, Labridae, Loricariidae and Serranidae – are all well represented in tropical waters (Nelson, 1984).

Data on the ecology of tropical fishes used to come mainly from studies of commercially important food fishes (the larger species) but now come increasingly from underwater observations of fish behaviour made while SCUBA-diving (i.e. using a self-contained underwater breathing apparatus) and by filming and television in the clear warm waters over coral reefs and in the shallows of some tropical lakes. Conditions here are often also good for experimental work, for example building artificial reefs to study colonization by fish. The creation of man-made lakes behind hydroelectric barrages has provided large-scale experiments for studying the changes from riverine to lacustrine fish communities. Fish growth has been investigated both in natural waters and in reservoirs and ponds. Aquarium studies have assisted analyses of fish behaviour. The whole adds up to a picture of how ecological relationships affect fish behaviour and how their behaviour affects their ecology, both aspects playing their part in the evolution of species and of the communities in which they live.

Fisheries research aims at determining the optimum sustainable economic yield of desirable fish species. This involves finding out what kinds of fish are present and understanding their population dynamics, how fast the fish grow and reproduce, the size and age at which they spawn, their mortality rates and causes, their movements and so on. Methods for determining such variables are described in the International Biological Programme (IBP) Handbook no. 3 (Bagenal, 1978) and by Gulland (1978, 1983), Pauly & Murphy (1982) and Pauly (1983). Research in tropical waters presents special problems. It is difficult for instance to determine fish ages and growth rates in equatorial waters where seasons are ill-defined. Also, the numerous species living together, many of them difficult to distinguish, complicate research; as they may grow at different rates and breed in different places, it is important to be able to identify them correctly. Identifications of marine fishes in tropical seas are now greatly assisted by a series of handbooks prepared through the Food & Agriculture Organization of the United Nations (FAO, 1974–84; Collette & Naven, 1983).

The plan here is to look at ecological studies, first in representative freshwaters and secondly in the sea, starting with direct observations of the fishes on coral reefs, then fisheries studies on continental shelves, in upwelling areas and the open oceans. In Part IV the data are assessed to see what they tell us of the life history strategies of the fishes, of how the nutrient supply is cropped, of how communities are maintained and evolve, and of the production and exploitation of tropical fish faunas.

Tropical marine environments

Sea temperature is the major factor in the division of marine faunas. The tropical fauna lies between the 20°C isotherms for surface water temperatures at the coolest time of year, flanked on either side by subtropical faunas living where the water temperature does not fall below 16–18°C. The 20°C isotherms lie between latitudes 20 and 30°N and between 15 and 30°S (Fig. 1.1), with some seasonal fluctuations. Marine fish faunas are also greatly affected by depth of sea, current systems, salinity, oxygenation, availability of food and many other factors.

Tropical oceans are divided by land masses into the extensive Indo-Pacific and the smaller Atlantic. The richest fish faunas of the world are in the Indonesian area of the Indo-Pacific; waters around New Guinea include over 1000 species belonging to *ca* 241 fish families. Parts of the Pacific date back to the Upper Jurassic (see Springer, 1982). The Atlantic, a very much younger ocean (dating back to the Cretaceous and perhaps only *ca* 75 million years old), has a much less varied fauna; this is richest in the western Atlantic, where the Caribbean region was in contact with the eastern Pacific before the formation of the Pleistocene landbridge connecting North and South America. The Bahamas and adjacent waters have over 500 species representing 89 fish families. The continental shelves have far richer faunas than those of the pelagic zone or deepwater habitats. These tropical shore faunas are now separated by four highly efficient zoogeographical barriers: (1) the New World land barrier; (2) the African land barrier; (3) the vast relatively landless stretch of deepwater across the eastern Pacific; and (4) the geologically more recent mid-Atlantic (Briggs, 1967). Despite these barriers, over 100 species of circumtropical fishes are known, of which 84% are pelagic; only about 14 (5 elasmobranchs and 9 teleosts) are shore species (Briggs, 1960). The possible influence of the Pacific lithospheric plate on the biogeography of 179 shorefish families has been examined by Springer (1982). Nelson (1984) cites areas of endemism in (1) the Indo-West Pacific, (2) the tropical West Atlantic, and (3) the tropical eastern Atlantic. The fauna of the tropical East Pacific is more like that of the western Atlantic than that of the Indo-West Pacific because of the mid-Pacific barrier and the relatively recent formation of the Isthmus of Panama.

Fish resources from tropical oceans come from a number of separate ecosystems: upwelling regions, offshore oceanic areas, continental shelves with hard or soft bottom deposits, coral reefs, mangrove swamps. The latter two ecosystems are found only within the tropics. Coral reefs only grow where the water is warm, clear and highly saline, whereas mangrove swamps line tropical estuaries and coasts where the seawater is diluted with freshwater and is turbid with mud.

Many fishes move from one ecosystem to another as they grow. Many reef fishes have oceanic larval stages. The larvae and juveniles of some oceanic fishes live in inshore waters, and some large oceanic fishes, such as tunas and marlins, cruise into continental waters. Recognized subdivisions of the ocean comprise the **Neritic Province**, which extends from the water's edge to the edge of the continental shelf, and beyond this the **Oceanic Province**. Neritic waters are well lit but turbulent, subject to

Fig. 1.1. Tropical seas, bounded by the 20°C isotherm (surface water temperature, shown as a heavier unbroken line) and their contained land masses, indicating main current systems, zoogeographical regions and some sites mentioned in text. Main reefs studied (R); trawl survey sites: T1 off Guyana, T2 off West Africa, T3 off Kenya, T4 Sind, T5 Indian West coast, T6 off Cape Comorin, T7 off Bangladesh, T8 Thailand, T9 West Malaysia, Sunda Shelf, T10 South China Sea. Rivers: *M* = Mekong (Asia); *K* = Kapuas (Borneo); *F* = Fly; and *S* = Sepik (Papua New Guinea).

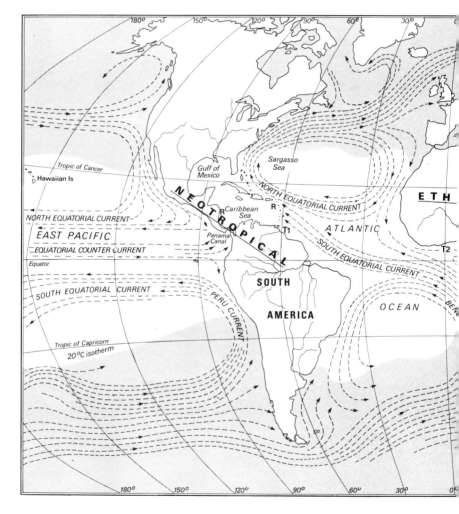

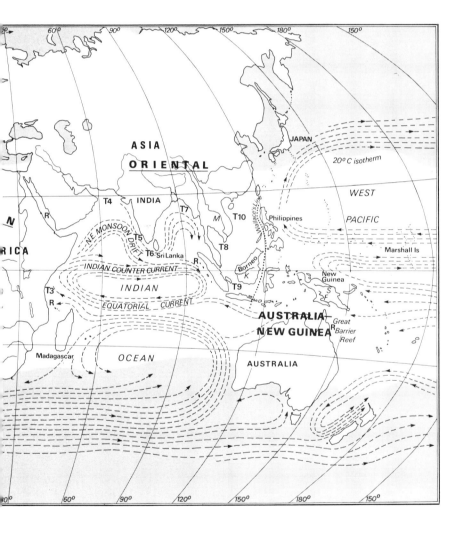

greater water movements (wave action, tides and currents), and with greater variation in salinity and water temperatures than the deeper waters, and they include many types of habitat. The **benthic division** (sea floor) includes several major zones: (1) the **littoral** or seashore intertidal zone from the wave- or surge-washed region above high-tide level (**supralittoral**) to (2) the **sublittoral** zone, extending from low water spring-tide level to the edge of the continental shelf in about 200 m of water; and (3) the **continental slope** and deeper zones beyond the edge of the shelf.

In the Oceanic Province, which extends from one continental shelf to another, conditions are relatively uniform: seasonal fluctuations only affect some areas and there are few types of habitat, but conditions do change with depth. The **epipelagic zone** (which extends into the Neritic Province) is 200 m thick and includes the warm well-lit euphotic zone where photosynthesis can occur. The epipelagic zone has sharp gradations of illumination and temperature with depth; it also has diurnal and seasonal changes in light intensity and temperature, and water movements here may be relatively rapid. Below this, the **mesopelagic zone** (200–1000 m) has very little light and a gradual temperature gradient with little seasonal variation; oxygen may be depleted, but this zone often has phosphate and nitrate present. The **bathypelagic zone** below 1000 m is characterized by perpetual darkness, low temperature (even in the tropics) and great pressure. In this book we are concerned only with warmwater fishes, those of the epipelagic zone.

The depth of the euphotic zone varies with conditions. The sun, which is overhead at low latitudes, may penetrate to 100 m if the water is fairly clear (compared with but 40–50 m at middle latitudes). Below 200 m the sea lacks light, though in clear tropical water a small amount of blue radiation may penetrate to at least 1000 m.

The heat absorbed at the sea surface at low latitudes produces a warm surface layer overlying colder, denser layers. In equatorial waters the surface temperature remains at 26–27°C throughout the year in most areas. In the open oceans there is often a distinct **thermocline** (temperature discontinuity) lying between 100 and 500 m, but its depth varies seasonally (with wind changes) and, because of the earth's eastward rotation, it dips deeper to the west. Off West Africa the thermocline is at 20 m or less. The continuity of colder water beneath the warm surface layers allows some temperate and subtropical species to traverse equatorial regions provided that they can live there at greater depths, a phenomenon known as 'tropical submergence'.

Most open ocean water has a salinity of 34–36‰. Open-sea fishes are generally **stenohaline**, unable to withstand significant salinity changes. Fishes of coastal and estuarine waters have to be **euryhaline**, able to withstand such changes. Many tropical river systems pour immense amounts of freshwater into the oceans: the Amazon, carrying much mud as well as freshwater, stains the sea for over 100 km from the river mouth.

Fish are also much affected by water movements, currents, and in the littoral and sublittoral zones, tides or surges. Ocean currents result from the effects of wind action on the ocean surface as the earth rotates eastwards, combined with density differences in various layers of the sea. Equatorial surface currents flow at about 8–14 km day^{-1}. The effect of the earth's rotation (Coriolis force) causes a clockwise deflection of currents (gyres) in the northern hemisphere, and an anticlockwise one in the southern hemisphere (Fig. 1.1). The seasonal monsoon winds (and associated rainfall) lead to changes of flow, even reversals, in some ocean currents.

Fish production in the sea depends ultimately on primary production by algae at the base of food webs; light and temperature values remain high throughout the year, so primary production is generally limited by availability of nutrient salts. As there is no winter surface cooling, the strongly developed thermocline often persists throughout the year and acts as a barrier to vertical mixing. Dead organisms and excreta rain down into the bottom waters and thus out of circulation, so that nutrients become depleted in the euphotic zone, limiting production to a low level except where upwellings of nutrient-rich water come to the surface. Upwelling in the sea is caused by offshore winds but it also occurs at convergences and divergences where adjacent currents flow in opposite directions, as between equatorial and counterequatorial currents in the Pacific (Fig. 1.1). At low latitudes, upwellings occur along the western coasts of continents to replace the westward-flowing surface water of the equatorial currents.

Throughout the tropics, wherever there is a permanent thermocline, production is generally low, as for example in the Sargasso Sea. However, although at a low level, production is continuous throughout the year and it may extend through a greater depth of water column than at higher latitudes. Moreover, the turnover rate is very high, since growth rates are very fast and life cycles short at the high temperatures. Thus the total annual production in many tropical areas is much higher (perhaps five to ten times higher) than in temperate seas. In some warm seas, production may be fairly constant throughout the year but where there are seasonal changes in winds (monsoons), causing great variations in water circulation or seasonal upwellings, the seasonal increase in nutrients is quickly followed by increased production. The seasonal increase in warm seas is, however, seldom more than tenfold, whereas it may be 50-fold in temperate seas.

Conditions in tropical freshwaters

Tropical regions are characterized by high temperatures where these are not modified by altitude, and by relatively little seasonal variation in temperature or daylength compared with temperate regions. In equatorial regions the daylength is practically constant at 12 h throughout the year; even at 10°N and 10°S annual variations in daylength are less than 1 h. Seasonal variations do, however, exist in most tropical waters, caused primarily by wind regimes and fluctuations in rainfall, which lead to regular flooding of immense tracts of country, expanding the freshwater environment seasonally on a scale rarely known outside the tropics.

As the tropical rain falls mainly when the sun is overhead, rainfall maxima occur around May to July north of the equator and November to January south of it. The rainy seasons are, however, modified by the shape of the continent, the presence of mountain masses, and wind systems. Equatorial regions have two rainfall peaks a year, centred on the March and September equinoxes as the sun passes overhead. Africa, bisected by the equator, has comparable areas of freshwater to the north and south of it; in Kenya such rains cause rivers to flood twice each year, though which flood is the main one may vary from year to year. The Asian tropics include India, lying well north of the equator, mainland Southeast Asia, reaching south into equatorial regions, and islands (such as Borneo) on the continental shelf straddling the equator. South America also straddles the equator, with both northern and southern hemisphere rainfall regimes affecting the Amazon system.

The equatorial regions have the heaviest rainfall (over 1500 mm yr^{-1}) and the rain is well distributed throughout the year. Seasonality of the climate, including rainfall and flooding, increases with increasing latitude both north and south. Except where cleared by man, the zones of heaviest rainfall carry dense forest through which large permanent rivers flow; this forest slows the runoff from the land. Bush, passing to wooded savanna, then desert succeed each other in areas of lower rainfall and the rivers diminish in volume and persistence, drying into seasonal pools. The mirror-image effect of decreasing rainfall with increasing latitude to north and south of the equator is very clearly shown in Africa, though the most arid regions are slewed to northeast and southwest. The largest tropical river systems, the immense drainages of the Amazon and the Zaire, flowing mainly through dense forest, receive tributaries from both sides of the equator; these flood at 'opposite' times of year, causing a bimodal flood or prolonged highwater season in the lower reaches. The more seasonal rivers at higher latitudes flood into seasonal swamps and over surrounding savanna grassland. Floodwater which takes a long time to travel downstream may arrive after local floods have subsided, causing a second flood (a 'black' flood as sediment has been deposited on route). The fishes respond to rise in water level rather than to local rains, moving into newly flooded aquatic habitats on otherwise arid plains.

The high temperatures throughout the year speed up development and growth processes. Aquatic plants grow throughout the year in the permanent waters, giving cover both to the fishes and to the aquatic invertebrates on which they feed; very varied food sources may remain available throughout the year in these waters. Large amounts of decomposing plant material in the water lead to oxygen depletion. As oxygen is less soluble in water as the temperature rises, deoxygenation is more of a problem in static waters at tropical temperatures than it is in the temperate zone (see Beadle, 1981). Many tropical fishes have special respiratory adaptations enabling them to use atmospheric air. Enormous areas of swamp dry out completely, either annually or during dry phases of climatic cycles, and certain fishes have evolved mechanisms which enable them to overcome this desiccation.

The latitude within the tropics determines the seasonality of the flood regime which in turn has a profound effect on the biology of the fishes. Tropical conditions grade gradually into subtropical and temperate ones. Some lakes technically within the tropics, such as Lakes Chad and Kariba in Africa, are at high enough latitudes for seasonal changes in temperature and illumination to affect fish behaviour and growth. In tropical lakes a seasonality may also be imposed by wind action bringing nutrient-rich bottom water into circulation, for example by the southeast trade winds which blow from April to September along the Great Lakes Tanganyika and Malawi (Nyasa).

Divisions of the freshwater environment

Freshwater environments are divided into two main groups: running water (**lotic** environments) such as streams and rivers, and standing waters (**lentic** or lenitic environments) such as lakes, ponds and swamps. Most river systems pre-date their associated lakes, which are formed when rivers are dammed. Many of the tropical river systems are immense and have a very long history during which their form may have changed radically, for example both Amazon and Zaire drainages had large lakes. Tectonic changes have led to great changes in river level, falls and rapids forming barriers isolating some fish populations, while river captures have introduced new elements into local faunas.

'Classical' temperate rivers are typically smaller and divided into three regions: torrential upper reaches with a steep profile and swift current flowing over rocky or stony bottoms, give way to a flatter profile as the river emerges from foothills, depositing its silt load, and finally the river broadens out and gains organic matter from plant and plankton production in the slower-flowing meandering lower reaches. Such temperate rivers are inhabited by a succession of fish communities, often trout (which need well-oxygenated water) in the upper reaches and 'coarse' fish such as cyprinids in the lower reaches. The picture is much more complex in the large tropical rivers: changes in land levels have led to some rivers originating in swampy regions, while falls and rapids at widely spaced intervals along the course of a river such as the Zaire allow an alternation of swift-water and calm-water species. Such alternations help to explain the high numbers of species in these tropical rivers. (Such zonations are discussed by Balon & Stewart, 1983.)

Conditions in headwater streams depend very much on slope, which dictates both current speed and nature of the bottom deposits. Whether a stream flows through dense forest or is open to sunlight affects the sources of fish food. Where sunlight penetrates to the water surface, algal growth or floating meadows of aquatic plants provide and support sources of fish food, but where the light is cut off by overhanging forest the fishes are highly dependent on allochthonous (exogenous) foods, such as plant debris and aerial insects produced in the forest. Gallery forest persists along the streams far into savanna areas, so the width of the stream affects light penetration. Extensive lateral flooding in the highwater season, a feature of tropical floodplain rivers, leads to temporary lacustrine conditions over vast areas annually. It is hard to draw distinctions between riverine and lacustrine species, as many fishes lead a lotic existence at low water when they retreat to the riverbed, and a lentic one in the flooded areas with their interconnected pools and swamps in the highwater season. Conversely, some fishes adapted for life in rushing waters in flooded streams have to withstand conditions in stagnant pools in the dry season. In years of low rainfall some fishes may be trapped in ponds throughout the year. The rivers themselves, with their anastomosing channels, quiet backwaters and islands that block the current, also provide some lentic biotopes. Lotic and lentic conditions may exist side by side on riverbeds, where the current is less than 20 cm s^{-1} on the inner bank but exceeds this on the opposite 'bounce' bank.

These laterally flooded areas are mainly flooded forest shaded from the sun in equatorial regions, or savanna open to the sun at higher latitudes. The forest waters remain cool and lack light for photosynthesis, and the penetration of light is often further reduced when the waters are darkly stained with humic acids. The flooded savannas on the other hand heat up rapidly in the sun and there is abundant light for plant growth. Decomposing tree debris makes the flooded forest water very acid and deoxygenated, and the trees protect the river surface from wind mixing. On the flooded savannas the oxygen falls sharply as the water rises and submerged vegetation rots, but the oxygen supply is replenished with the aid of wind mixing, diurnal temperature changes and photosynthesis. These floodplains carry important fisheries in many parts of the tropics.

Many physical and chemical changes accompany flooding. The violent increase in water velocity may be accompanied by a drop in water temperature and rise in turbidity, together with changes in water chemistry. Where these effects fluctuate seasonally, they are generally non-catastrophic and the aquatic life is adapted to take advantage of the flood conditions. In the smaller rivers and streams the dominant factor may be desiccation, partial or complete, which leads to spatial contraction, isolation, rise in temperature and deoxygenation. These are limiting factors to which some fishes have become adapted but which are avoided by the majority of them by active migration. Small streams in the tropics often rise suddenly overnight but cease to flow shortly after the rains and are then reduced to interconnected pools and swampy depressions. The fishes respond to the changing conditions in rivers by moving about a very great deal, often making very long and well-defined migrations up and down river. Some of the fishes living in lakes continue to migrate up inflowing rivers to spawn, but many lake fishes appear to be much less influenced by seasonal changes. Immense river systems such as the Amazon and Zaire receive tributaries which pass through both forested and open country and through many degrees of latitude.

Riverine fishes are very mobile, moving long distances up and down river. They respond to the floods, which may arrive after the local rains have ceased, by making lateral movements out over the floodplain and into its pools, returning to the main river channels as the floods subside (unless trapped in pools until the next flood season). The typical seasonal cycle of events is summarized in Fig. 1.2. Most riverine fishes breed at the start of the flood season. The highwater period is the main feeding and growing time, and fat stores are accumulated on which the fish subsist through the dry season when they eat little. Young fish are thus born at the time with most food, when abundant plant growth provides cover from enemies. Floodwater seeping over the plain is greatly enriched by nutrients from the rapid decomposition of grasses and animal dung, or from forest litter. This leads to an explosive growth of microorganisms, in turn followed by an explosion in numbers of larger invertebrates (e.g. insects, crustaceans, molluscs) used as fish food, and of aquatic vegetation. The biomass of fishes increases rapidly during the highwater season, a large proportion of it due to rapid growth of young-of-the-year. As the floodwater subsides, fish losses are often enormous, from strandings in drying pools where innumerable birds feast on the fish, and from falling prey to the larger fishes which feed heavily on the young as they move down the restricted channels back to the rivers.

In the equatorial forest rivers the fishes move out into the flooded forest to feed. Shade from overhanging trees limits primary production and

Fig. 1.2. The seasonal cycles of events in a floodplain river.

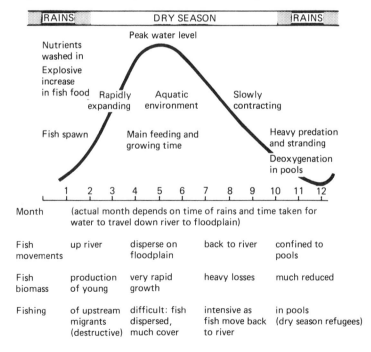

fishes are very dependent on forest products for food. Conditions in some forest rivers may be less markedly seasonal; spawning peaks may occur in both flood seasons.

Many tropical rivers have very long estuarine reaches, especially in Asia and South America where the land surface is lower than in most of Africa. Tropical estuaries are often vast, and the immense outpourings of freshwater dilute the sea for many kilometres, while in dry seasons or years saltwater runs long distances upriver where the land is low; being denser, the saltwater tends to move up beneath the freshwater, enabling bottom-dwelling marine fishes to move upriver more easily than surface dwellers. Above the brackish water zone, the freshwater may be backed up by the tides for long distances (over 160 km in Guyana). The fresh/salt interface moves long distances up- and downriver seasonally (over 200 km in the Amazon). This can ease the migration of marine species into freshwater, especially where primary and secondary freshwater fishes are lacking (as in Australia/New Guinea) and land is low as in South America.

Large swamps are characteristic of all the tropical regions. In swamps the limiting factors are generally deoxygenation or desiccation, restricting the fish fauna to species which have structural or behavioural adaptations enabling them to live through difficult periods. Conditions in the immense swamps of Africa are discussed by Beadle (1981); comparable areas in South America and Asia present the same basic problems to the fishes, though different faunas are involved.

Tropical lakes vary enormously in size from mere ponds to the 68 635 km^2 Lake Victoria in East Africa, about the size of Switzerland. The ratio of inflow to storage also varies greatly. Many reservoirs, e.g. on the Nile, and floodplain lakes on the Amazon, are flushed through with river water many times each year, whereas in the very deep Lake Tanganyika the inflow is only 1/1500 of the volume stored. In the flushed lakes new nutrients are brought in with the river water, in amounts and proportions depending very much on the geological formations over which the river flows. In lakes lacking a regular inflow of nutrients, production is governed by recycling processes within the lake.

The enlargement of the pelagic zone, with calm water suitable for plankton growth, is a feature of lakes compared with rivers. In tropical lakes the water temperature and light energy are high throughout the year, as in tropical seas, and the availability of nutrient salts for plankton growth limits production. Tropical lakes stratify (as temperate lakes do in summer); the surface water (**epilimnion**), the euphotic zone inhabited by the plankton, is separated from the deeper nutrient-rich but generally deoxygenated **hypolimnion** water by a relatively steep temperature gradient or discontinuity (thermocline). At the higher temperatures prevailing in the tropics, water has a greater density difference per degree of temperature change and in tropical lakes stability can be maintained despite very small temperature differences between surface and bottom water. Most tropical lakes lack the surface cooling which in temperate lakes leads to an 'overturn', bringing nutrient-rich bottom water into the euphotic zone seasonally. Very deep lakes, such as Tanganyika and Malawi, may remain more or less permanently stratified. Some nutrients are, however, mixed into the epilimnion by wind-induced currents (as discussed by Beadle, 1981). These winds are often seasonal, but in shallow equatorial lakes, such as Lake George in Uganda, winds throughout the year lead to a daily exchange of nutrients from the bottom mud to the lake water.

The Great Lakes of East and Central Africa have been much studied as they have valuable commercial fisheries; they also have spectacular faunas of endemic cichlids of great interest to students of evolution, as reviewed by Fryer & Iles (1972). The study of small crater lakes in Cameroon has helped to illuminate ways in which fish communities have evolved in the African Great Lakes (see Trewavas, Green & Corbet, 1972), and the new man-made lakes have provided large-scale field experiments in which to study the changes that occur when a riverine community becomes converted into a lacustrine one.

Studies on smaller lakes have included intensive work by an International Biological Programme team on Lake George in Uganda in an attempt to determine why this lake is so productive (Dunn *et al.*, 1969; Greenwood & Lund, 1973). Lack of space precludes discussion in this book of numerous studies on smaller lakes; many of these in Africa were considered by Beadle (1981). These include some extreme environments for fishes, such as Lake Magadi in the Kenya Rift Valley, a soda lake (pH 10.5) with hot springs and a salinity of 40‰, where the small endemic tilapia *Oreochromis grahami* lives in water up to 39°C, only 2 degrees below its lethal temperature (Coe, 1966).

South America has no comparable great lakes at the present time but the trumpet-shaped mouths of Amazon tributaries have many lake-like features and innumerable small lakes become cut off temporarily from the anastomosing river channels when the water level is low. Nicaragua, in Central America, has large lakes dominated by cichlids as in many African lakes. The world's highest lakes occur in the Andes, the largest being Lake Titicaca (*ca* 7600 km² at 3800 m). In Southeast Asia the Grand Lac of the Tonle Sap, a lateral lake of the Mekong system, used to support one of the world's largest freshwater fisheries, and many new lakes are being created in the Mekong system. In India and Sri Lanka many of the lakes and 'tanks' which support fisheries are man-made. Southeast Asia has a very long tradition of fish culture.

Tropical fish faunas

Fishes may be classified according to their way of life – pelagic fishes living in open water, benthic (demersal) fishes on or near the bottom. A close relationship exists between the evolutionary status of the fish, exemplified by its shape, body structure and physiology, and the environment in which it lives, its feeding ecology and behaviour. The systematic classification of the most important families of fishes found in tropical waters (Appendix) expresses evolutionary relationships (see also Lauder & Liem, 1983) and tells much about structure and probable way of life. Fishes with soft fin rays, such as herrings (family Clupeidae), are considered less highly evolved than those with spiny-rayed fins, such as percoid bass (Serranidae) (Fig. 1.3). Body changes which have occurred during the course of evolution include many which make the fish more manoeuvrable: the forward shift of pelvic fins, for instance, and the closure of the duct from swimbladder to gut (from 'physostomous' to 'physoclistous' condition) which has enabled the swimbladder to become a hydrostatic organ. Changes in jaw structure in which the maxilla is no longer part of the gape allow the more evolved fishes to protrude the jaws, and thus to exploit more diverse food sources. Teeth also vary greatly.

Many soft-rayed fishes, such as herrings, are built to swim through open water, unencumbered by heavy spines or scales, feeding as they swim. Many such fishes have simple breeding habits, scattering large numbers of eggs in the sea. The more highly evolved spiny-rayed fishes are able to take advantage of cover in reefs as they can manoeuvre into crevices while extracting the numerous types of food offered by reef environments. Some have pelagic larvae, but others (e.g. damselfishes) care for their eggs, producing relatively few young at a time. The territorial way of life of many spiny-rayed fishes can lead to isolation of local populations, itself conducive to speciation, and there have been spectacular adaptive radiations in some of these percomorph families.

Fish families are more numerous in the sea than in freshwaters, the majority of them percomorphs. The provisional numbers of genera and species in the families are indicated in the Appendix. The smaller territorial fishes, like gobies and blennies, and benthic dwellers such as serranids and pomacentrid damselfish, are more speciose than the larger far-ranging fish families such as tunas and billfish. Most of the marine

Fig. 1.3. Representative fishes important in tropical communities. Most grow to 10–60 cm except gobies and damselfishes (<10 cm), and groupers, scombroids and some carangids (60+ cm). (a) Clupeid, herring; (b) mormyrid; (c) characoid; (d) siluroid, catfish; (e) exocoetid, flyingfish; (f) aulostomid, trumpetfish; (g) pomacentrid, damselfish; (h) serranid, grouper; (i) lutjanid, snapper; (j) haemulid, grunt; (k) carangid, jack; (l) scarid, parrotfish; (m) acanthurid, surgeonfish; (n) scombrid, tuna; (o) gobiid, goby; (p) balistid, filefish; (q) tetraodontid, pufferfish.

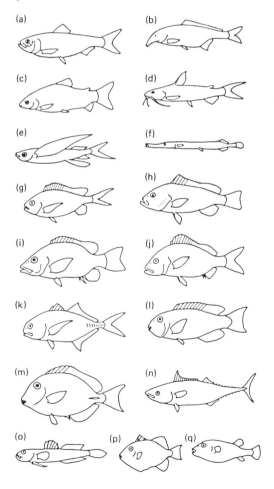

families are represented in both the Indo-Pacific and Atlantic Oceans, though their relative importance in these two areas may differ. We have little evidence of the relative ages of families, but unlike the freshwater family Cichlidae, which includes representatives of all trophic groups, trophic specialization in these marine fishes has generally proceeded to family level. Certain families, such as grey mullets (Mugilidae), are specialized detritivores; others, such as acanthurid surgeonfishes and siganid rabbitfishes are specialized algivores. Serranids and many other carnivores feed mainly on benthic invertebrates with some fish; many damselfish are specialized zooplanktivores; barracudas, large tunas and billfish are pelagic, mainly piscivorous carnivores.

The tropical belt includes four main, widely separated areas with their enclosed freshwaters (Fig. 1.1), making up the Neotropical, Ethiopian, Oriental and Australian zoogeographical regions. These comprise: South and Central America; Africa; tropical Asia and islands on the Sunda continental shelf west of Wallace's line, which runs down the Makassa Straits between Borneo and Sulawezi (Celebes); and New Guinea/ Australia and islands east of Wallace's line.

Freshwaters have their own faunas of primary freshwater fishes, those which have evolved in freshwater and are unable to tolerate brackish waters. Of the world total of *ca* 6650 freshwater species, about 93% are otophysans (see Appendix), characterized by the possession of Weberian ossicles, modified parts of anterior vertebrae which link swimbladder to ear and are involved in detection of underwater sound and, possibly, pressure (hence depth) changes. These fishes include carps (cyprinoids), characins (characoids) and catfishes (siluroids) of numerous families. Many primitive families, such as dipnoan lungfish, osteoglossids, and most of the endemic families of African fishes, are also primary freshwater families; most of these contain but few species. Secondary freshwater fishes, species evolved in freshwater but of marine groups and generally able to withstand slightly brackish water, include the cichlids which are so important in the lake faunas of Africa. Diadromous fishes, which make regular migrations between freshwater and the sea, include at least 115 species (*ca* 0.6% of living fish species (Cohen, 1970)). Representatives of marine families, peripheral fishes, complementary species or just sporadic visitors, also visit freshwater, especially in areas with few primary or secondary freshwater fishes, as in Australia/New Guinea.

Immigration of peripheral fishes from the sea has occurred independently up the rivers of each land mass. Most primary freshwater fishes are unable to tolerate sea crossings, and their distributions and relationships indicate former connections between the land masses. Characoid fishes, for example, are indigenous only to the Neotropics and Africa, regions now thought to have separated in the late Cretaceous about 75 million years ago, taking with them stocks of shared fishes. These regions both have cichlid fishes too. India and Australia were once part of this southern continent, Gondwanaland, but India and Madagascar separated from it at a much earlier date (probably in Jurassic times), too early for India to take many except primitive fishes with it on its journey across the ocean before it contacted Asia (probably in Eocene times, about 45 million years ago). Australia lacks primary freshwater fishes except the ancient lungfish (*Neoceratodus*) and osteoglossid *Scleropages*. Much later, in Pleistocene times (within the last 2 million years), faunal exchanges between Asia and Africa could have been facilitated through the Middle East when the sea level was lowered by about 100 m during glacial epochs; Africa and Asia now have certain genera in common. The large islands Borneo, Java and Sumatra, which lie on the Sunda shelf and were in contact with the Southeast Asian mainland when the Pleistocene sea level was lowered, have a fauna of Asian freshwater fishes. The systematic analyses that are needed to decipher the relationships of these African and Asian fishes are as yet in their infancy; Howes (1980) has illustrated possible dispersal tracks and discussed vicariant events (caused by geomorphological changes) which may have affected distributions of Asian and African bariliine cyprinids.

The net result of these dispersals (active or passive movements of individuals to new areas) and of vicariance (fragmentation of formerly continuous distributions through the appearance of a barrier, see Rosen, 1978) is that South America and Africa both have characoid and cichlid faunas, as well as lungfishes and some other primitive groups, but South America completely lacks cyprinoids. Africa and Asia both have cyprinoid fishes, but Asia lacks characoids and has very few cichlids. All three areas have many siluroid families, the relationships of which have yet to be unravelled (Howes, 1983). The proportional composition of South American, African and Asian freshwater fish faunas are summarized in Fig. 1.4, which also highlights the differences in lacustrine and riverine faunas in Africa.

Re-examination of the interrelationships of ostariophysan groups and their biogeography, the subject of lively discussion for many years, has led Fink & Fink (1981) to conclude that the group is a very old one and, contrary to 'traditional' belief that the characiforms were the most primitive, they consider the cypriniforms to be the sister group of the characiforms and siluriforms (Siluroidei and Gymnotoidei), and think that many of the extant characiform lineages evolved before the separation of Africa from South America. Cyprinoids dominate the freshwater faunas of tropical Asia, as they do in the temperate zone; they lack jaw teeth but pharyngeal teeth are well developed, and the body form of the African species is rather uniform. In South America characoids fill niches occupied by cyprinoids in the other two continents. Characoids show great variation in body form and have well-developed jaw teeth which have undergone evolutionary changes in relation to the food consumed. Nelson (1984) cites 10 families of characoids with *ca* 252 genera and at least 1335 species, of which some 176 live in Africa, the rest in the Neotropics; no genera or species are common to Africa and the Neotropics. Characoids are, together with siluroid catfishes, the dominant fishes in South American freshwaters. The siluroids include about 30 families from tropical waters. Two of these, Ariidae and Plotosidae, which have become secondarily marine (about the only ostariophysan fishes to do so) are widely distributed, the former in all tropical regions

Fig. 1.4. The proportional composition of the freshwater fish faunas of South America (Brazil), Southeast Asia (Thailand), African rivers and African lakes, based on the numbers of species indicated.

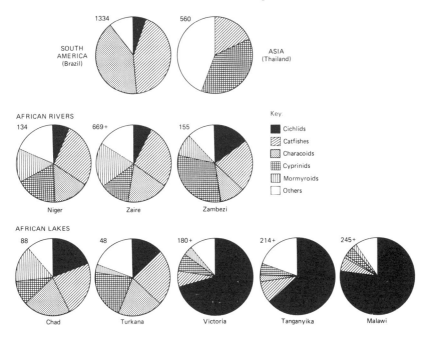

including Australia, the latter in countries bordering the Indo-Pacific. There are 14 endemic families of truly freshwater siluroids in South America, eight in Asia and three in Africa, and in addition to these the Bagridae, Schilbeidae and Clariidae are represented in both Asia and Africa.

The perch-like cichlids have a natural distribution restricted to South America and Africa, apart from a few species in Iran, Madagascar, Sri Lanka and southern India, but some species of tilapia are now circumtropical pond fish. Cichlids have undergone spectacular adaptive radiations in the African Great Lakes and parallel radiations in the Neotropics; over 70 species of the genus *Cichlasoma* have evolved in Central America (Miller, 1966). Tolerant of seawater, cichlids may have spread from South to Central America before the Pleistocene landbridge was formed.

Relict species of certain archaic groups of primary freshwater fishes are widely distributed. The living Dipnoi comprise *Lepidosiren paradoxa* in South America, four species of *Protopterus* in Africa, and *Neoceratodus forsteri* in Queensland, Australia. The Osteoglossidae include *Arapaima gigas* and two species of *Osteoglossum* in South America, *Heterotis niloticus* in Africa, and two species of *Scleropages* in Southest Asia and Queensland.

The small percomorph family Nandidae is represented in all three tropical areas; these are most numerous and include larger species (such as *Nandus nandus* and *Priolepis fasciatus*) in Asia, where they also move into brackish water. Two genera of small but voracious nandids (*Monocirrhus* and *Polycentrus*) occur in South America, another two (*Polycentropsis* and *Afronandus*) in Africa; their leaflike appearance enables them to drift near to unsuspecting small prey fish, as is well known to aquarists. Small cyprinodontiforms, toothed carps and killifishes of aquarists (revised by Parenti, 1981) occur in all tropical regions except Australia; many of the neotropical species are live-bearing. The families Notopteridae, Mastacembelidae, Anabantidae and Channidae (= Ophiocephalidae) are all represented in both Asia and Africa, but they are more conspicuous members of the fauna in Asia. The proportion of freshwater fishes of marine origin is much higher in Southeast Asia and South America, where low-lying land has permitted fishes to move far upriver, than in Africa, where much of the continental land mass is high, and falls or rapids near the coast obstruct the passage of fishes as, for example, in the Zaire River.

Remarkable convergences are shown by armoured catfishes of unrelated families which occur in stony streams of all three continents:

loricariids in South America, certain amphiliid genera in Africa, and *Sisor rhabdophorus* in India. Juveniles of these armoured fishes all live in rushing streams where the rapid flow might damage naked skins against the rocks. Possibly the armour, by its weight, helps these dorsoventrally flattened fishes to adhere to rock surfaces when the water flow is fast; the swimbladder is very reduced in both *Sisor* and loricariids. Adaptations for life in fast-flowing water include dorsoventrally flattened bodies, enlarged horizontal paired fins, and suctorial devices enabling the fishes to adhere to rock surfaces (modified pelvic fins in some Indian fishes, a suctorial mouth in others (Hora, 1930)); special modifications to enable the fish to respire are needed when the mouth adheres to the substratum. In a number of stream fishes the body is very elongated, with a whiplash tail which propels the fish forward in rapid jerks. Spectacular convergences are also shown by electric fishes – the African mormyroids with the unrelated Neotropical gymnotoids. No such electric fishes are known from Asian freshwaters.

PART II

Freshwater studies

2

The African fish fauna

Faunal composition

Africa has over 2000 known species of indigenous freshwater fishes. Table 2.1 lists the families they represent, giving approximate numbers of species, although these are only provisional because in some families, especially the Cichlidae, many new species have yet to be described. The table also shows whether families are endemic to Africa, or also found in the Neotropical or Asian tropics or in marine communities. Table 2.2 compares numbers of all species and of cichlids in the main river systems and their associated lakes. Examples of African fishes are shown in Fig. 2.1.

The most striking feature of the African freshwater fish fauna is the high degree of endemism. Eighteen of the families are endemic to Africa and this endemism at family level occurs amongst the less-advanced (pre-Acanthopterygian) fishes: lungfish Protopteridae, brachiopterygian Polypteridae, clupeomorph Denticipitidae and Congothrissidae, osteoglossomorph Pantodontidae, Mormyridae and Gymnarchidae, anotophysan Kneriidae (now taken to include Cromeriidae and Grasseichthyidae (Greenwood, 1975; Nelson, 1984)) and Phractolaemidae, and otophysan characiform Hepsetidae, Distichodontidae, Citharinidae, siluriform Amphiliidae, Mochokidae and Malapteruridae, and cyprinodontiform Aplocheilidae. At generic level almost all genera are endemic to Africa, with the exception of some euryhaline fishes of marine origin and seven genera shared with the Oriental region, *viz.* the cyprinid *Barbus*, *Garra*, *Labeo*, *Raiamas* and loach *Neomacheilus*, the catfish *Clarias*, also *Channa*.

The African freshwater fish fauna thus now consists of: (1) remnants of archaic elements of wide distribution, such as the dipnoan lungfish and osteoglossid *Heterotis niloticus*, which have living relatives in South America and Australia, respectively; (2) endemic families which have

Table 2.1. *The main fish families in African freshwaters, indicating whether they are endemic (E), or represented in the Neotropics (N), Asian tropics (A), or marine communities (M), with approximate numbers of known species in tropical waters of Africa*

Family	Range	Genera	Species
Protopteridae	E	1	4
Polypteridae	E	2	10
Osteoglossidae	N/A	1	1
Pantodontidae	E	1	1
Notopteridae	A	2	2
Mormyridae	E	18	198
Gymnarchidae	E	1	1
Anguillidae	M	1	6
Denticipitidae	E	1	1
Clupeidae	M	20	38
Congothrissidae	E	1	1
Chanidae	M	1	1
Kneriidae	E	2	24
Cromeriidae	E	1	1
Grassichthyidae	E	1	1
Phractolaemidae	E	1	1
Hepsetidae	E	1	1
Characidae	N	18	109
Distichodontidae	E	17	90
Citharinidae	E	3	8
Cyprinidae	A	23	475
Bagridae	A	16	100+
Schilbeidae	A	8	40+
Clariidae	A	14	100+
Amphiliidae	E	8	50
Malapteruridae	E	1	2
Mochokidae	E	9	155
Ariidae	M	1	1
Plotosidae	M	1	1
Aplocheilidae	E	6	255+
Poeciliidae	N	8	75+
Cyprinodontidae	A/N	1	2
Centropomidae	M	1	6
Nandidae	A/N	2	2
Cichlidae	N/A	100+	700+
Gobiidae	M	4	~26
Channidae	A	1	2
Anabantidae	A	2	~29
Mastacembelidae	A	2	~44
Synbranchidae	M	2	2
Cynoglossidae	M	1	1
Tetraodontidae	M	1	66

Table 2.2. *The approximate numbers of fish families and species in the principal river systems of Africa and their associated lakes*

River	Families	Species		% Cichlids
		Total	Cichlids	
Niger	26	134	10	7.0
Nile	17	115	10	8.7
Zaire	24	690	40	5.8
Zambezi	18	110	20	18.2

Lake	Families	Species		% Cichlids
		Total	Cichlids	
Chad	23	176	13	7.4
Turkana	15	39	7	17.9
Albert	15	46	9	19.6
Edward/George	8	57	40	70.2
Victoria	12	238+	200+	84.0
Tanganyika	19	247	136	55.1
Malawi	9	242+	200+	82.6

Fig. 2.1. Representative African freshwater fishes (indicating total lengths commonly encountered): (A) *Protopterus* lungfish (100 cm, Dipnoi); (B) *Polypterus* (40 cm, Polypteridae); (C) *Heterotis* (50 cm, Osteoglossidae); (D) *Pantodon* (6 cm, Pantodontidae); (E) *Phractolaemus* (6 cm, Phractolaemidae); (F) *Kneria* (6 cm, Kneriidae); (G) *Petrocephalus* (15 cm, Mormyridae); (H) *Pellonula* (14 cm, Clupeidae); (I) *Campylomormyrus* (12 cm, Mormyridae); (J) *Mormyrops* (100 cm, Mormyridae); (K) *Hepsetus* (30 cm, Hepsetidae); (L) *Hydrocynus* (50 cm, Characidae). (M) *Alestes* (12 cm, Characidae); (N) *Belonophago* (12 cm, Distichodontidae); (O) *Distichodus* (50 cm Distichodontidae);

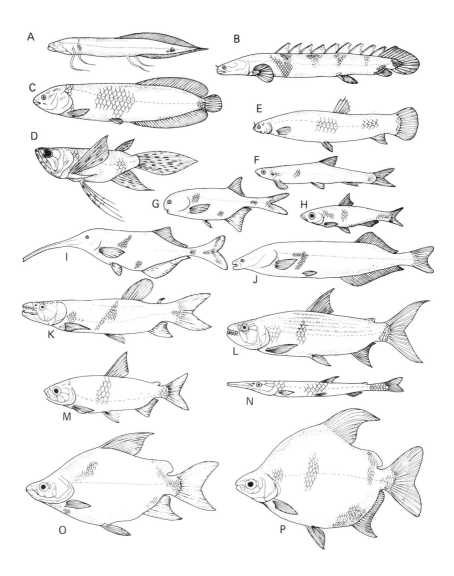

(P) *Citharinus* (70 cm, Citharinidae); (Q) *Labeo* (50 cm, Cyprinidae);
(R) *Barbus* (50 cm, Cyprinidae); (S) *Engraulicypris* (10 cm,
Cyprinidae); (T) *Schilbe* (20 cm, Schilbeidae); (U) *Malapterurus*
electric catfish (25 cm, Malapteruridae); (V) *Clarias* (50 cm, Clariidae);
(W) *Heterobranchus* (120 cm, Clariidae); (X) *Auchenoglanis* (80 cm,
Bagridae); (Y) *Bagrus* (100 cm, Bagridae); (Z) *Belonoglanis* (8 cm,
Amphiliidae); (AA) *Synodontis* (50 cm, Mochokidae); (AB) *Lates* Nile
perch (100 cm, Centropomidae); (AC) *Tilapia* (30 cm, Cichlidae);
(AD) *Haplochromis* (12 cm, Cichlidae); (AE) *Polycentropsis* (6 cm,
Nandidae).

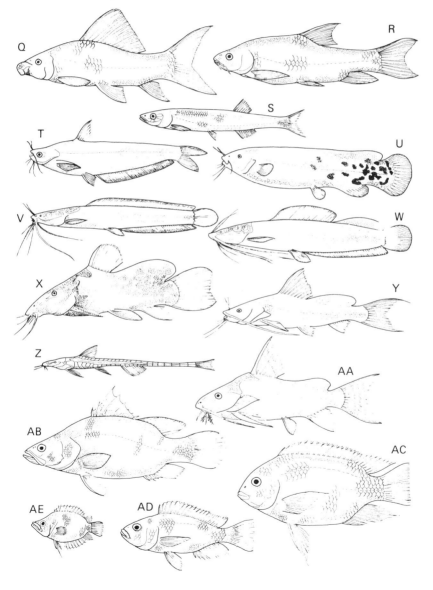

evidently evolved within the African land mass; (3) elements shared with South America, indicating their Gondwanaland origin; (4) elements shared with the Oriental region, some of which may have Gondwanaland ancestors, others, as indicated by the shared genera, resulting from much more recent land connections and possibly faunal exchanges; and (5) marine immigrants, from which freshwater species have evolved, and others moving from sea to freshwater either seasonally, during the life cycle or sporadically.

The families are represented very unequally. The Cichlidae and Cyprinidae have by far the most species, cichlids with at least 700, mainly in lakes, cyprinids with some 475 species (Daget, Gosse & Thys van den Audenaerde, 1984), mainly in rivers, followed by the endemic family Mormyridae with nearly 200 species and the endemic mochokid catfishes with *ca* 155 species. Some 176 characiform species are known from Africa and over 100 species each of bagrid and clariid catfishes, but many other families, including the endemic families and freshwater representatives of marine families, are represented by only one or two species.

Cichlids spawn in still waters and thrive in lakes. African rivers have only about 22 cichlid genera, but adaptive radiations have produced very numerous genera and species in the African Great Lakes. Another percoid family, Centropomidae, is represented by the large predatory *Lates niloticus*, distributed from West Africa to the Nile system. Offshore forms of *Lates* have evolved in Lakes Albert (Mobutu) and Turkana and Lake Tanganyika has a species flock of four species (Greenwood, 1976*a*).

African otophysans include *ca* 475 cyprinid species, mostly feeders on benthic invertebrates and vegetable debris. The genus *Barbus* (292 species) includes some large (to 90 cm total length (TL)) food and sport fish (see Banister & Clarke, 1980) and innumerable small species (*ca* 15 cm), very similar in appearance and thus hard to identify, often endemic to a particular river system. The 80 species of *Labeo* are detritus-feeders (see Reid, 1985). Bariliine cyprinids include 16 African species of *Raiamas* (see Howes, 1980) and small zooplanktivore lacustrine cyprinids (Howes, 1984), of which *Engraulicypris* in Malawi and *Rastrineobola* in Victoria are increasingly important as food fishes. Among some 20 other cyprinid genera, a dozen include only one or two species.

African characiforms fall into three groups but relationships between these groups and with the Neotropical characoids are not yet determined (Fink & Fink, 1981). The monotypic *Hepsetus odoe* may be close to the South American genera *Hoplias* and *Ctenolucius*. The Citharinidae includes three genera of deep-bodied mud-sucking fishes, the related Distichodontidae (*ca* 22 macrophyte-feeding *Distichodus* species and some small related forms) and genera formerly assigned to the Ichthyboridae, long slender fishes, found mainly in Zaire, specialized to feed on the fins of other fishes (Matthes, 1964). The characid assemblage of some 109 species includes large piscivorous *Hydrocynus* (6 species); *H. vittatus* in the Zambezi system and *H. brevis* and *H. forskahlii* in West Africa have been much studied, as these piscivores greatly affect the biology of other species in the fish communities (see Balon, 1971, 1974; Lewis, 1974*a,b*; Lauzanne, 1976). The omnivorous *Alestes*, several changing their main diets in wet and dry seasons, include the large species *Alestes ('Brycinus') baremose* and *A. dentex*, widely distributed across West Africa to the Nile, important food fishes which migrate up rivers to spawn (see Durand, 1978; Paugy, 1978). Africa also has about 50 species of smaller characids (Daget *et al.*, 1984).

The eight African families of siluriform catfishes include three also present in Asian freshwaters (Bagridae, Clariidae, Schilbeidae), three endemic ones (Mochokidae, Amphiliidae, Malapteruridae) and two occurring in the sea (Ariidae, Plotosidae). *Bagrus* is one of the main predators over a wide area, supporting important fisheries in lakes and rivers. Other bagrids include *Chrysichthys* species in the rivers of West Africa and Zaire and six species in Lake Tanganyika. *Auchenoglanis* bagrids are particularly abundant in equatorial West Africa, but the omnivorous *A. occidentalis* has a very wide distribution; small species of endemic bagrids live over sandy bottoms in swift water in the Zaire River (Poll, 1957, 1959*a*). Clariids, elongated catfishes with long dorsal and anal fins, include large *Heterobranchus* (*ca* 9 species) and 50 nominal species of *Clarias* in Africa (another 12 in Asia), mainly omnivores living over muddy bottoms. An arborescent respiratory organ in a cavity above the gills enables *Clarias* to wriggle through damp grass; clariids living in the open waters of the African Great Lakes (*Xenoclarias* and *Dinotopterus*) or in turbulent well-oxygenated rivers (*Gymnallabes* in the Lower Zaire) lack these accessory respiratory devices. Schilbeids include almost transparent, laterally compressed openwater species which have become very abundant in the pelagic zones of man-made lakes, and more-robust, widely distributed *Schilbe* and *Eutropius* species.

The endemic family Mochokidae includes over 80 *Synodontis* species (Daget *et al.*, 1984), catfishes with heads protected by bony armour and with formidable dorsal and pectoral spines; most explore bottom deposits with the ventral mouth surrounded by barbels, but a few species live inverted, taking food or oxygen from the water surface, and with reversed countershading. Amphiliids are slender-bodied stream catfishes, many armoured, living over hard bottoms in torrential streams. Electric catfish *Malapterurus*, widely distributed in rivers, are common in Lakes Tanganyika and Kariba; the dermal electric organ can give a formidable electric shock (see Howes, 1985 for catfish relationships).

Of the endemic families, the osteoglossomorph Mormyridae has the most species; these are especially abundant in Zaire and West Africa. Nocturnal fishes, they appear to use their electric signals to sense their way around and for communication. The mormyroid *Gymnarchus niloticus*, growing to over a metre long, makes a floating nest in swamps. Another endemic osteoglossomorph, *Pantodon*, a 'freshwater flyingfish', is well known to aquarists. The family Kneriidae includes small species living in swift mountain streams in Angola and West Africa, feeding mainly on aufwuchs (algae with contained microorganisms), and other species in pools in Zaire and East Africa. These and the microphagous *Phractolaemus* living in swampy regions of the Lower Niger and the Zaire, aided by an accessory respiratory organ, are recognized as primitive ostariophysans but, as they lack the Weberian ossicles present in Otophysi (the group previously known as Ostariophysi), they have been placed in the subgroup Anotophysi (Greenwood, 1975). Like the cyprinids, kneriids produce alarm pheromones, chemicals which warn other members of the school to take avoiding action if one is damaged (see p. 268). The endemic family Polypteridae, archaic fishes with brachiopterygian fins and shiny ganoid scales, includes 10 species of *Polypterus*; carnivores, the adults respire partly by a swimbladder lung and the young have external gills; a monospecific genus *Erpetoichthys*, an elongated form, occurs in West Africa. The dipnoan lungfish *Protopterus* (four species), well known for their ability to breathe air and to aestivate in a mucous cocoon in the mud in the dry season, are carnivorous with strong crushing jaws; their diet includes molluscs. They nest in swamps; their tadpole-like young have four pairs of external gills.

Of the non-endemic families, the osteoglossid *Heterotis niloticus*, a large microphagous fish, lives in swampy places where parents guard the young in a nest among aquatic plants. The Anabantidae, a very important family of food fish in Asia, is represented in Africa by a few small species of *Ctenopoma* and *Sandelia*; these have a labyrinthine accessory respiratory organ in a suprabrachial cavity, enabling them to live in deoxygenated water. The family Notopteridae includes elongate, laterally flattened fishes propelled by undulations of the long anal fin. Africa has but two species, *Papyrocranus afer* (which resembles the Asian *Notopterus*) and *Xenomystus nigri*; both live in West Africa and Zaire on the bottom amongst plants, feeding nocturnally on insects and small fishes; they have an epibranchial respiratory organ. The family Channidae (= Ophiocephalidae) of long cylindrical predatory fishes, common in Asia, has two African species; channids have accessory respiratory organs and *Parachanna obscura* is widely distributed in marshy places. Mastacembelids are rather anguilliform fishes living mainly among rocks, some partly buried; the African subfamily has two major lineages, *Caecomastacembelus*, a species flock of 22 species endemic to Lower Zaire rapids (some with reduced eyes, Roberts & Stewart, 1976), and *Afromastacembelus* of 16+ species including a group of species endemic to Lake Tanganyika which may be monophyletic (Travers, 1984). The presence of accessory respiratory organs in so many species of genera present in both Africa and Asia suggests that these fishes have had to survive deoxygenated conditions at some time in the past (Roberts, 1975; Roberts & Stewart, 1976).

Of the peripheral families of marine origin, the Clupeidae is represented by small species living in rivers; these have colonized the new man-made lakes, exploiting the zooplankton-feeding niche. Lake Tanganyika has two monospecific endemic genera which support the main fisheries in the lake; these were stocked into Lake Kivu (Dulnont, 1986) and Lake Kariba. Lake Mweru has three clupeid species. The Zaire has about six species of tetraodont puffer fishes. The Benue has a freshwater flatfish (*Dagetichthys*) and an endemic stingray (*Dasyatis garouaensis*) (see Reid & Sydenham, 1979). Anguillid eels of four species migrate up the rivers of eastern Africa from the sea (Frost, 1955; Balon, 1974).

The distribution of fishes in Africa

The effects of geomorphology, hydrographic history and Pleistocene climatic fluctuations on African fish distributions have been discussed by Roberts (1975) and Greenwood (1983c). They recognize about ten ichthyofaunal provinces, based largely on the present drainage systems. Much of present-day tropical Africa is drained by four great river systems – Niger, Nile, Zaire and Zambezi – together with smaller rivers flowing to east and west coasts (Fig. 2.2). There are also several inland drainage areas, such as Lake Chad lying between the Niger and the Nile, Lake Turkana (Rudolf) east of the Nile, and the Okavango swamps and Makarikari depression in Botswana. The fish fauna is richest and most varied in equatorial West Africa and Zaire; in Gondwanaland times this

Fig. 2.2. Ichthyofaunal regions of Africa and main sites discussed in text. (Lakes: A = Albert; Ch = Chad; E = Edward; G = George; Kb = Kariba; Kf = Kafue Flats; Kj = Kainji; M = Malawi; NN = Nasser/Nubia; Ok = Okavango; Tk = Turkana; Tn = Tanganyika; V = Victoria; Vt = Volta.)

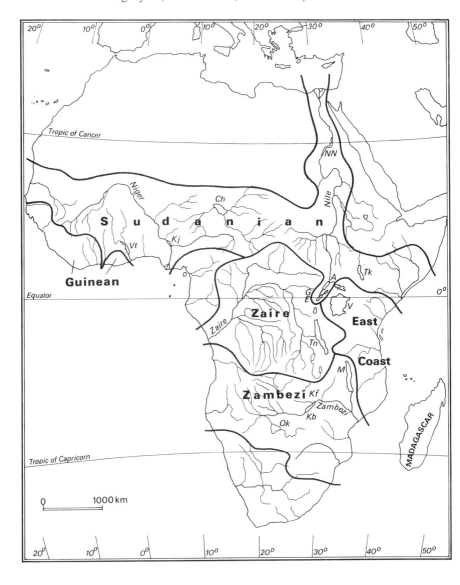

would have been the centre of the area connected with South America, probably drained by rivers flowing to the then westward-flowing proto-Amazon. The rivers of eastern Africa have numerous types of fish such as cyprinids which appear to have entered Africa from Asia and to have worked their way southwards down the east side of the continent. Much of this land is higher, with cooler waters which suit cyprinids well.

The Zaire basin has the richest fauna, 690 species (excluding Lake Tanganyika), 80% of which appear to be endemic, representing 23 families (Table 2.2). Each expedition collection produces new species. In addition to the otophysans which dominate the fauna (23% siluroids, 15% characoids, 16% cyprinids) the Zaire basin has 75 mormyrid species (18% of the fauna). The numerous endemic genera include the characoids *Mesoborus, Phagoborus, Belonophago, Eugnathichthys* and *Clupeopetersius*, the bagrid *Gnathobagrus*, the clariid *Channallabes*, and the cichlids *Teleogramma, Heterochromis* and *Steatocranus*. The west-flowing rivers of Cameroon and Gabon contain elements from Zaire and from West Africa, and also have a very rich fauna. Of the subtropical regions lying either side of the Zaire basin, the Nilo-Sudan or sudanian province to the north has a much more diverse fish fauna than the Zambezi province to the south. This sudanian region includes the savanna rivers south of the Sahara from Senegal on the Atlantic coast, across the Chad basin, to the Nile (Fig. 2.2). The sudanian fish fauna has been reviewed by Daget & Durand (1981), during a comprehensive survey of all the aquatic fauna and flora of this region with very useful identification keys (Durand & Levêque, 1981). The Niger and Nile share 74 fish species (some of them subspecifically distinct), 22 of which also occur in the Zaire (Greenwood, 1976*b*). The less rich and diversified fauna of the Nile compared with the Zaire probably reflects the different histories of the basins and also the less varied habitats and greater seasonality of conditions in the Nile.

The sudanian fauna is probably a very ancient one, formerly even more widely distributed than at present, a suggestion supported by the presence in Tanzania of a fossil denticipid and a fossil osteoglossid, both families now restricted to West Africa and the Nile. Relict status also seems to explain the five 'nilotic' species (*Protopterus aethiopicus, Polypterus bichir, P. senegalus, Ichthyborus besse* and *Oreochromis niloticus*) occurring in the upper Lualaba tributary of the Zaire system (Greenwood, 1976*b*; Banister & Bailey, 1979), which Poll (1973) had interpreted as indicating a former connection between the Nile and the Upper Zaire.

The almost pan-African distribution, from sudanian to Zambezi provinces, of taxa such as the catfishes *Malapterurus electricus, Schilbe mystus* and *Heterobranchus longifilis*, the characids *Hydrocynus vittatus, Alestes imberi, A. macrolepidotus* and *Hepsetus odoe*, and the cyprinid *Barbus paludinosus* strongly suggests interconnected waterways at times in the past. The break-up of early hydrographic patterns and consequent isolation of biotas, with differentiation in that isolation, are all elements of a classical vicariance pattern in historical biogeography (see Rosen, 1978), the details of which will emerge only when we know more about the phyletic relationships of the present-day fauna (as discussed by Greenwood, 1983*c*).

In West Africa, Daget & Durand (1981) distinguish between the 'sudanian' fishes living in the waters of the extensive savanna-covered peneplain, and 'guinean' species in forested streams. Ecological replacement species are well marked in some genera (of clariids, anabantids, cyprinodonts). The sudanian species often grow larger than their guinean counterparts and migrate long distances annually, so it is not surprising that these species have become widely distributed. The frequent river captures and variations in climate over long geological periods have resulted in guinean and sudanian species occurring within the same river system, separated neither by watersheds nor by waterfall barriers. Ecological conditions, particularly the presence or absence of forest, appear to account for their continued coexistence within one river system and it may be difficult for a species to extend its range when the river already holds a related species. Guinean species are found in the upper reaches of Niger tributaries in the higher more dissected country in Guinea, but in the Ivory Coast the situation is reversed as they occur in the lower reaches of the river, which are here forested, while the open country to the north carries sudanian species (Merona, 1981). This suggests that the overriding factor is shade or food from the forest. The Dahomey Gap, where a belt of savanna comes right down to the coast, breaks the guinean forest zone into western and eastern sections (Fig. 2.2), affecting the distribution of some guinean species.

The Zambezi fauna (*ca* 101 species) lacks many families of fishes present in West Africa and Zaire; it also has fewer kinds of mormyrid and characoid, but cyprinid numbers are high. The Upper Zambezi, above the Victoria Falls, a 'reservoir type' river, has a richer fish fauna than the Middle Zambezi (from the Falls to the Cabora Bassa rapids, now a hydroelectric barrage), probably for ecological reasons – the many swamps provide cover round the year for the fishes, whereas the Middle

Zambezi before the formation of Lake Kariba here was a 'sandbank river' shrinking into pools in the dry season. The Kafue tributary, which joins the Middle Zambezi below Lake Kariba, has a fish fauna more like that of the Upper Zambezi with certain Zaire elements (Bowmaker, Jackson & Jubb, 1978). The east-flowing rivers of eastern Africa also have attenuated faunas. As in southern Africa, cyprinids and cichlids are important; many families present in Zaire are absent and others have fewer species.

Fish faunas are continually receiving or losing species from other systems. Bell-Cross (1965) described the movements of six species of fish (of *Barbus, Clarias, Aplocheilichthys, Tilapia* and *Ctenopoma*) from the Zaire to the Zambezi system, the fish moving up the gentle slope from the Zaire tributary onto the watershed plain and back into tributary streams on both sides of the watershed at the end of the rains. In the man-made Lake Kariba, the Upper Zambezi species *Alestes lateralis* appeared and replaced the indigenous *Brachyalestes imberi*, as the new conditions suited the spawning habits of *A. lateralis* better than those of the indigenous species (see Balon, 1974; Jubb, 1977).

The lacustrine faunas are dominated by cichlids, with the exceptions of Lakes Albert and Turkana (Rudolf) which have retained nilotic riverine faunas (Table 2.2). Most of the cichlid species are endemic to a particular lake, having undergone adaptive radiations within that lake while exploiting the diverse feeding opportunities offered by lakes compared with the river. The non-cichlid faunas reflect those of the drainage system within which the lake lies: Lake Victoria draining to the Nile, Lake Tanganyika to the Zaire, and Lake Malawi to the Zambezi. Of Lake Victoria's 38 non-cichlid species, 16 (42%) are endemic; of Lake Tanganyika's 67 non-cichlids, 47 (70%) are endemic; and of Lake Malawi's 42 species, 26 (62%) are endemic. In Lakes Albert and Turkana, endemism among the non-cichlids is much lower (only 8% in Albert, 5% in Turkana).

The higher endemism of non-cichlids in Lakes Tanganyika and Malawi has been thought to reflect the probable greater ages of these deep lakes, though the greater diversity of Tanganyika's fauna must be related in part to the rich Zairean stocks which have contributed to it. Seven of the ten families which contribute endemics to the Lake Tanganyika fauna are also represented by endemics in the rapids of the Lower Zaire (*viz.* Characidae, Cyprinidae, Bagridae, Clariidae, Mochokidae, Cichlidae and Mastacembelidae); Roberts & Stewart (1976) suggested that evolution in these rocky habitats of the Zaire basin preadapted fishes such as the cichlid *Lamprologus* for life in the rocky littoral of Lake Tanganyika.

The Zaire: an equatorial forest river
The Zaire basin

The Zaire system is divisible into about six regions (Poll, 1957): (1) the Lower Zaire, below Boma and entering a mangrove-lined estuary, which has many euryhaline species (pristids, elopids, sphyraenids, mugilids), freshwater representatives of marine families (clupeids, gobies, tetraodonts), and secondary freshwater fishes (cichlids, cyprinodonts, anabantids); (2) a stretch with 32 falls and rapids in 350 km between Matadi and Pool Malebo (= Stanley Pool), one of the most extensive stretches of such rapid water in the tropical world – this bars access to marine fishes; (3) Pool Malebo, a 500 km^2 enlargement of the river rarely more than 10 m deep, the level varying about 3 m during the year, with an archipelago of islands and sandbanks; (4) the great central basin of the Zaire, cut off from estuarine reaches by the rapids below and from the Lualaba by falls just above the Kisangani (Stanley) Falls, over 1500 km upriver and almost on the equator; (5) the Upper Zaire basin or Lualaba above the Kisangani Falls; and (6) the Upemba lakes on the Lualaba floodplain.

The ecology of Zaire basin fishes has been studied at widely separated places: in and around Pool Malebo (Poll, 1959a); around Yangambi near Kisangani ca 1500 km upriver (Gosse, 1963); in the central basin on the large tributary Tschuapa River in the Ikela District; in Lake Tumba, a lateral lake below the Tschuapa/Zaire confluence (Matthes, 1964); in the rapids of the Lower Zaire (Roberts & Stewart, 1976); and in the Upemba National Park lakes (Poll, 1976). Banister & Bailey (1979) also visited the central basin. Only a brief summary of the ecology of this very complex fauna is possible here; more detailed accounts are given by Roberts (1972), Marlier (1973) and Lowe-McConnell (1975) when comparing the ecology of Congo and Amazon fishes.

In Pliocene times the Zaire basin contained a great lake or lakes which drained westwards when a coastal river broke through its western rim. The lower courses of streams flowing into this lake then extended across the lake floor to form the present Zaire system. The main Zaire River and most of its tributary rivers now flow in an anticlockwise direction; the affluent streams of the left bank cross the low-lying central basin and enter the river through swampy zones, whereas the right bank tributaries come from higher ground and have a different fauna of stream fishes above the 500 m contour.

This great river system straddles the equator where rain falls throughout the year. Tributaries from far to the north flood from August to

November, those from far to the south flood from May to June, bringing their waters into the lower reaches in a bimodal peak. Much of this central basin is cloaked in equatorial rainforest and the flooded river spills out, inundating huge areas of forest many kilometres from the main river.

The main biotopes and their fish faunas

Specific associations of fishes characterize the main river, the marginal waters along the shores and islands and in the bays and creeks, the inundated forest zone, which leads back into swamps and permanent pools, and affluent rivers and streams of various sizes. Most of the fishes change their habitat, and thus their association, as they grow and also with sexual activity and with the season. Fishes of many species move into the inundated forest at high water to breed and feed, and back to the low-water channels as the level falls. Some fishes, especially the young, stay behind in pools which become isolated as the level falls, and fishes with special adaptations to withstand deoxygenated conditions may remain in the swamps. The shallow marginal waters along the banks and islands, and over them when the river is high, carry more fishes than do the open waters.

The clupeids are found only in the main rivers. The mormyrids are principally fluviatile (many living in large schools near the bottom in Pool Malebo), though several species of each genus occur in streams and swamps. Among the characids, citharinids and cyprinids the large species are fluviatile, small species frequent streams, and none lives in swamps. Of the catfishes the bagrids and mochokids live mainly in the large rivers, but some species occur in streams. The schilbeids are typically openwater catfishes, all living in rivers, except for one species specialized for life in forest streams in each of the two principal genera *Schilbe* and *Eutropius*, the convergence between these two species being remarkable. The behaviour of the clariids allows them to colonize all three biotopes – river, stream and swamp – but in each they frequent calm muddy bottoms. The amphiliid catfishes are current-loving fishes of forest streams and river rapids. Cichlids in the rivers include various tilapias, *Tylochromis* and *Lamprologus mocquardi*, while most other species are confined to small streams. The cyprinodonts are specially adapted for life in streams and small water bodies. The anabantids occur only in the swamps. The Yangambi area is rich in forest-loving mormyrids, characids, bagrids and cyprinids absent from Pool Malebo. Some of these forestwater species also occur in Cameroon and Gabon.

The **rapids** below Kinshasa have a fish fauna specialized for life in very fast turbulent water, highly oxygenated, over rocks, and with poor visibility (see Roberts & Stewart, 1976). Here live many species adapted to fast-flowing water (cyprinids *Garra* and *Labeo*; catfishes *Atopochilus*, *Euchilichthys*, *Chiloglanis* and *Gymnallabes*; cichlids *Steatocranus*, *Teleogramma*, *Leptotilapia*), well shaped for adhering to the rocks or with anguilliform bodies for living in crevices between them. Both the accessory respiratory organs and the eyes are much reduced in the catfish *Gymnallabes*, and the eyes are no longer functional in some of the endemic *Caecomastacembelus*, of which there is here a species flock of 22+ species (Travers, 1984). Species of several families show convergent adaptations for a sedentary life with reduced vision. The endemic fishes here are representatives of lowland families (cyprinids, mochokids, cichlids), not of families (such as Kneriidae) typical of headwater streams in Africa (Poll, 1959*a*).

The **swamps** are either permanent, carrying a very specialized fauna adapted for life in deoxygenated waters, or temporary, carrying a much more varied fauna seasonally. The swamp water is very acid (pH 3.8–5.0), the water shallow (a few centimetres to 3 or 4 m) often over a leaf-carpeted bottom or organic mud. Almost all the fishes feed on insects or fish; primary production is very low. Cyprinodonts and *Hepsetus odoe* take the better-aerated water from just below the surface, but many species have accessory respiratory organs (Poll, 1959*b*): swimbladder lungs in *Polypterus* and *Protopterus*, epibranchial organs in *Papyrocranus* and *Phractalaemus*, arborescent organs in clariids, labyrinthine organs in anabantids, and pharyngeal diverticula in *Channa*.

In the **main rivers** the biotopes are much more numerous and diversified than in the streams and swamps, species are more numerous and the fish associations less well defined. Rivers more than 50 m wide, exposed to sun and wind, have higher temperatures (22.5–33°C) than the streams. The open waters have pelagic and benthic zones. In the pelagic zone live immense schools of small plankton-feeders, clupeids (*Microthrissa*) which migrate up- and downriver seasonally, *Clupeopetersius* and small *Barbus*, followed by predatory *Hydrocynus vittatus*. Also in the pelagic zone are large characoids and cyprinids, feeding mainly on aerial insects stranded on the water surface and themselves preyed on by *Hydrocynus* and *Lates*. The benthic fishes are mainly aquatic-insect-feeders, such as many mormyrids, *Chrysichthys* and *Synodontis, Tylochromis* and *Barbus*, but they also include mud-eating *Citharinus*, detritus-feeding *Chrysichthys*, some omnivores, and piscivorous *Polypterus, Mormyrops, Chrysichthys* and *Channa*. Rocky bottoms are rare here but the litter of fallen trees provides hiding places equivalent to those offered by rocks, and these are frequented by light-shy nocturnal mormyrids and siluroids. The **marginal waters** present a diversified series of biotopes, very important to the fishes, many of which reproduce and spend much of their lives here. These are transitional waters between the open water and the inundated forest zone. There is little current and the temperature is high (25–35°C) though the banks are usually shaded; the water is rarely more than 3 m deep (up to 10 m in some channels). Relatively stable habitats, they are nevertheless subject to greater seasonal changes than are the open waters.

The **inundation zones** carry juvenile fishes of many species: entomostracans and insect larvae are abundant here and there are rich bottom deposits for *Citharinus*, which become mud-feeders when about 3 cm long. At low water the young fish move into the floating meadows or marginal macrophytes, where an intensive basket fishery catches many species of mormyrid, the characoids *Hydrocynus, Distichodus* and *Citharinus*, and siluroids *Clarias, Synodontis* and *Schilbe*, and also tilapias.

Matthes (1964) recognized separate fish associations in: (1) the littoral zone, subdivided into areas with (i) rocky shores (wave-washed laterite cliffs), (ii) sandy or stony beaches, or (iii) marshy shores, shaded, sandy or muddy bottoms littered with vegetable debris; (2) channels, creeks and oxbows, with shaded calm pools, little current, sand or mud bottoms rich in vegetable debris; (3) floating meadows, along the shaded banks and heads of bays; and (4) inundation zones, inundated forest and pools adjacent to the rivers, unstable habitats rich in vegetation and fish foods.

The ecological factors shared by all these marginal waters are the still water, shallow depth, generally dense vegetation, abundant fish food, and bottoms littered with debris, often over organic mud. The aquatic flora is well developed, the invertebrate fauna particularly rich and diversified. The fish fauna is immensely varied, almost all species spending part of their lives in such waters, though most only come here periodically, some in the floods to reproduce, others in the dry season when the inundation forest is dry. Young fishes find shelter here from predators and current, as well as abundant food.

Near Yangambi the **streams** flow from the plateau; they are shaded, with very small variations in microclimate, the bottoms of mobile sand. The upper reaches are colonized by an association of insectivorous fishes. In a sample collected by chemofishing with rotenone, of 670 individuals representing 16 fish species, 84% had been feeding almost exclusively on terrestrial insects (compare Amazonian streams (p. 143) and Bornean streams (p. 165)); aquatic insect larvae, shrimps and fish formed the other foods (Gosse, 1963).

The open water of the streams presents a succession of biotopes, the extent of each governed mainly by the depth and current speed. Shallow zones less than 50 cm deep with mobile bottom deposits have few fishes, though some *Afromastacembelus, Amphilius* and *Auchenoglanis* hide in the little piles of branches in the sand. The stillwater zone along the bank is inhabited by juveniles of marginal species, exploiting aerial insects and organic debris from the water surface. Where the water is deeper (>1 m) the current is stronger and the stream physiognomy changes completely; the deeper sections are often littered with tree debris. Fishes here tend to be omnivores, staying near the bottom, with predators such as *Hepsetus odoe* and *Mesoborus crocodilus* lurking in holes in the banks, while light-shy silurids hide by day in the tangle of submerged branches.

The lower stretches of the streams carry the same associations along their margins and in open water, but as the stream widens and the current slows down the bottom omnivores become more prominent. The numbers of species, the size of individual fish, and the length of the section they colonize depend on many factors such as the breadth and depth, types of bank and water levels. Many riverine species move into streams at certain times, *Hydrocynus vittatus* and *Distichodus* only into the larger streams. Juvenile *Citharinus, Distichodus* and *Alestes* move upstream in floods and out into inundated forest where they make their early growth.

The fish fauna of these streams is a well characterized association, with the admixture of swamp and river species. Succession in these associations is governed principally by current speed. Upper reaches carry mainly insectivorous fishes living on aerial insects, and as the current slows bottom omnivores appear.

The Zaire system carries several large **lateral lakes**. The best studied of these, Lake Tumba, a 765 km² extent of water near the equator, is not a true lake but a permanent zone of river inundation, most organisms living here being dependent on materials brought in by inflowing streams (Matthes, 1964). The depth ranges from 3 to 10 m. The banks are covered with forest which is inundated twice each year, a little flood in May/June and a larger one in September to January. The water is stained brown with humic acids (pH 4.5–4.9) and is of low conductivity. Dissolved oxygen is, however, present all year as this shallow open water is stirred by violent winds. Temperatures range from 27 to 33°C.

Lake Tumba has a rich fish fauna with numerous mormyrids,

characoids, bagrids, clariids and cichlids, and a small endemic species of *Clupeopetersius*. In the 1960s a local fishery was catching mainly *Citharinus, Hydrocynus, Distichodus* and *Auchenoglanis*. Most species breed in August–September at the start of the main flood when the fish move into the inundated forest. The fish return to the lake as the water falls. The richness of the fish fauna, despite the chemical poverty of the water, appears to be due to a combination of the very extensive shoreline, with its rich littoral fauna and flora and allochthonous material (insects and vegetable debris) contributing to the food supply, and the floods allowing the fishes to find food far from the lake in the inundated forest. Should the banks be cleared and the fish denied access to the forest, the fishery would probably cease to be productive.

Trophic groups

The trophic groups recognized by Matthes (1964) are shown in Table 2.3. Plankton-feeders were not abundant as the river has so little plankton. The carnivores included a group taking aerial insects from the surface, another group using benthic aquatic insects and river-margin carnivores feeding mainly on aquatic insects and Crustacea. The 'mixed carnivores' group ate fishes and large invertebrates (shrimps and Odonata), the piscivores included those swallowing whole fishes and fin-biters, specialized to tweak pieces from the fins of living fishes. Each family and almost every genus had an omnivorous representative; this was often the most widely dispersed species.

Matthes listed the species representing these trophic groups in each of the main habitats in his field area, showing both the preferred habitat of the species and other habitats where it could also be found. His data are too detailed for reproduction here, but the summary of them in Table 2.3 indicates the preferred biotopes of the various genera. Most of these genera are represented by a number of species, each preferring a different biotope. Some species occur in several habitats and a biotope may have several species (in some cases of the same genus) occupying a particular trophic niche. For instance, two macrophyte-feeding *Distichodus* (*D. noboli* and *D. fasciatus)* occur in bays and pools, and another two omnivorous species (*D. sexfasciatus* and *D. atroventralis*) both live in the littoral zone.

Within each habitat as broadly defined here Matthes recognized that there are clearly a multitude of ecological niches differing in detail, and that there are continuous changes in the course of time, particularly with seasonal changes in water level, food available, and the numbers of

Table 2.3. *The preferred biotopes of fish genera representing various trophic groups in equatorial forestwaters of Central Zaire (data abridged from Matthes (1964) who listed the particular species and indicates all the biotopes where each occurs)*

Trophic groups	Open waters		Marginal waters
	Pelagic zone	Benthic zone	Littoral
Mud-feeders		*Citharinus* *Labeo*	*Synodontis*
Detritus-feeders	*Alestes*	*Gnathonemus* *Chrysichthys* *Auchenoglanis*	*Petrocephalus*
Omnivores	*Bryconae-thiops* *Barbus*	*Gnathonemus* *Chrysichthys*	*Gnathonemus* *Petrocephalus* *Alestes* *Micralestes* *Petersius* *Bathyaethiops* *Distichodus* *Parauchenoglanis*
Herbivores algal-feeders			
macrophyte-feeders			*Eutropius*
Plankton-feeders	*Microthrissa* *Clupeo-petersius*		*Barbus*
Carnivores using allochthonous material (surface insects)	*Petersius* *Barilius*		*Micralestes* *Barilius*
Bottom[b] insect-feeders		*Gnathonemus* *Barbus* *Gephyroglanis* *Synodontis*	*Petrocephalus* *Marcusenius* *Gnathonemus* *Chrysichthys* *Tylochromis*

Table 2.3. (continued)

Marginal waters		Swamps	Forest streams
Bays, pools, creeks, dead arms, channels	Floating prairies	Pools: seasonal or [a] permanent	
		Phractolaemus	
Auchenoglanis *Synodontis*		*Stomatorhinus* *Clarias* [a]*Clariallabes* *Channallabes*	*Clarias*
Alestes *Phenacogrammus* *Xenocharax* *Clarias*	*Distichodus* *Barbus*	*Stomatorhinus* [a]*Ctenopoma*	*Alestes* *Bryconaethiops* *Phenocogrammus* *Congocharax* *Neolebias* *Barbus* *Nannochromis* *Ctenopoma*
Hemigrammo-petersius *Pelmatochromis* *Distichodus* *Tilapia*	*Neolebias* *Distichodus* *Synodontis* *Aplocheilich-thys*		
Phenacogrammus *Bathyaethiops*		*Pantodon* [a]*Ctenopoma*	*Micralestes* *Phenacogrammus* *Epiplatys* *Aphyosemion* *Hypsopanchax*
Polypterus *Petrocephalus* *Gnathonemus* *Parauchenoglanis* *Microsynodontis*		*Polypterus* *Stomatorhinus* *Clarias* *Kribia* [a]*Ctenopoma*	*Barbus* *Auchenoglanis* *Clarias* *Eutropius* *Chiloglanis* *Amphilius* *Mastacembelus*

(continued)

Table 2.3. (*continued*)

Trophic groups	Open waters		Marginal waters
	Pelagic zone	Benthic zone	Littoral
River margin carnivores			*Mormyrops* *Microstomatich-thyoborus* *Eutropius*
Mixed carnivores	*Mesoborus*	*Mormyrops* *Chrysichthys*	*Polypterus* *Mormyrops* *Schilbe* *Eutropius* *Malapterurus*
Piscivores	*Odaxothrissa* *Hydrocynus* *Lates*		*Phagoborus*
Fin-biters	*Eugnathichthys*		*Phago*

[a] Genera found in permanent swamps.
[b] Over rocky bottoms or amongst tree debris genera include: *Gnathonemus, Chrysichthys, Dolichallabes, Synodontis, Nannochromis, Lamprologus, Mastacembelus.*

particular species present. Such an unstable medium is dynamic and resilient, and the diversity of niches in a relatively small stretch of water permits the coexistence of numerous closely related species with apparently identical ecological requirements, as is the case for certain mormyrids, characoids and siluroids. The seasonal instability explains the absence of very highly specialized trophic forms (as found in the African Great Lakes described below) but there is not as much seasonal variation in food abundance as in seasonal rivers such as the Niger. Zaire species which are apparently specialized can readily adapt to diverse conditions; for example, the long rostral probe in *Campylomormyrus tamandua* can seek insects in bottom mud or sand, or in organic debris or among rocks. Predators which specialize by biting fins of other fishes can live in any biotope where suitable prey fish are present. Under normal conditions most of the Zaire fish species show a marked preference for one particular biotope, for which they are probably slightly better adapted, but if conditions change suddenly they can survive. These Zaire species are also eurythermic, able to withstand temperature changes of the order of 5–10 Celsius degrees (Matthes, 1964).

Marginal waters		Swamps	Forest streams
Bays, pools, creeks, dead arms, channels	Floating prairies	Pools: seasonal or *a* permanent	
Polypterus	*Xenomystus* *Nannocharax* *Hemistichodus* *Heterochromis* *Ctenopoma*		*Phractura* *Nannocharax* *Trachyglanis* *Hemichromis*
Protopterus *Clarias* *Pelmatochromis*	*Ctenopoma*	*Clarias*	*Hemichromis* *Ctenopoma*
Parophiocephalus			*Hepsetus*
Belonophago			*Phago*

Breeding seasons and growth

Cichlids appear to breed more or less throughout the year in equatorial waters, but most of the Zaire basin fishes spawn at the start of the highwater season, September–October (continuing until December for some species in Lake Tumba), with another breeding season of less general importance during the high water in April–June (Matthes, 1964). The relative importance of the two seasons may vary in different parts of the basin and from year to year. Away to the south in the more seasonal Lualaba, and in the Luapula–Mweru tributary system, the principal breeding season for most species is at high water in January to March, with a secondary peak for cichlids in September–October (de Kimpe, 1964).

Most species migrate upstream as the waters start to rise, then laterally into the inundated forest. Around Lake Tumba fishes penetrate many tens of kilometres into the flooded forest to spawn and feed on the abundant and varied foods (Matthes, 1964). Exceptions to this include certain pelagic clupeids, cyprinids and schilbeids, and some benthic bagrids and mochokids tied to habitats in large rivers or lakes.

The fish eggs hatch very quickly, within several hours or 2 days at the most, and growth is fast in the inundation zone. The larger young leave this zone at the same time as the adults when the water starts to fall, but many juveniles remain behind in the numerous creeks, bays and ponds. Some species are well adapted for life in these still waters; others are merely trapped here and many perish in the shrinking pools. We know virtually nothing about the breeding behaviour of many of the species of economic importance.

Hardly anything is yet known about the growth rates of the Zaire fishes. The sizes of fishes taken in August–September suggest that sexual maturity is generally attained towards the age of 2 years in the large species, while certain small species complete their life cycles within one year (Matthes, 1964). The two periods of high water, during which the fish grow well, complicate attempts to determine ages from marks on skeletal structures. In residual pools and swamps fishes often mature at a smaller size than they do in the main river (nanism).

The Niger: a seasonal floodplain river

African seasonal floodplain rivers fall into two groups: those north of the equator across the sudanian region, the savanna zone south of the Sahara from the Senegal to the Nile; and south of the equator in the Zambezi and its tributary systems. The sudanian ones have been chosen for consideration here; for those of southern Africa see references in Bowmaker, Jackson & Jubb (1978).

Floodplain rivers are very dynamic environments, providing a whole mosaic of shifting types of habitat for the fishes. Welcomme (1979) has listed their major habitats, differing in morphology, chemical and physical conditions: in the flood season, the main river channels with rapid and turbulent flow and floating islands of sudd; tributary streams of many kinds such as rocky torrential streams and channels linking floodplain to subsidiary marsh or lakes above the main level of the floodplain; the floodplain itself with its floating meadows, open water and littoral fringe, its lagoons, depressions and lakes, with open water over sand or mud, and standing, floating or submersed vegetation; and areas of flooded forest of various types. In the dry season, the main river may dry up into semi-permanent channels, alternating pools and rocky riffles over various types of bottom, with or without vegetation and often much tree debris, meanders in the permanent channels producing a succession of habitats of varying depth and bottom type, deeps with slow or fast current, shaded by forest or open to the sun. On the floodplain, pools may dry out completely or become marshy and the larger lakes may stratify; waters open to the sun carry many different types of vegetation; backwaters connected to the main channel may be shaded.

Data on the ecology of sudanian fishes in the Niger and neighbouring rivers have been collated elsewhere (Lowe-McConnell, 1985; Welcomme in Levêque *et al.*, in press). The foundations were laid by Daget in his numerous studies of the fishes and their ecology in the Upper and Middle Niger inland delta (see Daget 1954, 1957; Daget & Durand, 1981). The building of the hydroelectric barrage across the Niger at Kainji stimulated pre- and postimpoundment studies in the lake area. In wet periods the Benue is linked via the Mayo Kebbi to the Lake Chad basin where fish and fisheries were studied by Blache *et al.* (1964) and later by workers from the ORSTOM laboratory in N'Djamena (Fort Lamy). The severe drought in the Sahelian region in 1972–74 stimulated special reports (see FAO, 1975) and led to the production of two volumes with identification keys for the aquatic flora and fauna of the Sahelo–Sudanian region (Durand & Levêque, 1981). In neighbouring rivers, the fishes of the Volta and Bandama and other rivers in the Ivory Coast have been much studied, again stimulated by the creation of man-made lakes.

Tributaries of the Upper Niger probably formed part of the Senegal system in Pliocene and early Pleistocene times, which explains the similarity in fauna of the Niger and Senegal. The Upper Niger, originating in hilly country, has a rapidly flowing section then spreads out over the immense Middle Niger floodplain, at high water contributing to a sheet of water up to 30 000 km^2 by October in wet years, shrinking to a 6 m wide river at low water (May), leaving shallow lakes, pools and swamps isolated on the floodplain (18 major lakes with a combined area of 2400 km^2 at low water (Welcomme, 1979). Below this central inland delta the Niger flows between well-defined banks and receives several large tributaries, such as the Sokoto from the northeast, entering the Niger *ca* 1200 km downriver from Timbouctou, and the Benue entering another *ca* 600 km downstream. Between these two tributaries the hydroelectric barrage at Kainji, closed in 1968, led to the formation of a 1280 km^2 lake, lying from 9° 50′ to 10° 57′N in northern Nigeria. Tributaries of the upper Benue are in close contact with rivers flowing into Lake Chad. In Late Quaternary times a lake here, Megachad, may have been five times the size of the present Lake Victoria and, at some time probably between 12 000 and 5000 years ago, channels linked the Chad system with the Nile (see Beadle, 1981). It is therefore not surprising to find so many fish species distributed right across Africa from the Atlantic to the Nile, forming this vast sudanian ichthyological province.

Most of these savanna-river fishes make longitudinal migrations up- and downriver for long distances and lateral migrations out onto the floodplain at high water, which helps to explain their wide distribution. Ecological differences are being revealed between the sudanian and smaller guinean species in the forested streams; for example, the sudanian species *Alestes nurse* has an annual spawning season in the floods, producing far more eggs than the comparable guinean species *Alestes imberi*, which produces small batches of eggs throughout the year (Paugy, 1979–80 *a*,*b*).

Biotopes of the sudanian area

The main biotopes of this area were described by Daget (1954, 1957) and Blache *et al.* (1964): large rivers with rocky, sandy or mud-bottomed stretches; temporary streams down which water floods and subsides annually; plains inundated at high water; residual pools, some permanent, many temporary; and lakes and swamps. In the Middle Niger 'delta' the latitude is high enough for the water to cool off during winter months, which affects fish growth; in January there is a minimum water surface temperature of 20°C compared with 29°C in June. The wet season is also the warm season, so eggs and young fish spawned as the water rises develop when temperatures are high, which speeds their development. Chemically the Niger water (pH 6–7) is not rich in nutrients and through-out the area the inundation zones are the main feeding grounds for the fish. These also provide the right conditions for species which make nests among water plants (such as the osteoglossid *Heterotis* and the mormyroid *Gymnarchus*). From this central delta, 40 000 tonnes of fresh fish were taken each year before the Sahelian drought. Traps in the Markala barrage above the central delta produced much information about fish movements (Daget, 1957).

Residual pools, depressions in the floodplains, have clay or mud bottoms. Turbidity varies greatly from pool to pool, fish such as *Clarias* stirring up the bottom water. Many pools lack vegetation, others have water lilies (*Nymphaea*), bladderwort (*Utricularia*) and many other kinds of widely distributed aquatic plants (see Durand & Levêque, 1981). Fish faunas vary greatly from pool to pool. Pools tend to keep the same types of fish from year to year, but conditions change as pools silt up, and new pools are formed as rivers change their courses. As the water level falls, fish populations from a wide area become concentrated in these pools and spectacular catches can be made from them. In the Lake Chad basin, Blache *et al.* (1964) quote 8 tonnes from a 0.2 ha (1 ha = 0.01 km^2) pool. Such catches are not due to production in the area of the pool (as they would be in a fish pond) but are the result of production from an unknown area draining into the pool. In Dahomey, special trenches (*whedos*) are dug across the floodplain to retain water and fishes through the dry season, a technique which could well be introduced to other floodplains (Welcomme, 1979).

Channels open to the sun merely provide seasonal passageways to and from the floodplain, but the streams shaded by gallery forest have a specialized fauna of small guinean species (*Barbus*, small *Alestes (Brycinus)* and mormyrids, cyprinodonts and *Amphilius* catfish). Leaves falling into the water make it rather acid but provide food for many higher crustaceans (*Caridina* and *Palaemon* prawns and *Potamon* crabs); molluscs are rare in the acid waters. The large lakes associated with the floodplain receive their water and fishes from the rivers in the flood season. The vegetation often includes dense plant stands (*Echinochloa, Oryza, Vossia, Paspalidium*). Fishes are very abundant in these lakes, restocked from rivers and acting as huge fish traps, but they have as yet been little studied. The lakes are shallow with swampy margins, populated with riverine fishes and, unlike the Great Lakes of eastern Africa, with few endemic species.

In these vast West African river systems, with their complex biotopes, faunal diversity appears to reflect the available biotopes, rather than increasing from source to mouth of river. The longitudinal zonation of fish species has however been examined in the smaller Bandama River in the Ivory Coast (Merona, 1981) and in the Ogun River in Nigeria (Sydenham, 1977).

Composition of the fauna

The Niger fauna, like that of most rivers, is dominated (60+%) by ostariophysan fishes. These include about 40 cyprinid, over 30 characoid, and about 47 catfish species. The cyprinids are mostly small *Barbus*, omnivorous little fishes living in schools, but two kinds of mud-sucking *Labeo, L. senegalensis* in sandy places and *L. coubie* in more openwater and rockier habitats, are important food fishes. The characoids include the deep-bodied mud-feeding *Citharinus* and the macrophyte-feeding *Distichodus* and among the characids are several large *Alestes*, such as the omnivorous *A. baremose* and *A. dentex*, and the highly piscivorous *Hydrocynus* species: all these are important in the fisheries. The catfishes also include many large species: *Bagrus, Chrysichthys, Clarias, Heterobranchus, Clarotes, Eutropius* and *Schilbe*, about 20 *Synodontis* species, *Arius gigas*, and *Malapterurus electricus*. Next to the ostariophysans, the endemic family Mormyridae is best represented. Mormyrids all have electric organs. The cichlids, so important in the Great Lakes, are poorly represented in the Niger (*ca* 13 species) as in most riverine faunas. Nor are there any endemic species of cichlid in Lake Chad; in this the Chad fauna resembles those of Lakes Albert and Turkana in East Africa which also have sudanian (Nilotic) faunas and a paucity of cichlids compared with the other African Great Lakes (see Table 2.2). Although Lake Chad has been a much larger lake at times in the past, drying out in arid periods probably counteracted evolution of endemic cichlids, and the presence of a fully differentiated riverine fauna, with numerous large piscivores, probably had an even greater role in this.

Two other important food fishes in the Niger include the detritus-feeding osteoglossid *Heterotis niloticus*, which lives in shallow water and does well in ponds, and the large piscivorous centropomid *Lates niloticus* (Nile or Niger Perch) at the top of the food webs. Unlike fisheries in the African Great Lakes which take mainly cichlids, these sudanian river fisheries take over 50 types of fish, mainly non-cichlids, in commercial catches.

The riverine faunas vary with type of bottom and current speed. Rocky stretches with rapids and falls occur mainly in the upper reaches. In the Upper Niger these harbour a specialized fauna, species such as the cyprinid *Garra waterloti* with its adhesive buccal disc, and the elongated *Gobiocichla wonderi*, the only endemic cichlid in the Niger system. Moving sands are rather barren, though frequented by small cyprinids (*Barbus* and '*Barilius*' *Leptocypris*). Where the sand is fine and hard it is colonized in the wet season by filamentous algae and a rich fauna of aquatic insects (Odonata, Hemiptera, Coleoptera). Such places are frequented by *Alestes, Labeo, Barbus*, tilapias and predatory *Hydrocynus*. Where the bottom is muddy with rooted aquatic plants in the shallows, many mormyrid and *Synodontis* species, *Heterotis*, large catfishes such as *Auchenoglanis* and *Heterobranchus*, and tilapias live.

A unique opportunity to census the fishes in a stretch of the River Niger occurred during the construction of the Kainji Dam (Motwani & Kanwai, 1970). Half the river, here divided by an island, was enclosed between two coffer dams in mid-May 1966 and the water pumped out in July. The enclosed area, a 1.7 km long, 18 ha lake with steep rocky walls over a rocky bottom with coarse sand, lacking vegetation, produced over 80 species of fish (Table 2.4). Fifty of these species had mature or ripe gonads; from the fry collected it was clear that many of them had spawned there. The channel appeared an inhospitable home for fish; it was probably used as a low-water refuge, and some fishes may have been moving up- or downriver when trapped between the coffer dams. The water level here normally starts to rise in August, but local rains start in April/May and are at a maximum in July.

In the Sokoto tributary, tagging experiments on 14 of the 27 species common in residual pools – pools of 0.07–4.7 ha cut off from the river for about 4 months each year – showed that the same individuals of *Hydrocynus vittatus* and *Lates niloticus* occupied a particular pool year after year. Some species (*Distichodus rostratus, Channa obscura, Alestes leuciscus, Hepsetus odoe*) were only caught in the pools in certain years; the cichlids and *Schilbe mystus* did not appear to be migratory (Holden, 1963). Sandy- or sandy/mud-bottomed pools carried higher biomasses of fish than mud-bottomed pools (see Table 14.2); the latter appeared to be nursery pools with few large fish. There was a direct relation between fish

Table 2.4. *Fishes caught from the coffer-dammed channel of the River Niger at Kainji (data summarized from Motwani & Kanwai, 1970, who recorded a total of 82 species representing 18 families)*

Family	Total no. species	Species mature or ripe	Individuals No.	%	Weight kg	%
Mormyridae	19	13	1198	20.7	219	19.5
Characidae	8	6	2103	36.3	136	12.1
Citharinidae and Distichodontidae	5	5	354	6.1	212	18.9
Cyprinidae	6	2	192	3.3	48	4.3
Schilbeidae	3	2	463	8.0	41	3.6
Bagridae	7	7	422	7.2	204	18.2
Mochokidae	18	11	1064	18.0	209	18.7
[a] Representatives of other families	16	present	–	–	–	–

[a] Including species of *Polypterus, Heterotis, Gymnarchus, Microthrissa, Malapterurus, Lates, Oreochromis, Tilapia, Tetraodon, Hemichromis, Pelmatochromis, Lutjanus.*

size and pool size. Biomasses varied considerably from year to year; good year classes appeared to be due to an early rise in river level. The mean biomass was 415 kg ha^{-1} for the dry season area of the pools but, as these occupied only 3–4% of the floodplain, this crop was estimated to represent only about 12–17 kg ha^{-1} from the floodplain as a whole.

In the Lower Niger down to the estuary and in the Benue tributary, *atalla* liftnets are used from canoes along the river during the six-month highwater season. Catches include 38 species of 13 families, small fishes dominated by clupeids, schilbeids and characids, mostly small species, with fry and fingerlings of large species (*Alestes, Hydrocynus, Citharinus*, bagrids, mormyrids) making up less than 20% of the catch.

In the Niger estuary, an estimated 8800 km^2, 80% of it mangrove swamps, salinities are very variable; freshwater reaches almost to the river mouths in heavy floods. Despite the low salinities, marine species predominate in catches throughout the year (Pillay, 1967). Pillay listed 128 fish species from these estuarine stretches. Castnets catch the West African shad (*Ethmalosa dorsalis*), other clupeids and grey mullets (*Mugil* spp). Set nets and long lines catch barracuda (*Sphyraena*), snappers (*Lutjanus*), croakers (*Otolithus, Sciaena*) and grunts (*Pomadasys*). Many types of trap take prawns and small fishes, also tilapias that can withstand brackish water and catfishes. These estuarine regions present great possibilities for fish culture.

The trophic groups of the main genera and species of the sudanian fish fauna have been studied most intensively in Lake Chad (see below). In the Niger system *Alestes* change their food seasonally, feeding on seeds and insects on the floodplain but subsisting on zooplankton in dry-season pools, losing weight and fat content while doing so (Daget, 1952). Within the genus *Citharinus*, all deep-bodied mud-feeding characoids, Daget (1962) found that three species coexist in the Chari–Logone and Benue (*C. citharus, C. latus, C. distichodoides*), and another three in the Zaire basin (*C. congicus, C. gibbosus, C. macrolepis*). In the first case the three species are known to share the same food, to have the same breeding seasons and comparable early growth rates, and the young live together; the three species in the Zaire basin are often caught together. Daget commented that these appear to provide exceptions to the competitive exclusion principle (Gause's law, that species with the same ecological requirements cannot coexist), but these only coexist in regions where the biotopes suitable for *Citharinus* are vast and permit ample migratory movement; smaller river systems each have only one species (*C. citharus* in Gambia, *C. latus* in southern Dahomey).

Lake Chad: a sudanian riverine lake

The ecology of sudanian fishes has been studied intensively in Lake Chad and rivers flowing into the lake. This work was carried out primarily by a team of scientists from the Office de la Recherche Scientifique et Technique Outre-Mer (ORSTOM) based at N'Djamena (Fort Lamy) south of the lake, as an International Biological Programme (IBP) project. Their results, published mainly in the *Cahiers d'Orstom* (now the *Revue d'Hydrobiologie Tropicale*) and listed in ORSTOM (1983), have been brought together in a monograph (Carmouze, Durand & Levêque, 1983). These and studies made on sudanian fishes in connection with the new man-made lakes have contributed greatly to our understanding of how riverine fish communities change as the fishes become adapted to lacustrine conditions. In the recent Sahelian droughts Lake Chad shrank and the later studies indicated how changing environmental conditions affected the fauna.

Lake Chad lies in very dry country, from 12°N to 14° 20'N, just south of the Sahara, an inland drainage area with large rivers flowing into the lake from the south and west. Historically Chad has oscillated greatly in size and during these studies the 20 000 km^2 lake of 1968 was reduced to 6000 km^2 in 1975. The lake has two basins, a southern one influenced greatly by the seasonal inflow of the large rivers of the Chari–Logone system draining wetter regions far to the south, and a northern more lacustrine one. In 1965 the open water was initially 4–7 m deep in the 4000 km northern basin and 2–4 m in the southern one, but during the 1965–72 study period the lake fell by over 2 m. The Sahelian drought began in 1968 and after the rivers failed to flow in 1973 the northern basin dried out completely, as it had done in 1907–8. Hit again by drought, the lake was reported to be almost dry in 1984. Lake Chad is far enough north for a winter fall in water temperature (to 18°C) to affect fish growth, particularly in the northern basin (Hopson, 1972). In the south, fish biology is governed mainly by the seasonal inflow of the great rivers. Daget regarded Chad as an extension of these river systems in which fluviatile species had decreased in importance while lacustrine species flourished; this appeared particularly true in the southern basin.

Before the drastic drought the lake had three main ecological zones: (1) open water; (2) the archipelago of islands, mainly along the east and southeast coasts; and (3) the south coast, with its deltas, which receives the inflowing Chari and Logone rivers. Many of the lake fishes migrated long distances (up to several hundred kilometres) up these rivers to spawn, the resulting juveniles feeding in the inundation zones before returning to the lake down the El Beid and other channels. Evaporation from the lake was very high and conductivity increased towards the north, from where this appeared to limit certain elements of the fauna, as oligochaetes, benthic snails and mormyrid fishes were not found in the northern basin. The fish fauna had many more species in the southeastern archipelago than in the northern basin, a trend which became further exaggerated during the drought.

Trophic relationships of Chad fishes
    Studies of the diets of the most abundant fishes in the southeastern archipelago and in the open waters of the southern basin, comparing foods used at high- and low-water levels (Lauzanne, 1976), led to the conclusion that fish diets were much more varied in the archipelago than in open waters, which lacked many food sources such as prawns, aquatic insects, plants and many small species of fish (*Barbus, Haplochromis*). They also showed that seasonal variations in foods eaten were most marked in terminal predators, which at low water fed on small species of fish (e.g. *Barbus, Haplochromis, Alestes (Brycinus)*), but utilized the fry and young stages of the larger species at high water. Terrestrial insects blown onto the water surface were also an important food source at high water, as were prawns within the archipelago at low water.
    The Lake Chad fishes fell into four main trophic groups: (1) detritivores (*Sarotherodon galilaeus, Labeo senegalensis, Citharinus citharus*); (2) carnivores taking mainly zooplankton (*Alestes baremose, A. dentex, Synodontis membranaceous, S. batensoda*); (3) carnivores taking mainly benthic invertebrates (*Heterotis niloticus, Hyperopisus bebe, S. schall*); and (4) terminal carnivores taking mainly fishes, either true predators feeding on live fish (*Hydrocynus forskahlii, H. brevis, Lates niloticus*) or predators with a tendency to scavenge dead fish (*Bagrus bayad, Eutropius niloticus, Schilbe mystus*).
    These Chad fish represented three trophic levels. In the archipelago, primary consumers included the detritivorous *Citharinus citharus* and *Labeo senegalensis*, the phytoplanktivorous tilapia *Sarotherodon galilaeus* and *Alestes macrophthalmus* (which also ate some macrophytes and terrestrial insects); in open water, primary consumers were all detritivores: *C. citharus* and *C. distichodoides, L. senegalensis* and *L. coubie, Distichodus rostratus*. In the archipelago, secondary consumers *Alestes baremose* and *Hemisynodontis membranaceous* were strict zooplanktivores, while *Brachysynodontis batensoda* fed mainly on zooplankton and *A. dentex* took it with other food items; the benthic-

invertebrate-feeders included *Synodontis schall*, the mormyrid *Hyperopisus bebe* (which also ate some macrophytes) and the osteoglossid *Heterotis niloticus*, which also took macrophytes, prawns and zooplankton. In the open water, *A. baremose, Hemisynodontis membranaceous* and *B. batensoda* were all strict zooplanktivores; the benthic invertebrates included many molluscs, fed on by *Hyperopisus bebe, Synodontis schall* and *S. clarias*. In the archipelago, where food webs were very complex, terminal consumers included the piscivorous *Lates niloticus*, feeding mainly on detritivorous fishes, and *Hydrocynus, H. brevis* feeding mainly on zooplanktivores and some tilapia (*Sarotherodon galilaeus*), while *H. forskahlii* fed on prawns and fish of many kinds; the less strictly piscivorous *Bagrus bayad, Schilbe uranoscopus* and *Eutropius niloticus* all included prawns in their diet, and *E. niloticus* took many terrestrial insects. In open water, where food chains were simpler, zooplanktivorous fishes were the main source of food, and terrestrial insects were important in the highwater season; few benthic detritivores were eaten. *E. niloticus* was itself preyed on by larger piscivores (Fig. 2.3).

Fig. 2.3. Food webs of top consumers in the open waters of Lake Chad, indicating the percentage composition of their diets (after Lauzanne, 1976). H. br = *Hydrocynus brevis*; H. fo = *H. forskahlii*; L. ni = *Lates niloticus*; B. ba = *Bagrus bayad*; S. ur = *Schilbe uranoscopus*; E. ni = *Eutropius niloticus*.

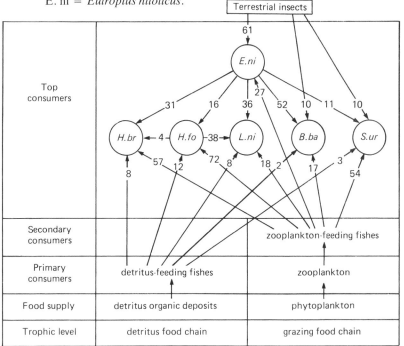

The quantitative contributions of the various food items to the diets of the principal fishes at the different trophic levels were determined by Lauzanne (1976). During IBP studies on production, attempts were made to quantify the transfer of energy from one trophic level to the next in various environments. These Lake Chad studies showed some of the difficulties in trying to do so. Trophic interrelationships proved to be very complex. Some fishes took foods from their own trophic level (for example *Lates* feeding on *Hydrocynus* and *Eutropius*), or from several levels. Calculations indicated that starting with 100 calories provided by the algae, the top piscivore *Lates niloticus* would accumulate 5.2 cal when feeding on the algivorous tilapia (*Sarotherodon galilaeus*), but only 1.7 cal if it cropped zooplanktivorous *Alestes baremose* (Levêque, 1979). Moreover, Hopson's (1972) study of *Lates* in Lake Chad indicated how much the food eaten varies with size of fish, biotope and time of year. Juveniles of many fish species start as zooplankton-feeders before turning to other foods, be these algae, benthos or fish.

The quantitative studies did, however, bring out that the main weight of fishes (44% of the ichthyomass) cropped from the archipelago waters was of zooplanktivores, kept going by the high and stable plankton biomass present throughout the year. In the open water, top consumers formed the main weight (64% of the ichthyomass) and these were supported by the great biomass of terrestrial insects blown onto the water surface and the abundance of small zooplankton-feeding fishes (such as *Micralestes acutidens* and *Pollymyrus isidori*) whose production is high.

The shallow depth and well-oxygenated water made Chad comparable to a vast littoral zone. Macrophytes were very abundant, and appeared to be underutilized by the fish. The benthic fauna dominated by molluscs, averaging 37 kg ha^{-1} (dry weight), was one of the richest known (compare with 7.4 kg ha^{-1} from the equatorial Lake George in Uganda, another IBP study site). Despite this rich fauna, the biomass of benthos-feeding fishes was not very high, nor were these much used by top consumers. Molluscs only appeared to be essential food in open water in the southeast where five fish species (*Hyperopisus bebe, Synodontis schall, S. frontosus, S. clarias* and the little pufferfish *Tetraodon fahaka*) fed on them. Although aquatic insects were found to be less important as food in the open water than nearer to the coast, they were, however, more important foods for the little mormyrids in the open water than in the southeastern archipelago. Here they were found in stomachs of numerous species but never formed the exclusive food of any one species. In Chad the mean zooplankton biomass, 8.1 kg ha^{-1} dry weight, was very close to that of

Lake George (8.3 kg ha^{-1} dry weight); this varied throughout the year. It was not possible to determine the mean biomass of fish in Chad but the fisheries yield was estimated to be 70 kg ha^{-1} yr^{-1} (fresh weight).

The numerous autecological studies in Chad brought out many interesting points. *Lates niloticus* is well adapted for lacustrine existence as it spawns in the lake and has pelagic eggs and larvae (Hopson, 1972). Winter checks in growth enabled age and growth rates to be determined; these showed that growth speeds up when *Lates* are about 75 cm standard length (SL) (6–7 years old), the size at which they turn to a diet of larger fishes (such as *Hydrocynus*). They grow to 140 cm SL. Juvenile *Lates* included prawns and other invertebrates in their diet. In Chad, *Lates* males matured at about 50 cm long, females at 60 cm (3–4 yr old), but females survived longer (to 15 yr compared with 12 yr for males). Durand's (1978) studies of the population dynamics of *Alestes baremose* supported earlier suggestions that this lake had distinct populations of this species. In *Citharinus citharus*, first-year growth was found to be very variable in the Chari–Logone; growth stopped November–March when the water temperature fell below 25°C (Benech, 1974). Effects of the drought on growth were shown in *Brachysynodontis batensoda* (Benech, 1975) (as also on *Hydrocynus* populations in the Middle Niger (Dansoko, Breman & Daget, 1976; Daget in FAO, 1975)).

During the drying-up period (1970–77), the small mormyrids *Petrocephalus bane* and *Pollimyrus isidori* disappeared from the southeast archipelago of Chad, which turned into a swamp (Lek & Lek, 1978). They were replaced by *Brienomyrus niger*, but *Petrocephalus bane* and *Pollimyrus isidori* became abundant again when lacustrine conditions returned. The 18 species of mormyrid known in the Chad basin were widely distributed and some were present in most biotopes. Lek & Lek (1977) also looked at the biology of the small characid *Micralestes acutidens*, a surface-dweller in open water where it feeds on Cladocera and terrestrial insects and is itself an important prey of *Lates* and *Hydrocynus*. These *Micralestes* spawned with the Chari floods when about 5 months old; they only lived for about a year and died after spawning, when 30–40 cm long.

In addition to the species studied intensively many other fish species were present. Zooplankton was also utilized by the endemic little *Brycinus* (*Alestes*) *dageti*, and the mormyrid *Pollimyrus isidori* ingested some as well as insects. In the open water, *Hydrocynus forskahlii's* main prey were *Micralestes* and *Pollimyrus*. Phytoplankton was used mainly by juveniles of many species and was also taken from the bottom by the numerous detritus-feeders. Oligochaete worms were scarcely cropped.

The ORSTOM team concluded that in Chad the search for food did not seem to be a determining factor in the distribution of the fishes; there was a great variety of available food, and the feeding regimes of most of the predatory species showed great plasticity. But where there were large concentrations of predatory fishes, these might have limited the density of prey organisms, such as certain benthic invertebrates.

3

Man-made lakes

Lake Chad provided an example of a natural lake in which many of the fish were under the influence of the inflowing rivers, and there was little distinction between riverine and lacustrine fish communities. The creation of a man-made lake behind a hydroelectric or irrigation barrage provides a large-scale experiment on how riverine species become adapted for lacustrine life, how faunas change and how lacustrine communities evolve. In the sudanian area, pre- and postimpoundment studies were made at Kainji, in Nigeria, where a hydroelectric barrage across the Niger, closed in 1968, led to the formation of a 1280 km^2 lake with an outflow:storage ratio of 4:1 and a large (10 cm) annual drawdown. Prior to this, the closure of the Volta Dam in Ghana in 1964 had created an 8800 km^2 lake, one of the world's largest man-made lakes; here the storage volume was four times that of the outflow and the drawdown only 3 m annually, providing more truly lacustrine conditions. Both lakes were populated with the same species of sudanian fishes from their inflowing rivers, the Niger and Volta, respectively. Lake Nasser/Nubia on the Nile, which formed behind the Aswan Dam closed in 1964, also has many of the same species of fish (Ali, 1984; Latif, 1984), as does Lake Kossou on the Bandama River in the Ivory Coast, formed more recently (Daget, Planquette & Planquette, 1973).

Within the tropics many other large lakes created in recent years have stimulated a great deal of ecological research and discussion (see symposia edited by Obeng, 1968, and Ackermann, White & Worthington, 1973). Information on Lake Kariba on the Zambezi River, one of the first new lakes to be studied in detail, was collated by Balon & Coche (1974), Bailey *et al.* (1978) studied changes in a Tanzanian lake, and several of the new African lakes were discussed by Beadle (1981). In the Neotropics, studies of new lakes include those of Brokopondo Lake in Surinam (Leentvaar in Ackermann *et al.*, 1973), of the numerous new lakes in

Brazil (Paiva, 1982; Ferreira, 1984; Beeckman & de Bont, 1985; Merona, 1985; Northcote, Arcifa & Froehlich, 1985); and of Gatun Lake in the Panama Canal (Zaret & Paine, 1973; Zaret, 1984a). Data from the many reservoirs and new lakes in the Asian tropics have been collated by Fernando & de Silva, 1984 (see p. 169). In Papua New Guinea the Purari River hydroelectric scheme (Petr, 1983) stimulated fish studies (Haines, 1983). Petr (1984) has listed species stocked in Indo-Pacific island lakes and reservoirs. In Hong Kong, where an arm of the sea was impounded to make a freshwater reservoir (Plover Cove), the fish population changes were monitored by Man & Hodgkiss (1977a,b).

In temperate regions the general sequence of events as a lake forms is well known. The new lake will be colonized by riverine fishes, unless steps are taken to prevent this and to stock with other fish. Some of these species will adapt themselves to the new conditions, others will drop out of the fauna. New lakes therefore tend to have less complex faunas than their river systems. But after the new lake has stabilized, species from elsewhere in the river system may gain access to it, and the numbers of lacustrine species will then increase. As the lake level rises, it floods the surrounding land and the rotting vegetation liberates abundant nutrients. This leads to an explosive development of fish food. The fishes that can take advantage of the abundant new foods offered at this time, and those which can spawn under the new conditions, will become very abundant. As these nutrients are used up, a 'trophic depression phase' sets in, the lake becoming much less productive. Eventually, as organic matter accumulates, the lake stabilizes at a new level, with fish production generally slightly raised again though not as high as in the initial stages. Fish spawning grounds may be reduced in area by fluctuations in lake level when the water is used for hydroelectric power. In the USSR it has been shown that lakes stabilize more rapidly at the higher temperatures in lower latitudes (6–10 yr south of 55°N, compared with 25–30 yr north of this). At tropical temperatures lakes may be expected to stabilize sooner. Lake Kariba, on the Zambezi, the first large man-made lake in the tropics to be studied in detail, stabilized about 10 yr after dam closure in 1958 (Balon, 1974).

A new lake offers expanded feeding grounds, but often with different types of food from those available in the river. Many riverine fishes feed on algae and benthic invertebrates from rocky and stony bottoms; as the new lake forms, these bottoms often lie in deoxygenated water so the fishes have to seek other foods.

Adaptations to spawn under the new conditions may be much more difficult. Many species will continue to attempt to reach affluent streams in which to spawn, and some of these may be lost to the fauna (as happened in Volta Lake). Finding their way to such streams through the still waters of the lake may present problems to the fish. Many species (characoids, cyprinids, catfishes) will continue to migrate up affluent rivers or streams to spawn and drop back to the lake to feed. On the other hand, populations of fishes able to spawn in still water amongst flooded vegetation may build up very rapidly as the new lake forms. In this way cichlids are eminently preadapted to take advantage of the new lacustrine conditions.

On an evolutionary scale, Corbet (1961) recognized four hypothetical stages in the colonization of lakes from rivers: (1) fish feeding and breeding only in rivers; (2) feeding in lakes and breeding in rivers; (3) feeding and breeding in both lakes and rivers; and (4) feeding and breeding only in lakes. He postulated that wave-washed rocky shores provide the nearest conditions in lakes to riverine spawning places, allowing riverine fishes to colonize lakes by this route, for eggs and young are least well adapted to new conditions. Man-made lake studies allow us to see these changes in action; for example, in Volta Lake *Leptotilapia irvinei*, previously known only from the Volta rapids, is now found in rocky places in the new lake.

Volta Lake

Volta Lake, in Ghana, lies from 6°N to 9°N, with a rainy season in May followed by floods until October. Little riverine forest was cleared and rapid deoxygenation led to fish deaths as the lake filled. In the deeper water there was virtually no oxygen. As the lake filled, which it did very gradually until about 1971, there was more extensive mixing and more oxygen available in deeper water, and the duration of deoxygenation when the lake was stratified in the dry season diminished. The Volta River flows in at the northern end and conditions there remained more riverine than in the south where the lake is broader.

The Volta Lake is low in nutrients, with a low standing crop of phytoplankton. Aufwuchs (algae with contained microorganisms) on the drowned trees formed the main basis of the food webs, supporting surprisingly high catches of fishes in the early years after impoundment. The fishes comprised potamodromous species (*Alestes, Citharinus, Distichodus, Labeo*), especially at the north end of the lake from where they moved up inflowing rivers in the wet season, and species which spawned within the basin, either annually (*Lates niloticus*) or throughout the year (tilapias). The potamodromous species relied heavily on the floodplains as feeding areas during the rainy season. Rotenone Sampling of 150 m^2 areas of the littoral zone between September 1969 and March 1971 produced an ichthyomass of about 155 kg ha^{-1} (unpublished studies by Ryder & Loiselle).

Apart from a fish kill just after dam closure, most of the Volta River fish thrived initially in the new lacustrine conditions (Petr, 1968, 1974). Exceptions were the mormyrids, which almost completely disappeared from the south end of the lake. The predominance of insectivorous fishes in the Volta River gave way to predominantly herbivorous and plankton-feeding fish in the lake. The mormyrids which are bottom-feeders may have vanished because their benthic insect food supply was submerged in deoxygenated water, but later work at Kainji (p. 70) questions this. After the first 2 yr, characid and cyprinid fishes (*Alestes* except for one species and *Labeo*) also disappeared from the south. These may have migrated up rivers to spawn and failed to return, or died out if they did not find suitable spawning grounds. Species common in Black Volta River samples became limited to the northern arm of the lake where the Volta Rivers enter the lake. Tilapias, on the other hand, flourished and fairly rapidly became the dominant fishes in most of the lake. Of the three species here *Sarotherodon galilaeus* was found to feed mainly on phyto-plankton, *Oreochromis niloticus* on periphytic algae, and *Tilapia zillii* on submerged grass and detritus.

The bark of the flooded trees developed a rich growth of periphyton, and the burrowing nymphs of the ephemeropteran *Povilla adusta* became exceedingly abundant in the dead trees. These provided a source of food for species which had been insectivorous bottom-feeders in the rivers, such as the catfishes *Schilbe mystus* and *Eutropius niloticus* – a good example of generalized feeders in the river taking to a more specialized source of food in the new environment. However, these same species also fed on the freshwater clupeid *Pellonula afzeliusi* which had become very abundant in the open waters of the lake where it formed the major food source of both the *Lates niloticus* and the much smaller common preda-tory cichlid *Hemichromis fasciatus*. In this new lake the flooded trees, with their flora of periphyton and fauna of *Povilla*, were a major element in the unexpectedly high production of fish. Volta Lake produced about ten times the weight of fish produced by Lake Kariba on the Zambezi in early years after impoundment.

Ecological studies of the indigenous clupeids in Volta Lake showed how pelagic fishes adapt to life in a new tropical lake (Reynolds, 1970,

1974). The pelagic zone is much enlarged as a lake forms, offering opportunities for plankton-feeders. Reynolds (1974) found that Volta Lake was colonized by five small pelagic species, all fractional spawners, producing batches of eggs in succession. Of these, two clupeid species and a cyprinid were dry-season spawners in the river and the two schilbeid catfishes were wet-season spawners. One clupeid, *Pellonula afzeliusi*, extended its breeding season to spawn throughout the year in the lake, where high water no longer presented a danger to its planktonic larvae. (*Pellonula* ova are relatively large and may be anchored to substrata but the larvae are pelagic.) The other, larger clupeid, *Cynothrissa mento* (growing to 170 mm TL), probably migrated up rivers to spawn; this species remained in the more riverine areas in the north of the lake. The wet-season-spawning schilbeids both extended their breeding seasons, throughout the year in *Physailia pellucida* and into the early dry season in *Siluranodon auritus*; for these species, lacustrine conditions provided a continued highwater season. The species which extended their breeding seasons rapidly became abundant in the lake.

Both Volta clupeid species made diel vertical migrations to and from the surface waters, appearing in small compact shoals at the surface at dusk. As the light failed, individuals dispersed in the surface waters, where they fed actively on emerging insects. They remained close to the surface in bright moonlight and were attracted to a light source. These feeding associations often contained other species, notably the transparent schilbeid catfish *Physailia pellucida* (up to 40% of the feeding shoal) and *Barbus macrops* (up to 10%), also the cyprinid *Leptocypris* (*'Barilius'*) *niloticus* and the three predators *Eutropius niloticus*, *Schilbe mystus* and *Cynothrissa mento*. In inshore waters two other predators, *Hemichromis fasciatus* and *Hepsetus odoe*, joined them. The clupeids re-formed their shoals at dawn and vanished from the surface. Thus these clupeids behaved very much as do those in Lake Tanganyika (p. 90). *Pellonula* is, however, essentially a riverine fish (present in West African rivers from Senegal to Niger) which can inhabit both fast-flowing and calm, muddy habitats; it is not known whether it carries out vertical migrations in rivers when conditions permit, and whether this is a legacy from a marine environment.

Pellonula is a facultative feeder taking mainly aquatic and terrestrial insects and small Crustacea (Ostracoda, Cladocera, Copepoda). *Pellonula* is in turn preyed on by a variety of fish – *Clarias*, *Hepsetus*, *Hydrocynus*, *Lates*, *Schilbe*, *Eutropius*, *Alestes*, *Synodontis*. *Cynothrissa* takes young stages of many fish species, including its own young. Both clupeids mature within one year. Reynolds commented that for these largely annual forms predation probably outweighs all other factors in determining population size each year, but the flexible and extended breeding season in *Pellonula* and *Physalia* is important in allowing rapid population growth and effective exploitation of the changed environment.

Lake Kainji

Preimpoundment studies in the River Niger at Kainji, Nigeria, included the census of fishes in the stretch of river enclosed by coffer dams, described above (p. 55, Table 2.4), and studies listed by Lelek & El-Zarka (in Ackermann *et al.*, 1973); many of the numerous postimpoundment studies were listed in Lelek (1973), Lewis (1974*a*,*b*), Blake (1977*a*) and Ita (1978).

About 90 species of fish representing 24 families occur in Lake Kainji, the majority of them the same species as in Volta Lake. Immediately after impoundment, catches rose far above the estimated 10 000 tonnes yr^{-1} to a maximum of 28 000 tonnes in 1970 and 1971, but the boom was short lived. Commercial catches declined rapidly, stabilizing by 1974–75. The 4500 tonnes in 1978 represented a catch of 35 $kg\,ha^{-1}$, very similar to that from the floodplain.

Studies of young fishes after impoundment showed that there were two spawning patterns among the lake fish: one an extended season (as in clupeids and tilapias), the other a restricted season with the fish spawning as the water level rose. Products of this restricted season predominated in catches in the first year after impoundment; young of these species were able to feed on detritus, not too affected by increased turbidity, and not so dependent on the formation of a littoral zone as are young tilapias.

Comparison of the foods eaten in different areas of the lake showed that the greatest weight of fish from the northern (lotic) section and channels over 15 m deep were primarily zooplanktivores, whereas in the shallow areas, both cleared and bush-covered, detritivores predominated, with some herbivorous and periphyton-feeding species. Predators were present in all zones.

The fish fauna started to change as soon as the barrage was closed in 1968. Mormyrid fishes, very abundant in the river, became scarce in the lake (as in Volta) and some species vanished. Cichlid populations had been expected to increase but they did not become important commercial species at first (as tilapia did in Kariba and Volta). The delay in build-up of their populations in Kainji Lake was perhaps related to the inhibiting effect of water level fluctuations in the littoral zone where they make their nests. *Sarotherodon galilaeus* populations did, however, build up in the flooded bush, where they were not fished, and by 1972 they were an important element of the fauna, caught in castnets though not common in experimental gillnets. In the first year *Citharinus* became the most important commercial fish, probably because of successful spawning as the lake filled. In the open waters there was a boom in small clupeids and schilbeid catfishes (as in Volta Lake), especially *Eutropius niloticus*. The clupeids were later caught in *atalla* liftnets and by a midwater trawl dragged between two boats, catches averaging *ca* 125 kg hr^{-1}. There was also a boom in predatory *Hydrocynus* and *Lates*. In Lake Kainji, *Hydrocynus forskahlii* fed on clupeids, and *H. brevis* fed on tilapia and by biting pieces from larger fishes such as *Citharinus* (Lewis, 1974*a*). *Bagrus docmac* took mainly bottom-living fish, particularly *Chrysichthys auratus* (Blake, 1977*a*). The two species of *Chrysichthys*, both omnivores, lived in different habitats: *C. auratus* in inshore waters, feeding on insect larvae, crustacea, fish and plants, and *C. nigrodigitatus* on the bottom in deeper water, taking bivalves, aquatic insect stages and detritus (Ita, 1978).

In the river, preimpoundment gillnet catches were mainly of mormyrids (14 species making up 40% of the total fish caught) and mochokid catfishes (11 species, 25% of total numbers), though characids predominated numerically in the rocky channel of the main riverbed (see Table 2.4). In the lake, in gillnet catches there was an immediate drop in percentage of mormyrids (to 1%) and a dramatic increase in *Citharinus citharus*. Catches of *Lates* also increased, following the increase in small clupeid prey, fish too small to be taken in gillnets. Schilbeid catfishes also increased and remained important in catches, as in Volta Lake (Olatunde, 1977).

By 1976 tilapia had become the main component in inshore waters less than 7 m deep, with a standing crop of 104 kg ha^{-1} (Ita, 1978). *Sarotherodon galilaeus*, *Oreochromis niloticus* and *Tilapia zillii* were caught in a ratio of 16:5:1. Preimpoundment castnet catches over sandbanks in the open river had taken mainly *S. galilaeus*. In this lake *S. galilaeus* was an algae- and detritus-feeder, *O. niloticus* took some worms and insects along with algae, and *T. zillii* was a macrophyte-feeder.

The species composition of the mormyrid and mochokid families which dominated riverine catches underwent substantial changes under the new conditions. In the lake *Hippopotamyrus pictus* was the dominant mormyrid in openwater gillnets, its young stages caught inshore. The larger species *Mormyrops deliciosus*, *Mormyrus rume* and *M. macrophthalmus* were still present in small numbers, but the marsh-loving *Brienomyrus niger* and rock-loving *Campylomormyrus tamandua* (which feeds by probing into rock and other crevices) both vanished. Of the species which feed mainly on insect larvae and nymphs, *H. pictus* was found to rely heavily on *Povilla adusta* nymphs for much of the year, *Mormyrus macrophthalmus* utilized small ephemeropteran nymphs other than *Povilla*, and *Marcusenius* took more conchostracans and ostracods. *Povilla* nymphs, which bore into wood or reed stems, emerge from their crevices to feed at night, when they are vulnerable to the mormyrids which feed nocturnally. Many mormyrid species utilized the same prey organisms (as Corbet (1961) had found amongst mormyrids in Lake Victoria, see p. 80). The paucity of benthos in Kainji Lake was reflected by the very limited range of prey organisms in stomachs. There was no evidence of lunar cycles of insect emergence which might affect food availability (as occurred in Lake Victoria. Blake (1977*b*) looked at feeding in the lake mormyrids whose populations had declined and concluded that the declines were unlikely to be the result of feeding limitations alone, even though the drawdown adversely affected the populations of benthic invertebrates, as submerged trees still offered extensive feeding grounds when the lake was at its lowest level. The need for riverine conditions for reproduction may have been an important factor in their decline (Blake, 1977*c*). We know very little as yet about where these mormyrids spawn; many of Lake Victoria's mormyrids migrated upriver (p. 97).

The species composition of the mochokid catfish fauna also changed. Willoughby's (1974) studies showed that these changes could be related to changes in available foods. There is as yet virtually no information on the breeding places and habits of these catfishes. Their distribution may also be affected by oxygenation of the water; Green (1977) found oxygen tolerances to vary greatly from species to species and that the blood haemoglobin content appears to be related to the species' normal habitat. The most tolerant species, *Synodontis schall*, has the widest distribution in Africa.

Mochokid catfishes fall into two main trophic groups: (1) pelagic feeders, taking zooplankton and insects from the water surface, feeding ventral side uppermost to do so and with reversed countershading (*Hemisynodontis membranaceous, Synodontis resupinatus*, with *Brachysynodontis batensoda* living like this part of the time); and (2) benthic-invertebrate-feeders, taking mainly juvenile insects and molluscs.

Among the 13 synodontid species caught in the coffer-dammed channel, *Synodontis gobroni* predominated (37%) with *S. gambiensis* (14% – an invertebrate-feeder greatly resembling *S. schall*), *S. budgetti* (12% – a robust species taking mud and humus with insect larvae), and the zooplankton-feeding *H. membranaceous* (11%). The castnets in the main river over sandy shallows took three species, all surface-feeders: *H. membranaceous, B. batensoda* and *S. resupinatus*. Gillnets along the river margins in channels and flooded areas took *S. gambiensis* (27%), *S. nigrita* (26% – a small species feeding on bottom deposits and the main species caught in traps set in grassy channels), *S. budgetti* (21%) and *S. clarias* (17% – a feeder on benthic insects and molluscs). The traps also took *S. ocellifer*, another small species and the most omnivorous one, eating plant material and insect larvae.

In the open lake the standard gillnet fleet took mainly *B. batensoda* (43%), *S. nigrita* (26%) and *S. schall* (9%); smaller nets inshore took juveniles of *S. nigrita, B. batensoda, S. schall* and *S. filamentosus* (Blake, 1977c). Of the 17 species in the lake studied by Willoughby, the three plankton-feeders had different habitat preferences: *H. membranaceous* in midlake, *S. resupinatus* close to shore where there was deep water and *B. batensoda* more evenly distributed at surface and bottom in deep water. *S. filamentosus*, which feeds mainly on chironomids, came from inshore. Rarer species preferred shallow and more sheltered biotopes. In the river below the dam, Sagua (1978) found *H. membranaceous* (32%), *S. gambiensis* (18%), *S. budgetti* (17%) and *S. violaceous* (15% – a species rare in preimpoundment samples from the river and not seen in the lake).

Thus, under lacustrine conditions the main increase was of the plankton-feeding *B. batensoda* and the bottom-feeding *S. schall* (instead of the very similar *S. gambiensis* commoner in the river). *S. filamentosus* and *S. nigrita* continued to be caught (mainly in inshore waters), while *S. violaceous* and *S. clarias* (a benthic-mollusc- and insect-feeder) were lost from samples.

Willoughby suggested that the lack of molluscs on the Kainji lake floor forced some synodontids into competition with the insectivorous synodontid species in the lake. He attributed the success of these catfishes to their stout dorsal and pectoral spines. When locked into erect position these prevent predators from swallowing even quite small individuals (as Hopson (1972) demonstrated for *Synodontis* from *Lates* stomachs in Lake Chad) (Fig. 12.2). It is interesting that the mormyrids, the other group with so many species in the sudanian fauna, also have a special device (the electrical system) which may help them to avoid predation, and that both these groups are nocturnal.

The Niger system below Kainji

The Kainji barrage greatly affected fishing in the river below the dam. The floodplains diminished and with them the fish stocks. The barrage provided an effective barrier to upstream movement but many kinds of fish congregated just below the dam; Sagua (1978) reported catches of *ca* 21 tonnes yr^{-1} here. *Lates* increased in numbers but catches were very seasonal, greatest when the outflow was highest; *Lates* appeared to do better in the river than in the lake. The proportions of predatory fishes generally rose; of the characids *Hydrocynus brevis* was common, but few *H. forskahlii* were caught. *Malapterurus electricus* became abundant, especially in April–July (Sagua, 1979). Of the eight bagrid species, *Bagrus bayad* and *Clarotes laticeps* superseded *Bagrus docmac* and *Chrysichthys auratus*, the commoner species in the coffer-dammed area. *Distichodus* were much commoner than *Citharinus citharus*, and of the cyprinids *Labeo coubie* compared with *L. senegalensis*. Five species of mormyrid made up less than 5% of the numbers of fish caught here, *Mormyrops deliciosus* and *Mormyrus rume* being the main species. Of the clariids, *Heterobranchus bidorsalis*, and of the cichlids, *S. galilaeus*, were the commonest species below the dam.

4

Lacustrine fish communities in the Great Lakes of eastern Africa

We have seen how West African riverine fish communities changed into lacustrine ones in the newly created man-made lakes, also how climatic changes during the 1970s droughts in the Sahelian zone affected fish populations in Lake Chad. How have fish populations responded to the much longer-lasting conditions in the Great Lakes?

Lakes in the Nile drainage with its Nilotic (sudanian) fishes fall into two groups: (1) those with little-modified faunas, showing a low level of endemicity and with few cichlids – Turkana, Albert and Tana; and (2) lakes with a high level of endemism – Victoria, Edward/George – lakes in which cichlids dominate the faunas, as they do in Kivu and Tanganyika in the Zaire drainage and Malawi in the Zambezi drainage (Table 2.2).

Lake Albert (Mobutu) in the western rift valley, a 150 km long, 35 km wide, 56 m deep lake, from which the White Nile flows, has a fish fauna most of which is shared with the Nile. The river has, however, a few species which have not colonized the lake and the lake has a few endemic species: about six endemic cichlid species, one cyprinid species and an offshore-living *Lates, L. macrophthalmus*, living over the deeper waters. The main fisheries in Lake Albert have been for tilapias, *Oreochromis niloticus* and *Sarotherodon galilaeus*, which live around the margins of the lake, together with *Oreochromis leucostictus* (not found in West Africa) and *Tilapia zillii*, and for *Citharinus citharus* in inshore waters, especially in the southern river deltas, *Alestes baremose* and *Lates niloticus*. The trophic relationships of Albert fishes were examined by a Belgian expedition (Verbeke, 1959) in comparison with those of Lake Edward, a 'cichlid' lake.

Lake Turkana

Lake Turkana, although no longer in contact with the Nile, evidently was so in the recent past and it has a very similar fish fauna to that of Lake Albert. A 7500 km² lake, up to 115 m deep, with an annual surface temperature range of 27.0–28.9°C, this lake has mainly soft deposits below 10 m deep, shores with rocky outcrops, sand, gravel and shingle beaches, sparse littoral vegetation except for beds of *Potamogeton pectoralis* down to 2 m along the more sheltered eastern coastline, and marsh in the delta of the Omo River, the main inflow which flows in at the north end of the lake. The fish fauna is basically Nilotic but the lake itself lacks mormyrids and has only one *Synodontis* species, *S. schall* – the species that Green (1977) found to have blood most adaptable to strong varying oxygen tensions. Both *Bagrus bayad* (common in the deeper water) and *B. docmac* (rare) occur, the latter in the rockier places. Endemic cichlids include three haplochromine species adapted for deep water, two pelagic *Alestes* (*A. minutus* and *A. ferox*) living in open water, a deepwater *Barbus* (*B. turkanae*), a pelagic cyprinid (*Neobola* ('*Engraulicypris*') *stellae*), and a dwarf offshore-living *Lates* (*L. longispinis*) comparable to the offshore *Lates* in Lake Albert. The endemic species show marked parallel evolution with Albert endemics in their adaptations to deepwater habitats (see Greenwood, 1976*a*). A few endemic subspecies among the non-cichlids reflect Lake Turkana's geographical isolation for a relatively short time. However, several Nile species present in Lake Albert (especially mormyrids) are very rare in Lake Turkana, probably due to the high ionic content of the water (*cf.* Lake Tanganyika).

The Lake Turkana Fisheries Survey (1972–75) provided much ecological information on the fishes (Bayley, 1977; Hopson, 1982). The lives of many of the fishes are governed by the inflowing rivers (as they are in Lake Chad). Of the 48 species, 36 occur in the lake and 12 are confined to the inflowing Omo River. The ten endemic species all live in the pelagic zone or deep water. A few species (*Neobola stellae, Barbus bynni, Synodontis schall, Lates niloticus*) are widespread in both inshore and offshore habitats but most species are restricted to a narrower range of conditions. Four fish communities were recognized: (1) **littoral**, from lake margin to about 4 m deep; (2) **inshore demersal**, from the littoral to some 10–15 m deep; (3) **offshore demersal**, extending out from some 8–20 m offshore according to the season and about 3–4 m above the lake bottom; and (4) **epipelagic**, subdivided into a superficial pelagic community above a midwater scattering layer and a deep pelagic community below this layer down to the demersal zone, extending down to some 60 m in the deepest part of the lake. The boundaries of the various communities shift seasonally, determined chiefly by the light climate in the lake; a decrease in light penetration in the flood season stimulates fish to move closer to the surface and inshore. At night, too, boundaries tend to break down as fishes move towards the surface and the shore. The pelagic fishes undergo extensive vertical migrations surfacewards at night.

In the **littoral community**, the tilapia *Oreochromis niloticus* and the catfish *Clarias lazera* occur throughout, whereas *Sarotherodon galilaeus*, *Alestes nurse*, *Micralestes acutidens* and *Chalaethiops bibie* live mainly over soft deposits, *Tilapia zillii* and *Leptocypris* (*'Barilius'*) *niloticus* live over rocky or stony bottoms, and *Haplochromis rudolfianus* and *Aplocheilichthys rudolfianus* are found among emergent or submerged macrophytes. In the inshore demersal zone, *Labeo horie, Citharinus citharus* and *Distichodus niloticus* are found over hard substrates, and in the few rocky areas *Bagrus docmac* is common. Offshore, in the deeper water, live *B. bayad*, the endemic *Haplochromis macconelli* (which has a hypertrophied laterosensory system, clearly an adaptation to life in deep water) and *Barbus turkanae*. In the pelagic zone the midwater scattering layer is composed of fishes which give a distinct echotrace in a band some 4 m thick, on average perhaps 10 m from the surface (ranging from 1 to 2 m deep in turbid water to 30 m deep in very clear water at the south end of the lake); the dominant fish here are two endemic *Alestes* – *A. minutus* and *A. ferox*. *A. minutus* forms the main biomass of fish in the lake and is an important link between zooplankton and the predatory fishes *Schilbe uranoscopus* and *Lates longispinus*. Above this scattering layer, the superficial pelagic layer from the fringe of the littoral to midlake is frequented by *Alestes baremose, Hydrocynus forskahlii*, postlarval *Neobola stellae* and juvenile lake prawns. Below the scattering layer, where larger fish are scarce, *N. stellae, Lates longispinus, Schilbe uranoscopus* and adult prawns are found.

The **midwater scattering layer** appears to be unique to this lake. The endemic species nearly all live in the pelagic zone or in relatively deep water. In the midwater scattering layer the endemic *Alestes minutus* (maximum size 3.7 cm fork length (FL)) feeds chiefly on zooplankton and occasional small insects, the larger *A. ferox* (maturing at 6–8 cm FL) chiefly on prawns with some zooplankton, insects and small fish. These sympatric endemic species were thought by Hopson (1982) to have evolved from *Alestes nurse*, a fish restricted to the littoral region of the lake. In this lake *Synodontis schall* is mainly pelagic, though some live demersally both inshore and offshore, feeding on ostracods and molluscs as they do in West African waters. In Lake Turkana *Bagrus bayad* is the dominant fish living demersally in the deeper water, whereas elsewhere it is a shallow riverine species.

Many of these species retain the habit of spawning in the **inflowing rivers**: in the inflowing Omo River only – *Alestes baremose, A. dentex, Citharinus citharus, Distichodus niloticus, Barbus bynni*; in the Omo and Kerio deltas – *Schilbe uranoscopus*; or in both rivers and ephemeral affluents when in spate – *A. nurse, Labeo horie, Clarias lazera, Synodontis schall*. Those spawning in the littoral region of the lake include the tilapias *T. zillii, O. niloticus, Sarotherodon galilaeus, Leptocypris niloticus, Aplocheilichthys rudolfianus*. In the open lake away from the shore spawn *Alestes ferox, A. minutus, N. stellae, Haplochromis macconnelli* and the predatory *Bagrus bayad, Lates niloticus* and *L. longispinus*. *Hydrocynus forskahlii* appears to have two populations, one migrating to the Omo to spawn and one spawning in the lake. Growth data were available for a number of species, the lake regime being seasonal enough to affect fish spawning seasons and in some cases to leave regular marks on skeletal structures which could be used for age and growth rate determinations.

The cichlid lakes: Victoria, Malawi, Tanganyika

Lake conditions

Cichlid-dominated faunas occur in Lake Victoria (the largest), Lake Tanganyika (the deepest) and Lake Malawi (Table 2.2). The spectacular endemic faunas of these lakes and the importance of their fisheries have stimulated an immense amount of ecological research. These lake faunas are very complex: in Lake Tanganyika *ca* 7000 fishes in a 20 m × 20 m quadrat represented 38 species, most of them cichlids (Hori *et al.*, 1983). The spectacular speciation among these cichlid fishes makes these lakes one of the prime sites in the world for the study of speciation and the roles of ecology and behaviour in evolution. The dramatic impact of piscivorous *Lates* on almost all of Lake Victoria's fishes, especially the haplochromine cichlids, since *Lates* gained access to this lake about 1960, stresses the urgency of making such studies while it is still possible to do so.

How old are these Great Lakes? It has been suggested that Lake Victoria is only 750 000 yr old in its present form (Greenwood, 1981), Malawi *ca* one million years old and Tanganyika slightly older, perhaps *ca* 2 million yr (Banister & Clarke, 1980). But the great accumulations of lake sediments found in recent drillings (more than 900 m in Lake Tanganyika) suggest an age for Lake Tanganyika 'no younger than the Lower Miocene' (i.e. some 20 million yr old) and Rosendahl & Livingstone (1983) think some parts of Lake Malawi may also be as old as this. Victoria, the largest in area of the three (69 635 km² – about the size of Switzerland), is a huge saucer-shaped lake a mere 93 m deep, lying astride the equator (0°21'N to 3°S) between the two East African rift valleys (see Fig. 2.2). Lakes Tanganyika and Malawi, lying to the south, are long narrow, very deep rift valley lakes. Tanganyika is 640 km long, 64 km wide and 1600 m (over a mile) deep, the second deepest lake in the world. Lake Malawi is 560 km long, 80 km wide and 722 m deep and lies at about 550 m above sea level, well to the south of the equator (9°29'S to 14°25'S), where the climate is much more seasonal than around Lake Victoria.

The faunas in these lakes reflect those in their present and former drainage systems (Table 2.2, Fig. 2.2). Victoria, which now flows to the Nile, has an attenuated Nilotic fauna from an earlier connection with the Nile (from which the indigenous *Lates* found in fossil beds near the present lake became extinct); Tanganyika is in the Zaire drainage basin; Lake Malawi connects via the Shire River with the Zambezi system, though many of its fishes have affinities with those in other east-coast rivers.

The deep lakes Tanganyika and Malawi are more or less permanently stratified (Malawi with most mixing between June and August). Although temperature differences between surface and bottom waters are very small compared with those in temperate lakes (*ca* 23.6–26.5°C surface, 23.1°C bottom in Tanganyika, and 23.5–27.5°C surface, 22.0°C bottom in Malawi), the difference in density per degree Celsius at the prevailing high temperatures is enough to ensure that these lakes remain stratified for long periods. This stratification has two main consequences for the fishes: (1) nutrients are locked away in the hypolimnion water except for local upwellings, so the lake is less productive than it might otherwise be; and (2) the bottom water is deoxygenated and so out of bounds to the fishes (see Beauchamp, 1964; Eccles, 1974; Beadle, 1981). Thus these deep lakes represent relatively shallow fish-living space (down to *ca* 200 m in Tanganyika, perhaps a little deeper in Malawi) on top of a huge volume of deoxygenated bottom water.

Lake Malawi has a deeper epilimnion than Tanganyika; plankton production tends to be highest between June and August (FAO, 1982) but Lake Malawi lacks the truly pelagic fish community found in Lake Tanganyika. In Lake Tanganyika the thermocline breaks down in certain years: the strong southeast Trade Winds funnelling up the lake in the cool season, May to September, combined with the Coriolis force (due to the earth's rotation), lead to turbulent conditions with strong offshore drift, upwelling and mixing, especially at the southern end of the lake and in bays (Coulter, 1963, 1966). Nutrient salts brought into circulation cause local phytoplankton blooms (in July and again in November–December in some years). These in turn affect the zooplankton on which the pelagic community depends. In Lake Tanganyika the important fisheries are based on this pelagic fish community, comprising endemic clupeids and their centropomid predators. In Lake Malawi the southeast Trade Wind, known locally as the *mwera*, often blows for days on end during the cool season, but stormier waters are caused by blows from the northeast during the well-defined rainy season from December to March.

Although Victoria is an equatorial lake, with two rainy seasons a year at the northern end, the relatively shallow open waters have a regular annual stratification and overturn as in a temperate lake, bringing the nutrient salts from bottom waters back into circulation; internal water movements (seiches) also spill hypolimnion water into various gulfs and bays seasonally. Fish may be caught at all depths in this lake.

The annual inflows into these lakes are very small compared with their volumes. In Lake Tanganyika, the inflow to volume ratio is only 1:1500, a great contrast with many man-made lakes in which annual inflows far exceed the volume. For long periods Tanganyika lacked any outlet, all water loss being due to evaporation. This led to concentrations of salts in the lake water, particularly of sodium, magnesium and sulphate ions (instead of the more usual calcium, potassium and bicarbonate ions found in other freshwaters). This concentration of salts, and of unusual ions, may present barriers to certain fishes. Only one species of mormyrid fish is found in Lake Tanganyika (*Hippopotamyrus discorhynchus*), though the affluent rivers contain at least two species (*Mormyrus longirostris* and *Mormyrops deliciosus*) which have colonized Lake Malawi. (Compare the lack of mormyrids from Lake Turkana.) It is interesting that the endemic clupeids and centropomids on which Lake Tanganyika's fisheries are based belong to marine families.

The numbers of families represented in these lakes are indicated in Table 2.2. Nearly all the cichlids are endemic to one particular lake. Non-cichlids are represented by endemic species too, notably clariids in Lake Malawi, bagrids, mochokids, mastacembelids, centropomids and clupeids in Lake Tanganyika.

Feeding adaptations
Lake Victoria
Non-cichlids

The most complete study of feeding habits of non-cichlids in the Great Lakes has been Corbet's (1961) analyses of gut contents of 26 non-cichlid species from Lake Victoria; more recently, research has been concentrated on the food of the introduced *Lates* (FAO, 1985). Corbet's fishes were caught over hard, soft and mixed bottom deposits, also from lakeside swamps, affluent rivers and the effluent Victoria Nile; for these he recorded occurrence of food items (in 35 categories) and main contents, i.e. those occupying half of the stomach content volume, for each species of fish, noting the size of the fish and its habitat when caught. Changes with time of year could not be demonstrated, but some species showed changes of food with phase of the moon, with which insect emergences were associated. Foods eaten were also found to change with size of fish in many cases and with feeding grounds on which the particular individuals were caught.

Insects, their larvae and nymphs formed by far the most important foods of non-cichlids in the Victoria basin. Nearly all species included some in their diet and for many (especially Mormyridae and juvenile fishes) they provided virtually the entire food. Surface insects were important food for *Alestes, Barbus, Schilbe, Synodontis afrofischeri* and *Aplocheilichthys*, whereas mormyrids, *Clariallabes* and *Afromastacembelus* fed almost exclusively on the bottom; yet others searched out insects from marginal vegetation. Insects were the most widespread of prey; they require of their predators the least specialization in feeding habits.

Molluscs formed the principal food of adult *Protopterus, Synodontis victoriae* and *Barbus altianalis*. *Protopterus* can crush shells of any size, *Barbus* only smaller ones; *S. victoriae* is able to remove opercula and flesh from gastropods. *S. afrofischeri* and *Clarias mossambicus*, which eat small amounts of gastropods, ingest entire shells. Gastropods were eaten more often than bivalves. None of the non-cichlid molluscivores showed any particular lacustrine adaptations, but certain molluscs were very abundant in the bottom mud of the lake, which provided an extended feeding ground.

Fishes, especially haplochromine cichlids, were preyed on by three indigenous non-cichlid species, as well as by many cichlids. Of the non-cichlids, *Bagrus docmac* was the main predator on haplochromines, while *Clarias mossambicus* ate relatively more large insects and arthropods, and *Schilbe mystus* fed more at the surface taking *Rastrineobola* and surface insects. Small *Bagrus docmac* lived among rocks and stones feeding on lithophilous insects, becoming piscivorous when about 15 cm long. Damaged fishes (including those caught in gillnets) were scavenged by *Protopterus* and *Alestes jacksoni*.

The small cyprinid *Rastrineobola argentea* appeared to be the only non-cichlid specialized as a pelagic feeder on **zooplankton**, though small surface-feeders, such as the cyprinodont *Aplocheilichthys pumilis*, ingested considerable amounts when taking insects from amongst marginal vegetation.

Plants and associated invertebrates were used as foods by the two endemic *Alestes* species, *A. jacksoni* and *A. sadleri*, *A. jacksoni* also taking larger insects and some fishes. *Barbus altianalis*, though primarily a mollusc-feeder, ingested some higher plant material, especially in rivers.

Most of these non-cichlids were facultative feeders, their diets varying according to the size of fish, its feeding grounds, and even the lunar phase when this affects the emergence and movements of insect prey. For instance, *A. jacksoni* fed on adults of the ephemeropteran *Povilla adusta* when these were emerging in the nights after the full moon, and on chironomids (pupae, exuviae or adults) at other times. The size of an insect or other organism and its position in the aquatic environment are important in determining its value as prey.

Many haplochromine cichlids utilized these same foods, as discussed below, so the amount of interspecific competition for food cannot be assessed without taking these into account too; there is as yet too little precise information on the ecology of all these fishes, and of the invertebrates on which so many of them feed, to discuss competition in any detail. The overlap in foods used by non-cichlids was well illustrated by food preferences amongst the five species of mormyrid and the six species of siluroid catfishes in the lake. Most of these species are unspecialized feeders, but no two had exactly the same diet, except when food was locally abundant as when surface-feeders were gorging on emerging *Povilla*. Greater knowledge of the taxonomy of prey organisms might reveal wider differences in prey than was shown by the broad categories used in this study. Where two species did eat the same food they had caught it or utilized it in different ways, but on the whole the non-cichlid fish were facultative feeders, able to range over feeding grounds of great variety and extent. Juveniles and adults of one species often lived in different habitats and ate different foods.

Lake Victoria's mormyrids provided clear examples of how misleading food preference studies can be if based on few specimens of uncertain origin. Within one species, *Mormyrus kannume*, foods utilized varied more according to where the fish were feeding (hard or soft bottom), size of fish and lunar phase, than between this and other mormyrid species. Mormyrids feed mainly at night. Over mud *M. kannume* ate mainly chironomid and chaoborid larvae; over sand and rock the diet was more

varied and included greater numbers of Trichoptera larvae, Ephemeroptera nymphs (especially *Povilla*) and *Caridina* shrimps. Over hard bottoms *M. kannume* ate larger kinds and individuals of prey species as their
size increased. Over soft bottoms chironomid (and some chaoborid)
larvae were eaten throughout life. The larger these fishes became, the
more they fed over mud, supporting the view that *M. kannume* of riverine
origin has become well adapted to the most extreme of lake environments, soft mud. In this species, food was less varied in older fish.
Amongst the smaller mormyrids, diets were more varied in the affluent
rivers, where they migrate to spawn, and there included large numbers of
swamp-living worms (*Alma*), fruits and dytiscid beetles (Okedi, 1971). In
contrast with the great flexibility in trophic behaviour shown by the
mormyrids, *Afromastacembelus* was found to be one of the few
stenophagous insectivores in the lake, a specialization possible as this
species remains in one biotope, amongst rocks, throughout its life.

Corbet concluded that interspecific competition for food amongst these
Victoria non-cichlids plays a very minor part in their ecology. The few
species that share specialized feeding habits appear to have a superabun-
dance of food, and the others cope with overlap by remaining mobile and
facultative.

Lates, which first appeared in the lake about 1960, has since spread all
over the lake and to almost every habitat. Fingerlings and first-year fish
feed primarily on prawns (*Caridina*) and fish fry, with some molluscs; they
then move to deeper water and turn to fishes, especially haplochromines,
Rastrineobola and now small *Lates* (FAO, 1985). The impact of *Lates* on
these other species is shown clearly by changes in the species composition
of catches (see Fig. 14.3).

Cichlids

Trophic specialization is the basis of the spectacular adaptive
radiations of the cichlids in these lakes, providing a striking contrast with
the rather general diets of the non-cichlids. Greenwood's studies of the
haplochromine cichlids showed that species had evolved which utilized
almost every available food source in the lake. However, in contrast with
the great variety of foods eaten by the haplochromines, the tilapias in the
lake were primarily algivorous. Of the two indigenous tilapia species,
Oreochromis esculentus lived mainly in sheltered bays where it fed mainly
on flocculent bottom deposits, while *O. variabilis* lived in clearer waters,
where the bottoms are harder, and in water lily swamps round the lake.
Detailed comparisons of their foods were never made, however, nor were
the foods used by the introduced tilapia species *O. niloticus, O. leucostic-*

tus and *Tilapia zillii* compared. Elsewhere, these *Oreochromis* tilapias are algivorous and *T. zillii* utilizes some macrophytes.

Greenwood (1981) now recognizes 20 distinct genera (many of them new) among the haplochromines in Lake Victoria. In this book the term *'Haplochromis'* is used in quotes when citing authors who have not recognized these new generic names, and also for the numerous (100+) haplochromine species in Lake Malawi – see below (instead of *Cyrtocara*, suggested by Greenwood (1979) as a temporary 'generic' name for the Malawi haplochromines, as these differ from those in Lake Victoria and appear to be a polyphyletic group). In Lake Victoria's haplochromines the general body form varies little but morphological differences in the feeding apparatus (mouth, teeth, guts) are associated with trophic groupings. These include **generalized bottom-feeders** (*Astatotilapia*, 10 species) which feed mainly on bottom debris with some insects, algae, plant material and small fish, and **benthic-invertebrate-feeders**, a superlineage of forms which include insectivores (*Paralabidochromis*, 8 species), sand-sifters (*Psammochromis*, 5 species), a substrate-rasper (*Platytaeniodus*, one species), and insect and mollusc-feeders (*Hoplotilapia*, one species and *Macropleurodus*, one species). **Molluscivores** include specialized 'oral shell-crushers' which spit out the shells (*Labrochromis*, 6 species and *Ptyochromis*, 4 species) and 'pharyngeal shell-crushers' which crush and swallow shells (*Gaurochromis*, 7 species and *Astatoreochromis*, one species). **Herbivores** form a superlineage of four genera with long coiled guts – these include phytoplankton-feeders (*Enterochromis*, 3 species) and algal grazers taking epiphytic or epilithic algae (*Haplochromis s.s.* as redescribed, 3 species and *Neochromis*, one species) while higher plants and algae are utilized by *Xystichromis* (one species). The **piscivores**, of which about 50 species have been described, contain the largest species (growing to *ca* 28 cm compared with but 13 cm SL in the other haplochromine groups. *Prognathochromis* (30 species) and the larger-growing *Harpagochromis* (17 species) feed on live fish; scale-scraping is practised by *Allochromis welcommei* (evidently related to the substratum-rasping *Platytaeniodus*). Paedophagy, obtaining eggs and embryos from brooding female cichlids, is practised by *Lipochromis* (6 species); this specialization is possible as some cichlids can be found brooding young at all times of year in this lake. Different ways in which the predators manage to extract the eggs and young from the brooding female, mainly by engulfing her snout or ramming her, have been described by Wilhelm (1980) and by McKaye & Kocher (1983) for three paedophagous species in Malawi.

The *Haplochromis* Ecology Survey Team (**HEST**) from the University of Leiden, operating at the south end of the lake, have added greatly to our knowledge of the ecology of these haplochromines (van Oijen, Witte and Witte-Maas, 1981; van Oijen, 1982; Witte, 1981, 1984*a*). More than 140 species live in habitats fished by the bottom trawls in the Mwanza Gulf (i.e. all habitats except over rocky bottoms, in dense plant stands or in very shallow water). These catches include many trophic groups: epilithic and epiphytic algal-scrapers, detritus-feeders, zooplanktivores, insectivores (which feed mainly on Trichoptera or Anisoptera or a whole range of invertebrates), molluscivores (shellers, oral crushers, pharyngeal crushers), crustacean-feeders (including cleaner fishes which remove leeches and crustacean parasites from other fish – a new discovery here – Witte & Witte-Maas, 1981), piscivores taking *Rastrineobola* or smaller haplochromines, and scale-eaters. The zooplankton-feeding cichlids, at least 12 species, were also a new discovery in this lake. The HEST team concluded that, on the whole, these Victoria haplochromines seem specialized in certain predation techniques rather than specializing on particular prey organisms; the team suggested that a new trophic classification is needed to take this into account, rather than classifying predators according to the systematic position of the prey.

In the Mwanza Gulf over 90% of the trawl-caught haplochromines were detritus/phytoplankton feeders, zooplanktivores or insectivores. Piscivores made up less than 1% of the trawl catch, though they included the largest number of species (*ca* 40%). A well-defined specialization on prey types may help to account for the large number of piscivorous species (as van Oijen (1982) showed) but, as the larger fish were selectively removed by the fishing, the low number of piscivores may have been a sampling effect.

Habitat partitioning was also apparent. The distribution of the various species is determined by factors such as substrate type, depth and water movement. A community of 16+ species of rock-dwellers was discovered in the Mwanza Gulf with representatives of the following trophic groups: epilithic algae-scrapers (4 species), zooplanktivores (one species), insectivores (6 species), molluscivores (2 species), a crab-eater (one species), a paedophage (one species), and a piscivore (one species). Few large haplochromines preyed on these fish: their main predators appeared to be *Bagrus*, cormorants and kingfishers. These rockfish showed convergence in squamation, teeth and jaws with the *mbuna* rockfish of Lake Malawi described below. Some other species were also restricted by type of substratum, for example '*H.*' *plagiodon* over sand and '*H.*' *teegeleeri* over mud, though some species occupied varying types of substrate or changed substratum as they grew (from hard in the young to mud in adults). Every species was found to have its own depth range, though the width of range may vary considerably (for example, from 8 to 25 m in one piscivore). Bathymetric segregation and its possible role in speciation have been discussed by Hoogerhoud, Witte & Barel (1983), who compared gill morphology and overlaps in diet and breeding in two molluscivorous species living at different depths. Depth distributions of adult piscivores were illustrated by van Oijen (1982). In some species adults live in deeper water than the juveniles. Faunal changes were marked at the 10 m zone; a thermocline develops at between 7 and 10 m in the rainy season, below which there is relatively little oxygen (2–3 ppm) (see Beadle, 1981). The decline in light penetration with depth may also affect a fish's ability to catch zooplankton. Distributions are also affected by exposure, whether the site is open to the wind or sheltered, and by the species composition. Every type of habitat has its own types of haplochromines, and several representatives of most trophic groups are present in most habitats. Despite this, there was no evidence of interspecific competition; subtle preferences for different prey mean that each species has its own unique combination of requirements.

Insectivorous species dominated in shallow littoral water over sand, detritus/phytoplankton-feeders in deeper open water over mud. These studies made it quite clear that these haplochromines are not evenly distributed over the various habitats, as had been assumed in the stock estimates for fishery development; most species are strongly habitat-restricted, so could be fished to extinction. Furthermore, each habitat has its own group of species exploiting a particular food source. Juveniles of piscivores living in the central waters of the Gulf were found in shallow areas in bays, where they may feed on planktonic Crustacea and haplochromine fry. The juveniles of deepwater species were also taken in the Gulf. Certain insectivores, molluscivores and piscivores brood young throughout the year, but in the transect samples the detrititus/phytoplankton-feeders and zooplanktivores had a distinct breeding season with a peak at the end of the rainy season.

Lake Malawi
Cichlids
The Lake Malawi cichlids show great specializations of mouth and teeth for using a particular type of food (Fig. 4.1). These cichlids

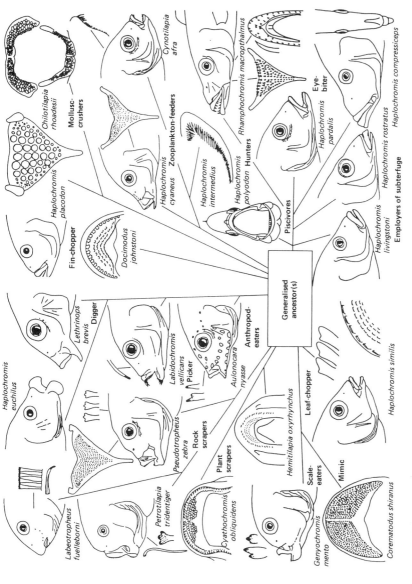

Fig. 4.1. Examples of adaptive radiation in Lake Malawi's cichlids (after Fryer & Iles, 1972).

have diverged to become plant and rock scrapers, zooplankton- or arthropod-feeders, or piscivores which hunt their prey in diverse ways, including mimicking their prey to rasp off scales as food. Specializations to use a particular type of food, such as aufwuchs, have occurred many times, and within one community numerous species appear to share the same resources. This was particularly marked amongst the aufwuchs-feeders, suggesting to Fryer (1959) that this source of food is superabundant; if so, factors other than food must control the fish numbers. In some cases competition appeared to be minimized by fishes feeding at slightly different depths, or cropping the food in slightly different ways, later confirmed by observations using SCUBA. Photographs of these very colourful Malawi cichlids are given in Axelrod & Burgess (1976) and Mayland (1982).

In his classic studies at Nkhata Bay on the northwestern shore of Lake Malawi, Fryer (1959) contrasted food webs of the fish fauna living on rocky and sandy shores. On a 300 m-long rocky shore he recognized 30 fish species and on an adjacent sandy shore of equivalent length 23 species. The food webs were based on few items, particularly in the rocky

Fig. 4.2. The food web on a rocky shore at Nkhata Bay, Lake Malawi (after Fryer, 1959).

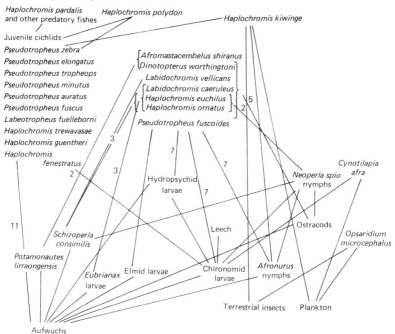

habitat where aufwuchs was cropped by more than ten species; fish here also took some plankton and insect larvae (Fig. 4.2). On the sandy shore the basis of the food webs was broader (plankton, terrestrial insects, *Vallisneria*, aufwuchs on *Vallisneria*, bottom debris and algae). In a nearby swampy creek mouth, bottom algae and detritus formed the main basis of the food webs, with higher plants and terrestrial insects supporting a few species. Many more fish colour morphs, possibly species, are now recognized from this area (Ribbink *et al.*, 1983*a*).

Ribbink *et al.* (1983*b*) made detailed analyses of resource utilization of 196 species/taxa of the rock-dwelling fishes known in Malawi as **mbuna**, mainly in the southern part of the lake and around Likoma Island, noting for each the depth range, preferred substratum and degree of territoriality. These analyses (of which an example is shown in Fig. 4.3) led them to conclude that almost all the *mbuna* are stenotopic and they also have a

Fig. 4.3. Resource utilization by the *mbuna* cichlid community at Nkhata Bay, Lake Malawi (after Ribbink *et al.*, 1983*b*). (Substratum S indicates sand, SR sand-rock, WR weed-rock).

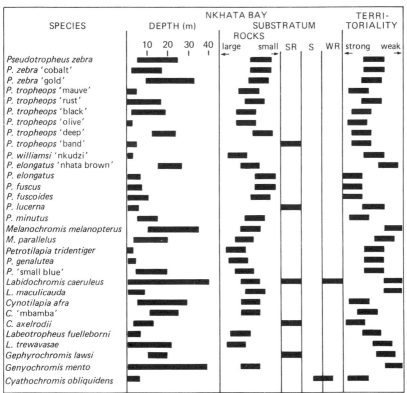

narrow natural distribution, many of them being restricted to tiny rock outcrops. Consequently the 'species' assemblage in each area is unique, and even widely distributed species are subject to interactions with different species at each locality. The *mbuna*'s stenotopy is due to their sedentary nature, demonstrated by tagging and also by the fact that species transferred within the lake have remained near the point of introduction. Furthermore, they appear to remain in one biotope throughout life, the large yolked egg enabling the young to hatch at a size large enough to use more or less the same food as the parent, and cutting out the need for a planktonic openwater stage (as found in most reef fishes, p. 199). Numbers of *mbuna* fall off with depth, being highest in the 3–10 m zone (more than 10 fish m^{-2}), though about 26 species live down to 40 m depth. Analyses of overlaps in resource use by three sympatric *Petrotilapia* species at Monkey Bay showed that the males utilized only 6% of the area defended as breeding territory for feeding and, though overlap of feeding territories was considerable, only 2.5% of the feeding space used by territorial males was shared. These males utilized food patches in the territories of the highly aggressive *Pseudotropheus* species, but the females fed mainly in undefended areas (Marsh & Ribbink, 1985).

Mbuna trophic groups recognized by Ribbink *et al.* (1983*b*) include: plankton-feeders; fishes using invertebrates other than plankton; scale-eaters (lepidophages); fishes taking loose aufwuchs; filamentous-algae-feeders; egg- and parasite-eaters; and piscivores. But almost all *mbuna* were found to contain a mixture of food substances. Virtually all *mbuna* will feed on zooplankton when this is abundant, regardless of other trophic specializations. Very little food partitioning was evident from analyses of stomach contents, and there was considerable intraspecific variation in foods used. Contrary to earlier belief that they are rigid specialists, *mbuna* appear to be opportunist feeders. But despite the overlap in resources used, Ribbink *et al.* (1983*b*) concluded that the *mbuna* do not contradict the competitive exclusion hypothesis, as how and where they feed (zonation, depth and microhabitat) appear more important for resource partitioning than the items on which they feed (as Smith & Tyler (1972) had found for coral reef fishes, see p. 188). How and why then have the *mbuna*'s trophic specializations evolved? Ribbink *et al.* (1983*b*) suggested that these come into play at times when resources are scarce, possibly at different lake levels (see Wiens (1977) on 'ecological crunches' and p. 303).

Food switching has now been observed among many Malawi cichlids: in the aufwuchs-feeding *Pseudotropheus zebra* and *Petrotilapia tri-*

dentiger and in species of *Melanochromis*, *Labidochromis*, *Cyathochromis*, *Docimodus*, '*Haplochromis*', *Labeotropheus* and *Oreochromis*. The periphyton-feeder *Hemitilapia oxyrhynchos* may take zooplankton particles in the water column, algae off rocks, fish, detritus and invertebrates from the sand (McKaye & Marsh, 1983).

A group of specialized planktivorous haplochromine cichlid species, known locally as **utaka** (Iles, 1960) usually occurs most abundantly in certain places (*virundu*) where upwellings lead to plankton abundance; *utaka* are not uniformly distributed in the lake and most of them come to sandy shores to breed or, possibly, breed in open water near rocks.

Non-cichlids

The zooplankton in this lake is also utilized by an endemic cyprinid, the little usipa (*Engraulicypris sardella*), as well as by the juvenile stages of numerous other species. *E. sardella* is, however, the only pelagic fish to venture far from shore in this lake; its numbers may be food-limited, particularly by shortage of suitable food for the postyolk-sac young in May–July, which appears to affect the size of year classes (FAO, 1982).

In Lake Malawi several large endemic clariid catfishes also live in the more open waters of the lake. These species, with much reduced accessory respiratory organs and lateral eyes, were formerly considered an endemic genus '*Bathyclarias*' (Jackson, 1959) but are now considered by Greenwood (1983c) to be representatives of *Dinotopterus*, a genus also found (one species) in Lake Tanganyika. Of the ten endemic species in Lake Malawi, several of the large *Dinotopterus* are piscivores: *D. gigas*, which grows to over 1.5 m and 30+ kg; *D. foveolatus*, a massive sluggish fish which skulks among rocks and mud at great depths; and *D. longibarbis*, a more active fish-eater fairly abundant at the north end of the lake. *D. worthingtoni*, a smaller species, lives among rocks and feeds on the *Potamonautes* crabs confined to this habitat. Several of the species live more pelagic lives, including *D. loweae*, a heavy species of open and deep waters where it may eat fish as well as plankton and *nkungu* lakeflies (chaoborids). The commonest species, *D. nyasensis*, has long close-set gillrakers adapted for eating plankton and *nkungu* flies; the fish is caught in gillnets at all depths down to the limits of dissolved oxygen and enters river estuaries and flooded areas during the rains (Jackson *et al.*, 1963).

Lake Tanganyika

Non-cichlids

Only *ca* 10% of Lake Tanganyika's coasts have yet been explored for fish. On rock slopes the catfishes are very varied, ranging from the small endemic bagrids *Phylonemus* and *Lophiobagrus* which hide in deep, dark crevices, to the big species such as *Chrysichthys, Bagrus, Heterobranchus*, the electric *Malapterurus* and *ca* 10 species (most endemic) of *Synodontis*. The catfishes are generally the largest fishes in the area but are rarely seen as they are all mainly nocturnal. At greater depths, schools of *Synodontis* wander on the lake floor. Most *Afromastacembelus* live in rock crevices; some seem to have a very restricted distribution. The large endemic cyprinodont *Lamprichthys tanganicus*, though pelagic, comes to rock shelves to spawn its transparent eggs. Some species of *Varicorhinus* and juvenile *Labeo* also live among surf-washed rocks.

In the **pelagic zone** of Lake Tanganyika, the endemic clupeids are planktivores which migrate surfacewards at night (*cf.* marine clupeids, p. 227). Both species feed on both phyto- and zooplankton when young. Adult *Stolothrissa tanganicae* (maturing at *ca* 70 mm SL) live in more offshore waters, feeding mainly on pelagic zooplankton (to a small extent on insect larvae, fish ova and diatoms), while the slightly larger (*ca* 95 mm SL) *Limnothrissa miodon* occupies the more inshore zone where it has a more general diet, including prawns, insects and their larvae, and young clupeids. Since both species can occupy the same habitat at the same time they may then be in competition, but *Limnothrissa* with its less specialized feeding habits appears better adapted to inshore life where the food is more varied, while *Stolothrissa* has the advantage in the pelagic zone. Lake Tanganyika lacks the dense swarms of 'lakeflies' (chaoborid dipterans and small Ephemeroptera) which rise like smoke from the surface of many of the other African Great Lakes such as Malawi, Victoria and George. Is their absence perhaps linked with the presence of pelagic clupeids in Lake Tanganyika?

Of the **centropomid predators** of the clupeids, young *Lates* of three species, *L. mariae, L. microlepis* and *L. angustifrons*, up to 3 cm SL, are common in the plankton. These then live among littoral macrophytes until about 35 cm TL, dispersing away from the shores as they grow. In the weed beds they feed mainly on prawns, small cichlids, and insects and their larvae. Length frequency data suggest that they stay here for about a year, growing at about 1 cm per month. Soon after leaving the weed beds, *L. microlepis* adopts a diet of clupeids; the others continue to take prawns and other invertebrates until maturity is reached. *L. mariae* is pelagic whilst *L. angustifrons* feeds on prawns and benthic fishes but migrates up to the surface at night to feed on clupeids in the months when these are most abundant. *Lates* (*Luciolates*) *stappersi* seems to be completely pelagic all its life, living near the surface in the upper 30 m, both young and adults feeding on clupeids. The trophic niches of these pelagic predators are more clearly separated than are those of their clupeid prey. While all feed on clupeids, *L. mariae* and *L. angustifrons* do so mainly when these are plentiful. Only *L. mariae* and *L. stappersi* appear to compete continually for the same food. None occupies the role of climax predator (Coulter, 1976).

In Lake Tanganyika endemic bagrid catfishes, *Chrysichthys*, fill the deep-water scavenging role that the clariid *Dinotopterus* catfishes have in Lake Malawi. Coulter's (1968) experimental gillnet catches showed that in spite of wide depth distributions the various species had clear depth preferences and food specializations were greatest between members of the pairs of species sharing a depth zone. In the sublittoral zone *C. brachynema* took mainly crabs and *C. graueri* mainly cichlids; in deep water *C. stappersi* took mainly crabs and *C. grandis* fish (especially cichlids but also other *Chrysichthys* and crabs). *C. sianenna* and *C. platycephalus* predominated in the littoral zone but were not common enough in catches to reveal diet preferences. *C. stappersi*, the most numerous species in catches at the south end of the lake (most abundant from 80 to 120 m), was not found in the northern basin. In addition to crabs, its very generalized diet included other invertebrates (insects, gastropods, lamellibranchs) and it also scavenged dead fish (cichlids).

Deepwater community

Lake Tanganyika also has a deepwater fish fauna. With increased pressure, decreased light and decreased temperature (24°C at 100 m), the deepwater habitat is the most specialized the lake has to offer. Below 60 m the dissolved oxygen falls rapidly – from 85% at 60 m, to only 20% at 70 m, 10% at 100 m, 4% at 140 m, and 2% at 170 m in the north basin. In the south basin oxygen sometimes extends down to 250 m. Below 20 m the bottom is generally of sand or mud and mollusc shells; rocky areas are rare. Below 100 m it is mainly of mud (with mud extending into shallower water in the river estuaries). These muddy bottoms in deep water present very special conditions for freshwater fishes and a remarkable fauna has evolved here. Cichlids, all endemic species (*Trematocara* with most species, *Xenotilapia* and *Limnochromis*), and the bagrid *Chrysichthys* predominate. The species vary with the depth and nature of the bottom. Many of these deep-living fishes move up to the surface at night. Deepwater fishes caught in bottom-set gillnets at 120 m, where there was no recordable oxygen, included five cichlid species (*Hemibates stenosoma, Bathybates ferox, B. fasciatus, Xeno-*

chromis hecqui, Limnochromis permaxillaris), two *Chrysichthys* (*C. stappersi* and *C. grandis*), *Dinotopterus cunningtoni* and *Lates mariae*. Coulter (1966, 1968) suggested that these species might be capable of some form of anaerobic respiration; this calls for further investigation. Not all the species found at these depths are restricted to the very deep water; *Hemibates stenosoma* is caught from 20 m to the oxygen limit, though most abundantly from 80 to 120 m; it feeds on cichlids (including its own young), clupeids and prawns, and is itself the main cichlid prey of *Lates mariae* and *L. angustifrons*.

Cichlids

In Lake Tanganyika tilapias live mainly in the river deltas and estuaries. The largest known cichlid, the endemic *Boulengerochromis microlepis*, growing to 80 cm SL and 3.5 kg, common and widespread in the lake, is mainly a bathypelagic coastal predator, hunting in schools, chasing the large schools of clupeids into bays and shallows. Individuals occupy different biotopes in the lake according to their size, feeding behaviour and gonad state. The fry are microphagous bottom-feeders. Juveniles live in shallower water than adults. Omnivorous juveniles less than 15 cm long hunt in schools of 100 to 5000 in water 3–10 m deep. As they grow, the schools become smaller, the diet includes more fishes (clupeids and small cichlids) and they hunt at greater depth; juveniles over 15 cm roam the lake at depths of 15–35 m in groups of several dozen individuals. These fish make diurnal vertical movements, rising to the surface and coming inshore at night. Mature fish live in pairs amongst rocks in the littoral zone, utilizing a more varied diet of crabs, molluscs, insect larvae and plant debris as well as fishes. Matthes (1961) concluded that they mature when 2.5 yr old (about 35 cm, 500 g).

Among the huge variety of cichlids in the **littoral zone**, an 'infauna' living amongst the rocks includes *Eretmodus*, *Spathodus* and *Tanganicodus* in the rubble-bottomed surf zone, stocky fishes able to withstand the pounding of the surf, pickers of food from off and between the rocks (Fig. 4.4). Immediately below this lives a community of aufwuchs-eaters (17 species of cichlid and one cyprinid) together with an infauna of carnivorous *Lamprologus* (many species), *Julidochromis* and *Telmatochromis*. 'Hoverers' living above the slope include *Cyphotilapia*, *Limnotilapia*, *Ophthalmochromis*, *Cyathopharynx*, *Perissodus* and various *Lamprologus* species. The infauna fishes tend to lead solitary lives, sedentary and territorial, the substratum-spawning cichlids being monogamous. Hoverers live gregariously in schools, wandering within the limits of the habitat, only territorial (if at all) when spawning, the cichlids mostly polygamous mouth-brooders (Brichard, 1978).

Over **sandy patches** typical sand-dwellers, such as *Callochromis,
Xenotilapia, Cardiopharynx*, are gregarious, pale-coloured fishes living
in schools of several hundred, feeding on particles sifted from the sand
with shovel-like teeth. On the **rocky shores** the very varied geological formations provide a wide variety of underwater rock shelters. Each rock
slope presents a different biotope depending on its structure, shape and
the size of the rocks, which all influence the biocover, and on variations
in the oxygen levels at different depths, which, like the influx of food,
depend on the waves and currents. Some species are restricted to shallow
surf-washed shores (with high oxygen content and generally a rubbly bottom), others occur only in deep water; a few live all the way from the surface to 60 m or so depth. Those species living over open bottom tend to
be streamlined for speed; they form schools more often than do those living in shelters, and some have teeth and jaws for shovelling sand. Rock-

Fig. 4.4. Examples of Lake Tanganyika's endemic cichlids from various
zones: (A) *Spathodus*, a surf-zone rubble-dweller; (B) *Tropheus*, an
algal-grazer of the rocky littoral; (C) *Perissodus*, a hovering scale-eater;
(D) *Boulengerochromis*, a pelagic piscivore. (L1–L3): *Lamprologus*
species showing variations in body form: (L1) *L. brichardi*, a gregarious
planktivore; (L2) *L. modestus*, a stone-turning benthic-
invertebrate-feeder; (L3) *L. elongatus*, a midwater predator.

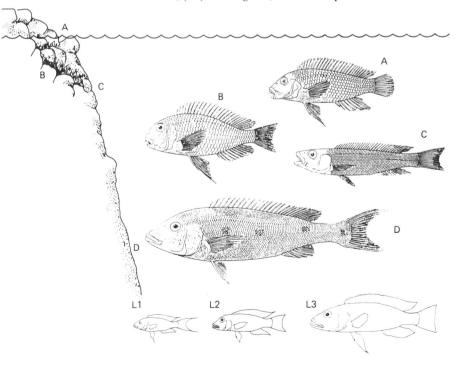

dwellers living in shallow surf-washed shores have short muscular bodies, those living among crevices have slim bodies; teeth and jaws are shaped to graze or to pick food from the rock cover.

The **trophic groups** in rocky biotopes include: grazers, such as *Tropheus, Petrochromis* and *Simochromis*; pickers, such as *Eretmodus, Spathodus, Tanganicodus* and various *Lamprologus* and *Julidochromis* species; carnivores with powerful canine teeth, such as *Lamprologus compressiceps*; omnivores living off particles suspended in the water, such as *Ophthalmochromis, Cyathopharynx* and *Limnochromis nigripinnis*; and scale-eaters with hooked teeth, such as *Perissodus*.

The abundance, microdistribution and trophic groups among the cichlids in a 20 m × 20 m quadrat at Luhanga on the northwest shore of Lake Tanganyika were studied in considerable detail by Hori, Yamaoka & Takamura (1983), Kawanabe (1981, 1983) and their colleagues. Here, a predominantly rocky substratum had patches of gravel or sand. Among the *ca* 7000 fishes of 38 species counted in the quadrat, planktivores (2 species) were most numerous (56%), omnivores (7 species) and aufwuchs-eaters (15 species) made up 21% and 18%, respectively, with zoobenthos-feeders (8 species) and piscivores including scale-eaters (6 species) each about 4%. Only the planktivorous *Lamprologus savoryi* and *L. brichardi* were gregarious. About half the other species were either aufwuchs-eaters, which preferred rocks and predominated in water less than 1 m deep, or zoobenthos-feeders, including six species of *Lamprologus* and *Telmatochromis caninus*, distributed in the different habitats. Piscivores included four *Lamprologus* species and the scale-eating *Perissodus microlepis* and *P. straelini*. Omnivores included *Limnotilapia dardennii* and *Lobochilotes labiatus*, a pair which appeared to have a symbiotic feeding relationship, and *Telmatochromis bifrenatus, Julidochromis marlieri, J. transcriptus, Gnathochromis pfeffereri* and *Xenotilapia* sp. Takamura (1984) noticed seven cases of associated feeding between pairs of carnivorous species, but in all cases small but clear differences in diet or feeding behaviour were observed which he thought could have been effective for the evolution of these symbiotic relationships. Three species were restricted to the shallow water. Ubiquitous species, which included piscivores (4 species), most omnivores (4 species) and two species of zoobenthivores, were represented by only a few individuals. Most species kept a distance from the piscivores. The feeding behaviour of the scale-eating *Perissodus microlepis* and its changes of food with development were studied by Nschombo (in Kawanabe, 1983); this semipelagic species takes scales from whatever fishes it can approach while they are engaged in their daily activities of feeding, breeding and so on.

The **aufwuchs-eating community** was dominated by two territorial species living symbiotically – *Tropheus moorei*, which eats mostly filamentous algae, and *Petrochromis polyodon*, eating mostly unicellular algae (Takamura, 1984). These two, which shared intensive grazing sites and temporal grazing patterns, chased away the other algae-feeders: *Petrochromis* (4 species), *Simochromis* (3 species), *Telmatochromis* (3 species), *Ophthalmochromis* (2 species), *Asprotilapia*, *Eretmodus* and *Tanganicodus*. *Tropheus moorei* and its related species resembled each other in food composition most clearly among the species taking mostly filamentous algae, and the *Petrochromis* species did so among the species taking mostly unicellular algae. Yamaoka (1982) looked at specific adaptations among the five algae-grazing *Petrochromis* species, particularly the relations between structural morphology and rate of feeding. All the algae-feeders take zooplankton in midwater when they first start to feed (Mbomba in Kawanabe, 1983).

Carnivorous *Lamprologus* **species** dominate the cichlid fauna in the lake as a whole (35 species, *ca* 25% of the cichlid fauna). At Luhunga they formed 75% of the cichlid population. scuba observations of feeding ecology and intra- and interspecific interactions, combined with gut content analyses, revealed the interrelationships of 13 species of carnivorous *Lamprologus* coexisting on the rocky shores at Luhunga (Hori, 1983); this study demonstrated very clearly how so many closely related species manage to coexist. Of the 13 *Lamprologus* species here, two were primarily planktivores; these were spatially divided as *L. savoryi* lives in small groups just above the stones (picking up some filamentous algae), whereas *L. brichardi* floats in shoals more than 1 m above the bottom in open water. Of the seven species which feed on zoobenthos, one is nocturnal (*L. toae*), one takes molluscs (*L. tretocephalus*) and the other five take shrimps and aquatic insects but obtain them from different substrates, using species-specific techniques to do so. Of the two species found mainly over stony rubble, *L. callipterus* shovels up bottom sand, while *L. compressiceps* feeds mainly in crevices, aided by its compressed body. Of two found among rocks, *L. leleupi* feeds on the rock surfaces, *L. furcifer* below the rocks. The gravel-dweller *L. modestus* turns over stones to find food. The zoobenthos-feeders are territorial but vary in their aggressiveness to other species: *L. furcifer* is very aggressive, *L. compressiceps* and *L. leleupi* are more tolerant. Of the four piscivorous species, the nocturnal *L. lemairi* is an ambusher; of the others, mainly diurnal, *L. elongatus* lunges at midwater prey, *L. fasciatus* stalks prey over stony bottoms, and *L. profundicola* (the largest species) cruises more widely, generally over sand. The piscivores feed mainly on juvenile *L. brichardi*, counterattacked by the guarding parents as they do so. Prey fish, though wary, are unable to guard against several hunting behaviours at once, and the piscivores benefit by complementing one another's hunting techniques. Facultative commensalism is also extended to other species present, piscivores ambushing prey from under cover of other cichlids, such as algae-grazers (compare the use of 'stalking horses' by aulostomid reef fish, p. 196). Hori also examined the daytime cruising areas, nocturnal resting places, inter- and intraspecific relationships and dominance hierarchies of these *Lamprologus* and the other cichlids. His studies highlight the roles of diversity in foraging behaviour and of facultative commensalism in permitting so many closely related species to coexist. Body forms and sizes of these various *Lamprologus* species also vary in accordance with their feeding habits.

Nagoshi (1983, 1985), who looked at growth and survival of early stages of *Lamprologus* of various species, concluded that interspecific differences in growth rate were related to the area in which the young lived while being guarded by the parent, growth being retarded in species such as *L. toae* and *L. tretocephalus* where the guarded fry were unable to expand their habitat (see p. 106), so a density effect operated. The young *Lamprologus* of many species take plankton, feeding on copepods while guarded by their parents and changing their food as they become independent: *L. toea* and *L. brichardi* to shrimps, *L. tretocephalus* to insects and gastropods, *L. elongatus* and *L. lemairi* to fishes (Gashagaza & Nagoshi in Kawanabe, 1983).

Breeding adaptations

Many of the non-cichlids continue to migrate up affluent rivers to spawn, and the young fish may remain for some time in the rivers, dropping down back to the lake gradually as they grow. Some species appear to have riverine and lacustrine populations. The fishes that continue to migrate up rivers generally retain seasonality of spawning. Some species have become truly lacustrine, spawning only in the lake. The cichlids nearly all spawn in the lakes and present a fascinating array of spawning behaviour.

Among the lake-spawning populations and species, some have distinct breeding seasons, but amongst many of them ripe fish may be found at any time of year. Such protracted spawning, an adaptation common in tropical species, enables the fishes to take full advantage of availability of foods throughout the year.

Non-cichlids
Lake Victoria

In this equatorial lake, affluent rivers at the northern end flood twice a year; in the south they flood in accordance with one well-marked annual rainy season, December to March. Among the non-cichlid fishes ascending rivers from the lake, Whitehead (1959) recognized three distinct patterns of migration: (1) long duration, fishes such as the large cyprinid *Barbus altianalis* entering the river over an extended period and ascending 80 km or more to spawn in swift rocky upper reaches; (2) medium duration, exhibited by most of the potamodromous fishes such as *Labeo victorianus* and *Schilbe mystus*, which enter the rivers in fairly compact schools and run for 8–25 km upriver before moving laterally into floodwater pools to spawn; and (3) short duration, species such as the small *Alestes* (*Brycinus*) ascending streams in enormous numbers when these temporary streams flood. *Clarias mossambicus* and various small *Barbus* species made very sudden ascents up Bugungu stream near Jinja, which often flooded at irregular times of year, and there took advantage of the short-lived flood conditions (Greenwood, 1955; Welcomme, 1969).

Labeo migrations were studied by Cadwalladr (1965) and by Balirwa & Bugenyi (1980). The small mormyrids (*Marcusenius victoriae, Gnathonemus longibarbis, Hippopotamyrus grahami, Pollimyrus nigricans, Petrocephalus catostoma*) move up the northern affluents which flood twice a year (April–May and September–December) on both floods, though whether the same fish move up and down twice a year is not clear. The ripe mormyrids remain near the river mouth until the flood comes, then migrate upriver at night, with peak runs at dawn and dusk. They spawn in pools 8–24 km upriver; after the eggs hatch the young remain in river pools for 3–7 months (Okedi, 1969, 1970). These mormyrids are not very fecund fish, egg numbers ranging from a mean of less than 500 eggs in *P. nigricans* to 6300 in *M. victoriae* (with wide variations, see Table 11.3). The larger *Mormyrus kannume* breeds within the lake, generally on rocky bottoms. Ripe fish are found throughout the year. Among the siluroids spawning in the lake, *Bagrus docmac, Clarias mossambicus* and *Schilbe mystus* displayed continuous protracted fractional spawning activity with at least half the population in spawning condition, but all species showing either January or August pulses; among the *Synodontis* populations, from 45% (*S. victoriae*) to 65% (*S.*

afrofischeri) were in constant spawning readiness (Rinne & Wanjala, 1983).

Lates niloticus produce pelagic eggs and in Lake Victoria ripe fish may be found at any time of year. Both sexes were mature when 2 yr old (70+ cm TL) (Acere in FAO, 1985). They exhibit sexual dimorphism in size and unequal sex ratios: the smallest ripe male was 53 cm TL (age > 1 yr), smallest ripe female 67 cm (age > 1 yr); males predominated up to 120 cm TL, females among the larger, fish (compare *Lates calcarifer*, p. 173).

Lake Malawi

In Lake Malawi the large salmon-like *mpasa* (*Opsaridium microlepis*) and the smaller *sanjika* (*O. microcephalus*) swim long distances up rivers from the northern end of the lake (North Rukuru and Bua Rivers); *Labeo mesops* and certain other species move up medium-sized rivers further south, and *Clarias mossambicus* short distances up spate streams to spawn in the flooded grassland (Lowe, 1952). Tweddle (1983) found that *mpasa* have an extended spawning run during and after the rains; he described spawning behaviour over a gravel bottom, often in very shallow water during the night and early morning, the males patrolling a small (1.5–2 m wide) territory until joined by a female. Eggs remain in the gravel until the yolksac is absorbed; juveniles live in the river for several months, then in the lake close to the river mouth until their second year. North Rukura fish generally mature in their third year (*ca* 28 cm TL). *Mpasa* ovaries contained from 1381 eggs in a 19 cm fish to 22 077 eggs in a 55 cm fish. *Mpasa* need clean gravel for spawning.

Labeo mesops migrates up medium-sized rivers to spawn, but *L. cylindricus* spawns around rocks in the lake. *Bagrus meridionalis* moves into shallow water to spawn where it makes a nest on a sandy bottom against a rock where both sexes guard the young. Cichlid young have also been found in association with young *Bagrus* being guarded by the parent *Bagrus* (McKaye & Oliver, 1980). *Bagrus* females grow to a larger size than the males and the males stay on the breeding grounds for longer periods (Jackson *et al.*, 1963). In the pelagic zone, larval young of the small cyprinid *Engraulicyrpis sardella* are found throughout the year, though peak numbers occur in June–August, when water mixing is greatest (FAO, 1982). Much remains to be learnt about the breeding habits of many of these non-cichlids.

Lake Tanganyika

In Lake Tanganyika species returning to affluent rivers to spawn include three endemic cyprinids, *Bagrus tropidolepis*, *Varicorhinus tan-*

ganicae and *Barilius moorei*, even though these have to pass from the lake waters with their high ionic content to river waters of very different chemical composition. The characoids *Hydrocynus vittatus, Alestes macrophthalmus* and *Citharinus gibbosus* migrate up the affluent Lufuba River to spawn in the November–March rains. Ruzizi River lagoons are also important spawning areas (Kwetuende in Kawanabe, 1983).

Fishes in the pelagic zone generally spawn within the lake. The clupeid life cycles appear to be related to the annual cycle of hydrological events and plankton production (Coulter, 1970). The strong winds in June to August are funnelled up the lake, causing offshore drift, upwelling and turbulence; the lake is most unstable at this time and offshore plankton increases during these months. September to December is the calmest time of year and plankton production in open water is generally high in November–December. Inshore phytoplankton blooms occur in July and in November or December.

Stolothrissa, which depends on plankton for food in both the larval and adult stages, is more affected by these seasonal changes than is the more inshore-living *Limnothrissa*, and responds with seasonal production of young. This is greatest in November–December, the calm months when water turbulence which might be damaging to eggs and larvae has died down (Coulter, 1970). Chapman & Well (1978a) considered that the lake population could not be considered one unit and found that peak spawning varied in different parts of the lake – August–December in the south, January–April in northern waters – the lag probably adapted to time of upwelling and plankton production. Most *Stolothrissa* only live one year.

Among their centropomid predators, spawning continues throughout the year but with the certain maxima (August–December according to Coulter, 1976) probably corresponding with the seasonal maxima in their prey; periodic spawning is suggested by well-differentiated size distribution in the young *Lates* found in the littoral weed. For *Lates* (*Luciolates*) *stappersi* the main spawning period is between January and April in Burundian waters (Ellis, 1978); Chapman & Well (1978b) say these spawn when 2 yr old (*ca* 25 cm), the time when they enter purse seine catches; few spawn again the next season so they may not have very long life cycles, but Ellis (1978) says cohorts of *L. stappersi* contribute to the fishery for 3–5 yr. Their eggs and fry are planktonic, found only in the upper layers of offshore water; young (to 13 cm long) feed on zoo-plankton, initial breeding size is *ca* 21 cm, and gonads are at maximum development from January to April. *Lates mariae* males and females mature at 44 and 49 cm TL, respectively, *L. microlepis* at about 47 and 51 cm and *L. angustifrons* at 50 and 57 cm; females grow larger than males in these species (Coulter, 1976).

Cichlids
Lake Victoria
Cichlids, like other fishes with well-developed parental care, produce relatively few young at a time, but may produce many broods in quick succession. In Lakes Victoria and Malawi and in most other African lakes, most cichlids are maternal mouth-brooders, the female brooding eggs and young in her mouth, which keeps them well oxygenated; Lake Tanganyika is exceptional in having groups of substratum-spawning cichlids, in which both parents guard the eggs and young (as in most Neotropical cichlids), as well as mouth-brooders. Among the cichlids seasonality of breeding varies very much from lake to lake and from species to species within one lake. In the equatorial Lake Victoria ripe individuals of most cichlids may be found at any time of year but seasonal peaks in numbers do occur; these peaks may be more marked in some years than others and peak time may vary with prevailing climatic conditions. Peak spawning in the indigenous tilapia *Oreochromis esculentus* occurs in the wet seasons, twice a year at the north end of the lake, once at the southern end, in accordance with the rainfall patterns in these areas; ovary conditions suggested that several broods are produced in quick succession (Lowe-McConnell, 1959). Fryer (1961) thought that the indigenous *O. variabilis* produced three, possibly five, broods in 8 months, but numbers spawning and survival of the young might vary acyclically from year to year. The introduced tilapias *O. niloticus, O. leucostictus* and *Tilapia zillii* spawn throughout the year in this lake (Welcomme, 1967*a*); these ousted the indigenous species (see p. 302).

Among the trawl-caught haplochromine cichlids in the Mwanza Gulf, the HEST team found that certain insectivores, molluscivores and piscivores brood young throughout the year, but in the transect samples the detrititus/phytoplankton-feeders and zooplankton-feeders had a distinct breeding season, with a peak at the end of the rainy season. Sudden invasions of brightly coloured males of other species normally found in deeper water were also indicative of seasonality in breeding and migrating to spawning areas. The factors controlling these breeding seasons and movements are not yet known, but nutrients injected into the Gulf from the land during the rainy seasons lead to phytoplankton blooms at this time. The peak spawning was 2 months earlier in 1980 (May–June) than in 1979 (July–September), though the reason for this was not clear. The spawning peak made it possible to trace the growth of a cohort of '*H.' nigrofasciatum*, one of the most abundant phytoplankton/detritus-feeders, throughout the year, which indicated that these reach adult size within one year. Nothing is known about the longevity of these fishes in the wild, nor of how the fishes manage to nest on the soft flocculent bottom mud and are able to see breeding displays in the very turbid water.

Brooding females, sexually active males and juveniles are found in the same areas as non-reproductive adults and are therefore affected by the trawl fishery. At peak spawning time, between 20% and 30% of the females in the catches appeared to be brooding young in the mouth. With each brooding female caught, its young are destroyed too. Such a drastic reduction in recruitment will be noticed in the next year's yield, as most of these fishes mature within a year. There is obviously an urgent need to stop fishing over the peak spawning period. Conditions may, however, differ in other parts of the lake, as they do amongst the tilapias.

Lake Malawi

In Lake Malawi the endemic tilapias have well-defined breeding seasons. Among the tilapias at the south end of the lake, *Oreochromis saka* and *O. lidole* were observed spawning in shallow water before the rains (October–November), the hottest time of year, *O. squamipinnis* in deeper water during the rains (December–February), while the most inshore-living species *O. shiranus* had the least restricted season (September–January) (Fig. 4.5). Time and depth of spawning, reinforced by

Fig. 4.5. Spawning grounds and seasons of tilapias at the southern end of Lake Malawi.

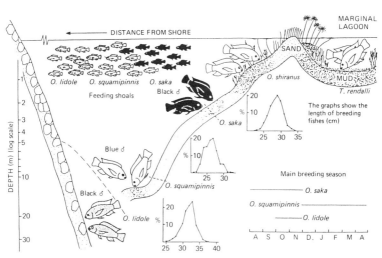

differences in male breeding colour, keep separate *O. saka* and *O. squamipinnis*, the most similar in appearance and ecology (Lowe, 1952, 1953; Lowe-McConnell, 1959, 1969*a*). The shallow-water spawners had predominantly black breeding males; in the deeper-spawning *O. squamipinnis* the breeding male became sky blue with a white top to the head – which should make it more readily visible to females approaching the nesting area from above. In these oreochromine tilapias, as in many other cichlids, the males congregate to display and make nests on certain special spawning grounds, 'lek' display arenas. After spawning, females pick up eggs and sperm in the mouth and move off to brooding grounds, generally in more sheltered waters. From aquarium studies it has been estimated that *O. squamipinnis* has a 6-week 'caring period' under natural conditions (mouth-brooding for 4 weeks, then guarding young for 2 weeks), but that females may be able to produce a second brood in the 4-month breeding season (Berns, Chave & Peters, 1978). These open-water-living Malawi tilapias brood their young to a larger size than do other tilapias, to 50 mm TL in the most openwater-living species *O. lidole*. The larger young are left in nursery areas and live in shallow waters, moving offshore as they grow.

Among the **zooplanktivorous** *utaka* group of cichlids living in the open waters of Malawi, some have more restricted spawning seasons than others. Several species spawn between March and July (Iles, 1960, 1971), the time when water mixing is greatest. Most of them probably have to come close inshore to spawn, where suitable places may be limited, though Eccles & Lewis (1981) have described midwater spawning in at least one species ('*H.*' *chrysonotus*). McKaye (1983) described a 4 km-long breeding arena, in 3–9 m-deep water off Cape Maclear, of the mouth-brooding *Cyrtocara* (= '*Haplochromis*') *eucinostomus*. This arena served only as a mating ground and at times over 50 000 males displayed here. Some males were present throughout the year from October 1977 through March 1981, but numbers fluctuated. Courtship took place in the morning and most of the males left in the afternoon to forage on zooplankton in deeper water, returning at dusk. All parental care was provided by the females which left the arena carrying the eggs. The natural history data suggested that the occurrence of arenas in shallow water helps *C.* ('*H.*') *eucinostomus* to avoid predation by catfish which move in from deeper water to feed on cichlids by night. Cormorant predation probably keeps the males from nesting at high densities in shallower water. Of 120 tagged males, 40% immediately returned and defended the nest from which they were taken. Between 18 and 46 eggs are laid at a time in this species. Data from another species in the area, *C.* (*'H.'*) *argyrosoma*, indicated that individual males remained on the arena for about 2 weeks. Some of the other species were found to have more restricted breeding seasons, whereas in *C.* (*'H.'*) *eucinostomus* some breeding activity could be seen at all times of year, though numbers increased at times which appeared to be correlated with high plankton densities.

Among the *mbuna* **rock-dwelling cichlids** many species breed through-out the year, with peaks when food is most abundant: August–October when plankton blooms, February–March when epilithic algae abound (Ribbink *et al.*, 1983*b*; Marsh *et al.*, 1986). The choice of spawning sites appears to be species-specific. Species living where sand occurs dig saucer-shaped nests; some favour areas alongside rocks. Most lithophil-ous species hide among rocks to spawn, though they may dig sand nests in aquaria. Ribbink *et al.* (1983*b*) noted that unlike behaviour patterns associated with feeding, which vary very greatly between species, courtship and spawning activities are very similar in most cichlids and are therefore of little value in species determination. However, degrees of territoriality range from non-territorial to very strongly territorial. Most *Labidochromis* species are non-territorial; weakly territorial species include members of the *Melanochromis melanopterus* complex in which territories are used for spawning only (vacated for feeding) and are only defended against conspecifics. Most *Pseudotropheus* and *Petrotilapia* species defend territories fiercely against conspecifics; the *Pseudo-tropheus elongatus* group and some *P. tropheops* are highly aggressive, both sexes defending from all intruders territories in which algal gardens may develop (compare certain coral reef pomacentrids, p. 195). *Mbuna* never guard their young once these become free-swimming; many *mbuna* will eat small fish and the young get straight into holes and crevices amongst the rocks. Some other rock-dwelling cichlids do guard their young after they have become free-swimming, and in Malawi a great many fishes adapted for life over sand visit rocky places to release their fry.

Ribbink, Marsh & Marsh (1981) have recorded at least 13 cichlid species in Lake Malawi hosting foreign fry, and at least 15 species guarding mixed broods of two or more species. Are these accidentally incorporated with their own young? Or are some species 'cuckoos' leaving their young with other species to rear, as Ribbink (1977) had suggested for *'H.' chrysonotus*, a surface phytoplankton-feeder which has at least three species incubating its young? Do mixed broods benefit the host fish, as has been suggested for *Cichlasoma* species in a Nicaraguan lake, or are the foreign fry undesirable additions? McKaye & McKaye (1977) had found the herbivorous *Cichlasoma nicaraguense* caring for the fry of one of its predators, *C. dovii*; they suggested that this 'altruistic' behaviour might be repaid later when *C. dovii* preyed on another species of herbivorous cichlid which competed with *C. nicaraguense* for the same natural resources. More recently, Ribbink *et al.* (1980) found that the mixed broods in Malawi came from one of two categories: those such as the *utaka* '*H.*' *chrysonotus*, which do not guard their free-swimming fry (taking them to the substratum or leaving them among the rocks), and those with well-entrenched protective care. They observed a brood of '*H.*' *kiwinge* fragmented by intruding predators, also the merging of two broods of different species in which the parental mouth-brooding fish could not accommodate all the mixed brood, so that some of its own fry were left out. This suggested that even if substratum-spawners benefit from 'kidnapping' other young (as McKaye & McKaye (1977) had suggested), foreign fry could be unwelcome guests in a mouth-brooder as they could lead to losses of its own young.

Lake Tanganyika

In Lake Tanganyika, cichlids have also been found hosting foreign fry. Yanagisawa (in Kawanabe, 1983) observed the substratum-spawning *Lamprologus elongatus* guarding *Perissodus microlepis* young as well as its own, and *P. microlepis* guarding *Hoplotaxodon microlepis* young. A parent *P. microlepis* whose mate was experimentally removed (Yanagisawa in Kawanabe, 1983; Yanagisawa, 1985) was seen to transport its fry and put them in the care of another *P. microlepis*, indicating that this species can behave as a 'cuckoo'. Circumstances also suggested that *H. microlepis* may leave its fry with *P. microlepis*, since *H. microlepis* guarding large fry are rarely found. Clearly there is much variation in cichlid breeding behaviour and much remains to be discovered. The variability of behaviour between species is well illustrated in the substratum-spawning *Lamprologus* studied by Nagoshi (1983) (see p. 106).

Lake Tanganyika cichlids show a whole range of breeding behaviour and parental care, well described by Brichard (1978, 1979) and investigated by a Japanese/Zaire team (Kawanabe, 1983). Unlike the other African Great Lakes, in which most species are maternal mouth-brooders, over 40 species (30% of Tanganyika's cichlid fauna) are substratum-spawners. These include 34 species of *Lamprologus* (a genus also present in the Zaire River) and the related endemic monotypic genera *Julidochromis* and *Telmatochromis*, all mostly rock-dwellers, and of the sand-dwellers the huge *Boulengerochromis microlepis* and some *Lamprologus* species, mostly small territorial species which choose sheltered places in which to spawn, some spawning as deep as 25 m. Twenty genera are known, or suspected with good reason, to be mouth-brooders: *Astatotilapia, Aulocranus, Bathybates, Callochromis, Cardiopharynx, Cyathopharynx, Ectodus, Grammatotria, Hemibates, Lestradea, Limnochromis, Limnotilapia, Petrochromis, Simochromis, Tanganicodus, Trematocara, Tylochromis, Tropheus, Oreochromis, Xenotilapia* (Poll, 1956; Coulter 1968; Oppenheimer, 1970; Brichard, 1978). Recently the maternal mouth-brooder *Perissodus microlepis* has been found to have features intermediate between mouth-brooders and substratum-spawners; the female mouth-broods eggs and embryos but the male stays near her and helps to guard the free-swimming young until they become independent of the parents (Yanagisawa & Nshombo, 1983).

From his extensive underwater observations of cichlids in this lake, Brichard (1978) commented that there appears to have been a gradual evolution from those which spawn in nests built and defended by the male, to direct buccal incubation without the use of any spawning site. Substratum-spawners are generally monogamous; mouth-brooders are polygynous and may be polyandrous too. The schooling fishes are generally polygynous mouth-brooders. In the mouth-brooding *Xenotilapia* there appears to be synchronous spawning (thought by Brichard to be an antipredation measure), in contrast to the substratum-spawning *Julidochromis* which spawns almost continuously (for details see Brichard, 1978). Polygynous spawning may occur without use of a nest site for contact between the sexes as in *Ophthalmochromis* and *Ophthalmotilapia*, species with elongated ventral fins ending in a blob of coloured tissue, and *Cyathopharynx*, or on a makeshift site in the rubble in *Simochromis, Eretmodus, Spathodus* and *Tanganicodus*, spawning occurring when ripe adults happen to meet. Direct buccal incubation occurs without the use of any spawning site in most *Tropheus* species; eggs laid one at a time are picked up by the female and fertilized immediately as her head bumps the male near his sperm duct. The release of fry to the substratum by *Tropheus* and other rock species has never been observed in the wild. Fry less than 15 mm long must live in deep concealment.

The rock-dwelling mouth-brooders are usually solitary wanderers within the limits of their habitat, territorial only (and then not always) when they spawn. This behaviour contrasts with the individualistic, territorial sedentary species where egg-laying and fry-raising occur on the spawning site. The general rule seems to be that where fishes have a territory and are sedentary they are substratum-spawners and when they are wanderers they take the eggs along in the mouth.

In the sand-dwellers, buccal incubation is known for all species living in very large schools and roaming over large areas. Group stimuli may synchronize spawning and perhaps the release of fry (an antipredation measure?) as in *Xenotilapia*. Buccal incubation also occurs in some large species (such as *Limnotilapia*). Open crater nests are made by large species of substratum-spawners, such as *Boulengerochromis microlepis* which comes up from deep water to spawn, both parents defending eggs and young, and large *Lamprologus*, which spawn on their usual grounds. Small species of *Lamprologus* have sheltered nests. Matthes (1961) suggested that most species spawn throughout the year and that the largest cichlid *B. microlepis* has three spawning peaks a year. Annual oscillations in lake level (of *ca* 1 m only) create lakeside pools wherein *Astatotilapia burtoni* spawns undisturbed by larger species (Fernald & Hirata, 1977).

Variations in mating patterns and parental care are well shown in the genus *Lamprologus*. Among ten species studied by Nagoshi (1983, 1985) in the 20 m × 20 m quadrat at Luhunga, each species showed a sub-stratum preference for feeding, spawning and parental care. Breeding territories were very small and never overlapped. All species were substratum-spawners, but they showed three types of parental care, each associated with a particular type of mating system: (1) five species (*L. elongatus, L. lemairei, L. savoryi, L. toae* and *L. tretocephalus*) were monogamous, both parents guarding eggs and young; (2) two species, *L. modestus*, the commonest and most ubiquitous species, and *L. furcifer*, a frequenter of dark places, were polygynous with a harem system – the male patrolling a territory in which several females guarded eggs and young; and (3) in *L. brichardi*, the most gregarious species, fry were guarded by several adults in a communal nursery. The other species (*L. compressiceps* and *L. leleupi*) did not spawn in the quadrat (probably spawning deeper than 12 m). Characteristics of the species-specific breed-ing territory determined whether the school of fry could expand horizon-tally or vertically as the fry grew; the growth of the fry was evidently retarded where space was restricted. Yanagisawa (in Kawanabe, 1983) suggested that among these Tanganyika *Lamprologus* defensibility of the fry in the habitat where the particular species spawns may be one of the important factors in the development of different parental care patterns and mating systems.

In *Lamprologus brichardi* and a number of other Tanganyika *Lamprologus* and *Julidochromis* species, conspecific helpers join in territorial defence and brood care maintenance, a phenomenon studied in field and aquaria by Taborsky & Limberger (1981). In *L. brichardi* families generally consist of the reproducing pair and several broods of their offspring all defending a common territory round the shelter site (Limberger, 1983). Costs and benefits of this brood care help have been analysed by Taborsky (1984). The helpers' defence is mainly (83%) against space (hole) competitors (such as *Telmatochromis* species), rather than against predators on the young (*L. elongatus* and *Afromastacembelus*), and in digging away drifted sand from eggs and larvae. In the sea, subdominant anemone fish (*Amphiprion alcallopisos*) help the breeding pair defend the anemone, but have not been seen to help with brood care; as their larval stages are pelagic, these subdominants are unlikely to be related to the breeding pair.

Thus, increasingly detailed studies emphasize how structured these lacustrine cichlid communities are, but differences are not so much in foods used by different species as in utilization of space to obtain this food and for spawning sites. Interactions with predators as they obtain food and spawn have yet to be explored.

5

Speciation: the African Great Lakes as laboratories of evolution

The high endemism in the African Great Lakes makes it clear that many fish populations respond to the lacustrine conditions in such a way as to become recognizable as distinct species, or even distinct genera. A species is generally taken to be a group of interbreeding natural populations which is reproductively isolated from other such groups. Generic distinctions rely more on the degree of difference as interpreted by various taxonomists. Within these lakes morphological differences within the family Cichlidae far exceed those between whole families of fishes in the sea. These lakes provide some of the best natural laboratories in which to study the ecological and behavioural changes accompanying the evolution of new species. A great deal has been written about speciation in these cichlids, much of it speculative, some of it controversial; see, for example, papers prepared for a recent symposium on the evolution of fish species flocks (Echelle & Kornfield, 1984). Earlier reviews by Fryer & Iles (1972) and Greenwood (1974, 1981) provided detailed analyses.

Controversies aroused include the following: (1) Can changes in habits and behaviour, engendered by changed ecological conditions, lead to speciation (as man-made lake studies seem to suggest and as Fryer (1977) held), or do genetically induced mutations, generally accompanied by slight morphological changes, generally occur first on which selection can than act (as held by Liem (1974) and Liem & Osse (1975))? (2) Are the barriers to gene flow leading to speciation always physical, albeit microgeographical, leading to allopatric speciation, or can changes in breeding behaviour (or colour) lead to sympatric speciation without any physical barriers to interbreeding, as discussed by McKaye *et al.* (1984)? (3) What is the role of predators in speciation? Do they inhibit it (as suggested from early studies of lakes with and without *Lates*), or promote it by contributing to the barriers between populations (as suggested for some riverine

fishes (Lowe-McConnell, 1969a))? (4) What is a species flock? Is this term
only applicable for species of monophyletic origin (as discussed in Echelle
& Kornfield, 1984)? (5) What is the role of polymorphism, as in the
Mexican cichlid *Cichlasoma minckleyi*, in speciation (and as discussed by
Hoogerhoud, Witte & Barel (1983) and Liem & Kaufman (1984))? (6)
What are the roles of sexual and social selection in cichlid speciation (as
suggested by Dominey (1984))?

Discussions bearing on whether some of the numerous sympatric lake
cichlids should be considered species, incipient species or polymorphic
forms (see Lewis, 1982) are included in recent papers on haplochromine
hybridization experiments by Crapon de Caprona & Fritzsch (1984) (who
also list known cichlid hybrids), and on the consistency of morphological
differences between wild-caught and tank-kept haplochromines by Witte
(1984b). (Another relevant paper is on the significance of microgeographical diversity in a South American gymnotid, *Eigenmannia macrops*, by Lundberg & Stager (1985).)

Among the **non-cichlids**, groups of endemic species apparently evolved
within one lake (though not in every case of monophyletic origin) include:
the small pelagic, openwater-living *Alestes* in Lake Turkana (2 species)
(Hopson, 1982); the *Dinotopterus* clariid catfishes (10 species) in Lake
Malawi; in Lake Tanganyika the bagrid *Chrysichthys* (6 species),
Synodontis (5 species), *Afromastacembelus* (7 species), *Lates* (4 species),
as well as the two species of clupeid considered generically distinct and
several other endemic genera and species (Poll, 1953). We have already
seen how some of these, such as *Dinotopterus* in Lake Malawi (p. 89) and
Chrysichthys in Lake Tanganyika (p. 91), share resources within one lake.

It is, however, the **cichlids** which have speciated most abundantly in
these lakes. Each of the three African Great Lakes now hold more species
of endemic cichlids than the number of species (192) making up the whole
of the European freshwater fish fauna, and far more forms of cichlid have
been recognized in the field than have yet been described as species.
Greenwood (1984) cites 200 cichlid species for each of lakes Malawi and
Victoria, whereas Ribbink *et al.* (1983b) consider that Lake Malawi
probably supports 400–500 species of cichlid fish, and Witte (1984a)
reports that the HEST research group has distinguished more than 250
different forms of haplochromines in the Mwanza Gulf (of Lake Victoria)
alone.

Until recently it was thought that the haplochromines in Victoria had
radiated within the lake basin from a generalized ancestral form which
might have resembled '*Haplochromis*' *Astatotilapia brownae* (a close

relative of *'H.' Astatotilapia nubilis*, a widespread species in Ugandan rivers). More recent phylogenetic analyses led Greenwood (1980, 1981) to distinguish lineages, many of which have representatives in lakes west of the Victoria basin (Edward/George and Kivu), and Greenwood now recognizes 20 distinct genera among the Victoria haplochromines (see p. 82), though their validities are not universally accepted (Hoogerhoud, Witte & Barel, 1983; Hoogerhoud, 1984). Why have tilapias not undergone the same explosive speciation as haplochromines? Fryer & Iles (1969) suggested that their 'generalized' state allowed tilapias to adapt to new conditions without diversification, but Trewavas (1983) pointed out that tilapias are probably too trophically specialized to do so.

Why cichlids? Liem (1974) pointed out the role of the pharyngeal jaws in freeing the mouth for food collection and manipulation, and in later papers (Liem, 1978, 1980) he pointed out that many of the same types of change occur repeatedly, which helps to explain the many examples of parallelism and convergence in body form and function between the cichlids of the various lakes. Greenwood (1981) considered that it is the suitability of the basic body plan, neither too generalized nor too specialized, for simple gene changes, such as those affecting relative growth rates and dental specializations, to effect morphological changes fitting the cichlids for very diverse trophic behaviour and also their ability to breed throughout the year and, for many of them, short generation times, permitting greater opportunities for genetic reshuffling than in annually breeding fishes. Fryer (1977) stressed behavioural attributes, the role of parental care, social communication and competition as creative forces acting on cichlids. Poll (1980) pointed out that cichlids, being relatively euryhaline, can withstand the higher ionic contents of lake waters compared with rivers, and that their physoclistous state (closed swimbladder) enables them to specialize for life at a particular depth in the lake (as Ribbink & Hill's (1979) experimental work has shown), in contrast to the stenohaline, physostome otophysan riverine fishes. Ribbink *et al.* (1983*b*) considered that cichlid adaptability, resilience and dietary flexibility all contribute to their success. Dominey (1984) focused on how changes in mating systems, brought about by sexual and social selection, could produce rapid reproductive isolation between diverging populations. He suggested that social selection (see West-Eberhard, 1979, 1983) rather than sexual competition might be responsible both for the rate of speciation and for maintenance and divergence of monomorphically brilliant species-typical colorations, for example among the biparental cichlids in Lake Tanganyika where both members of a pair defend a spawning site which may be a limiting resource.

Fryer has tabulated the numerous morphological and/or **ecologically equivalent cichlid species** in the three Great Lakes (Fryer & Iles, 1972). For example, in rocky habitats: the algae-feeding rock-scrapers with file-like teeth – *Tropheus* species in Tanganyika, *Pseudotropheus* in Malawi (see Fig. 4.2) and *Neochromis nigricans* in Victoria; those with long slender mobile teeth with spatulate tips – *Petrochromis* in Tanganyika, *Petrotilapia* in Malawi; invertebrate-pickers with forceps teeth – *Tanganicodus irsacae*, so like *Labidochromis velicans* in Malawi and *Paralapidochromis victoriae*; and invertebrate-eaters with fleshy-lobed lips – *Lobochilotes labiatus* in Tanganyika, '*Haplochromis*' *euchilus* in Malawi, *Paralapidochromis chilotes* in Victoria. In sandy habitats there are: the invertebrate-eating sand-sifters – *Xenochromis* species in Tanganyika, with *Lethrinops* species in Malawi; and mollusc-crushers with massive pharyngeal bones – *Lamprologus tretocephalus* in Tanganyika with '*Haplochromis*' *placodon* in Malawi (Fig. 4.1) and *Labrochromis* species in Victoria. In open water there are: zooplanktivores – such as *Limnochromis permaxillaris* in Tanganyika with the *utaka* group of '*Haplochromis*' in Malawi; and elongate slender-bodied piscivores – *Bathybates* in Tanganyika, *Rhamphochromis* species in Malawi (Fig. 4.1), *Harpagochromis* and *Prognathochromis* species in Victoria. Deepwater-dwellers include those with conspicuous acousticolateralis pits on the head – *Aulocranus* in Tanganyika, *Aulonocara* in Malawi (Fig. 4.1) – and scale-scrapers with similar habits but different dentitions – *Plecodus* and *Perissodus* in Tanganyika, *Corematodus* and *Genyochromis* in Malawi (Fig. 4.1), *Allochromis welcommei* in Victoria. Whether these similarities are due to parallel evolution in related forms or to convergences in unrelated ones is still under investigation (see Stiassny, 1980; Greenwood, 1983*a,b*). Convergence in feeding habits is also shown by feeders on argulid parasites – *Melanochromis crabro* in Lake Malawi and some newly discovered haplochromines in Lake Victoria (Ribbink & Lewis, 1982).

Among these lacustrine cichlids **reproductive isolation** leading to the evolution of sibling species often occurs with very little morphological change. For example, among the Malawi tilapias the morphologically similar *Oreochromis squamipinnis* and *O. saka* can be most easily distinguished by the male breeding colours. In cichlids the colour of the breeding male appears to be of particular significance for successful spawning (as a good deal of experimental work in aquaria has shown). Colour polymorphism is known amongst certain cichlid species in all three Great Lakes; orange-blotched and normal monochromatic forms occur in the tilapia *O. variabilis* in Victoria, nearly always in females, and also in a number of haplochromine species; in Malawi, Fryer (1959) noted colour polymorphism in five species of *mbuna* (species of *Pseudotropheus, Labeotropheus, Genyochromis*). Very many colour forms are now known (see Ribbink *et al.*, 1983*b*). Whether these different colour forms represent intraspecific polymorphism, or should be regarded as incipient species, or are reproductively isolated sibling species, is very difficult to determine. The key question is whether there is any evidence of assortive mating between the colour forms. Dominey (1984) has recently made a very good case for the important roles that sexual and social selection have probably had in the evolution of these cichlid 'species flocks'.

Where two or more species occur sympatrically, underwater SCUBA observations of their ecology and breeding behaviour can be made to assess the likelihood of complete reproductive isolation and, if so, whether this is likely to be due to microgeographical differences in where and when they spawn initiating their differentiation (as in allopatric speciation), or in any form of behaviour keeping the forms apart (as in sympatric speciation).

The significance of **colour morphs** in allopatric populations is much more difficult to determine. Starch-gel electrophoresis can however be used to indicate the amount of gene flow between the various populations (Kornfield, 1978). Such analyses have been spearheaded amongst the polymorphic complex of *Pseudotropheus zebra* and *Petrotilapia tridentiger* in Malawi (McKaye *et al.*, 1982, 1984). Fryer (1959) had reported colour morphs of these two species. Field studies of four colour morphs of *Pseudotropheus zebra* led Holzberg (1978) to conclude that these must be considered two sympatric species with complete reproductive isolation, probably based on an ethological mechanism, the males of one population discriminating against the females of the other, or females choosing a particular male. They form two mating groups: 'BB/OB' (blueblack males and mostly orange-blotch females) and 'B/W' (blue males, white females), respectively. B males have been observed to prefer steep rock surfaces where they occupy slightly larger territories and feed on plankton more frequently than the BB males (which show no preference for steepness of slope). Holzberg hypothesized that the B/W gene pool evolved from the BB/OB gene pool by several stepped mutational events, with ethological mechanisms enforcing the separation into two distinct sympatric populations. Comparative studies of the morphology and agonistic behaviour of the colour morphs in aquaria (Schroder, 1980) then showed that the BB/OB fishes were shorter, more compact and less aggressive than the B/W fishes. Schroder concluded that all the behavioural, ecological and breeding data justify the systematic division of *P. zebra* into two species. McKaye *et al.* (1984) found populations of the BB/OB group at several rocky localities from the north to the south end of Lake Malawi but only the Nkhata Bay population was accompanied by B/W pairs. The slight electrophoretic differences between the two species were found to be of the same nature, though of greater magnitude, as those between separate populations of the BB/OB pairs. They concluded that although the two species are now sympatric at Nkhata Bay, the possibility that at least some of their distinctive features were developed in microgeographical isolation (allopatry) cannot be ruled out, and that cichlid differentiation can be caused both by extrinsic geographical isolation of populations and by intrinsic behavioural factors which could lead to sympatric speciation. Trewavas (1983) has suggested that sympatric divergence may follow allopatric change. Ribbink *et al.* (1983*b*) recognized 27 species, or members of the species complex, as well as numerous colour forms which might be races or subspecies, or perhaps even additional species, among the *P. zebra* populations in Lake Malawi. They also recognized 17 colour forms of *Petrotilapia tridentiger*, at one time considered to be monospecific. They considered these colour forms to be species, of which three have now been described on the basis of adult coloration, maturation size, distribution, depth preference, degree of overlap of their territories, aggression and courtship behaviour (Marsh, Ribbink & Marsh, 1981). A genetic analysis of three of this complex suggested that these taxa are isolated sibling species which diverged recently, or are incipient species with minimal gene flow between the morphs (McKaye *et al.*, 1982).

Ribbink *et al.* (1983*b*) encountered what they believed to be sibling species at virtually every site studied; for example, at Nkhata Bay, seven of the *Pseudotropheus tropheops* complex, three of the *P. zebra* complex, three of *Petrotilapia*, two of *Labeotropheus*. Though these occur sympatrically, Ribbink *et al.* hypothesized that changes in the lake level could have divided a single gene pool into two or more gene pools (as already suggested by Fryer (1977) which might then have diverged during periods of isolation, each becoming adapted to a particular set of environmental conditions, remaining reproductively isolated when changing lake levels brought populations together again. So these sympatric species, could, in their view, have evolved allopatrically. *Mbuna* are markedly limited to particular habitats; studies in pressure tanks have revealed their stenotopy to a particular depth range (Marsh & Ribbink, 1981), suggesting that most would be unlikely to cross deep troughs separating rocky shores. Furthermore, their mouth-brooding habits and leaving the young amongst rock crevices means that some *mbuna* may spend their whole life within a few square metres (in striking contrast to most coral reef fishes, see p. 203).

Colonization studies of eight artificial cement-block reefs placed on a sandy bottom in 6 m- and 9 m-deep water off Cape Maclear, 1 km away from the nearest rock outcrop, showed that certain species of *mbuna* can move to the new habitats across stretches of sand; from 1978–83, 75 species were recorded from these artificial reefs, including nine resident *mbuna* species with territorial males or brooding females (McKaye & Gray, 1984). The early *mbuna* colonizers belonged to three groups which feed extensively on zooplankton, members of the *Pseudotropheus zebra*, *P. tropheops* and *Petrotilapia* complexes, which are also ones which Liem (1978, 1980) suggested are facultative in their feeding repertoire. Such species groups could provide founders for new populations. Interestingly, the *mbuna* groups which feed extensively on zooplankton and colonized the experimental reefs would appear to be the least likely to have restricted gene flow, yet they are amongst the most species-rich groups of *mbuna*, so clearly much further work is required to understand the evolutionary processes involved. Ribbink (personal communication) reported that the migrants to their reefs and to newly drowned rocky shores when the lake rose (as it did by 7 m between 1965 and 1980) were the widely distributed *mbuna* and not, in their view, the most speciose.

When considering speciation amongst Malawi's ***mbuna***, Ribbink *et al.* (1983*b*) stressed that changes in lake level have probably been one of the most powerful factors. They attributed the high rate of evolution among *mbuna* to be due to: (1) geographical isolation of small patches of rocky habitat; (2) the effect of changing lake levels; (3) *mbuna*'s philopatric tendencies, staying put in one place; (4) their stenotopic nature, and specific adaptations to microhabitats; (5) their retention of the ability to feed on alternative food sources; and (6) their ability to adapt to new environments. Climatic evidence suggests that the last 2 million yr have seen perhaps 20 interglacial cycles, each causing profound expansions and contractions of Africa's vegetational zones and affecting lake levels.

Ribbink *et al.* (1983*a,b*) considered that evidence for intralacustrine allopatric speciation among the *mbuna* in Malawi is excellent and that there is no evidence for non-allopatric methods of speciation among them. On the other hand, McKaye, who studied species dwelling on sand where physical barriers are less evident, considered that sympatric speciation may occur; he was also influenced by his earlier studies on assortive mating between gold and grey morphs of *Cichlasoma citrinellum* in Lake Jiloa, Nicaragua (McKaye, 1980; Barlow, 1983). Among the Malawi tilapias studied personally (Lowe, 1953), differences in depth and time of spawning reinforced by differences in male breeding colours appeared to keep apart the two most closely related species, *Oreochromis saka* and *O. squamipinnis*. Here it would seem that an extended breeding season could have become divided into two distinct ones, and differences in prevailing hydrological conditions before and during the rains could perhaps account for differences in depths of spawning; also, deeper spawning could have selected for the lighter coloration (especially the white head) of the deeper- and later-spawning *O. squamipinnis*, by making the males on their nests on the lek spawning arena more readily visible to the females circling above them; it was also suggested that the long-term changes in lake level affected spawning places. In Lake Victoria, Greenwood (1981) considered that the geomorphological and hydrological history of the lake suggests that allopatric speciation, through actual geographical isolation, played the prime role in the establishment and diversification of the haplochromine species flocks, but that sympatric speciation may also have had a role to play when one considers the immense numbers of species present, differing mainly in the colours of the breeding males.

Colour morphs are not, however, always to be regarded as incipient species, for example among many polychromatic cichlids in Malawi and Victoria. Marsh (1983) drew attention to marine studies on the serranid *Hoploplectrus* in which different colour forms do not mate assortively and electrophoretic analyses have indicated that this is probably one highly polymorphic species. In the Mexican cichlid *Cichlasoma minckleyi* electrophoretic studies and observations on the lack of assortative mating suggested that forms with molariform or papilliform pharyngeal teeth are conspecific (Kornfield *et al.*, 1982; Kornfield & Taylor, 1983), confirming an earlier hypothesis by Sage & Selander (1975) who had found these fishes using a variety of foods (algae and detritus, molluscs, fish). Liem & Kaufman (1984), who studied the functional morphology of these forms, concluded that the extremely specialized morphology of the molariform morph does not increase efficiency on preferred food but probably enhances exploitation of the secondary, less preferred food during eco- logical 'crunches' or 'bottlenecks' (Maynard Smith, 1966; Wiens, 1977). Greenwood (1981) suggested that if the genes controlling these features were to become linked with some reproductive feature (such as male breeding colour), then through the effects of assortative mating the morphs could become true species (a state which the Mexican species has evidently not yet reached). But he concluded that there is no evidence to suggest that balanced polymorphism could explain the situation in African lake 'flocks', although it might have played a part in their origins. Maynard Smith (1966) had suggested that stable polymorphism could be the first stage in sympatric speciation.

The controversial **role of predators** in speciation was discussed by Fryer & Iles (1972). Impressed by the paucity of cichlid faunas in Lakes Turkana and Albert, Worthington (1937) had suggested that the presence of large piscivores, such as *Lates*, controlled cichlid speciation in these lakes. Fryer & Iles had countered this by emphasizing the large numbers and kinds of piscivores in Lake Malawi; moreover, Lake Tanganyika has *Lates*, and Lakes Edward and George once had them (witness the fossil *Lates* found near Lake Edward). Lowe-McConnell (1975) considered that data from Lake Tanganyika suggested that predators quite probably inhibited speciation amongst the openwater-living fishes but may have promoted speciation among the rock-dwelling ones by increasing isolation between the various populations. The effects of introduced *Lates* on Lake Victoria cichlids (see p. 314) would seem to support Worthington's original suggestion.

The **rates** at which new species evolve are hard to determine. However, in Lake Nabugabo, a small (30 km^2) lake cut off from the northwestern shore of Lake Victoria by a narrow sandbar some 4000 yr ago, five of its seven cichlid species are endemic, differing in male coloration and some morphological characters from their closest relatives in Lake Victoria (Greenwood, 1965).

The rivers of Uganda have few cichlid species, which suggests that the early colonizers of the embryo Lake Victoria would have been mostly non-cichlid species, with which the initial cichlid colonizers would have had to compete for resources. Dominey (1984) suggested that the failure of the non-cichlid competitors to adapt to lacustrine conditions (which he attributed to continued gene flow with fluvial-adapted populations as they continued to breed in the rivers), was an important factor in the success of cichlids colonizing the lakes.

Greenwood (1984) considered that the haplochromines in Lake Victoria present an outstanding example of an extant punctuational evolutionary phase (as discussed by Fryer, Greenwood & Peake, 1983, 1985). During the 750 000-yr history of the lake these endemic fishes have come to occupy a dominant position in the lake's ecology as a result of their extensive and diverse trophic specializations. Greenwood has been able to trace clearcut trends in the development of different anatomical specializations associated with various feeding habits, anatomical features which appear to have been brought about mainly through simple changes in relative growth, affecting mainly syncranial characters and oral and pharyngeal dentition. The stages of derivation in any one morphocline are represented by distinct species, often with more than one species at a particular level of specialization still extant. This suggested to him that natural selection has not been of prime importance in the origin and evolution of these species flocks.

Thus it seems that both adaptive changes, in response to the changing ecological conditions in these lakes, and changes due to sexual and social selection may have contributed to the outstandingly splendid species 'flocks' in the African Great Lakes. We have a great responsibility to see that these are preserved. In Lake Victoria the destruction of the haplochromines by the introduced piscivorous *Lates* is now proceeding with alarming rapidity, a process which it will almost certainly be impossible to control. This tragic case provides a dire warning and stresses the need to guard the species flocks in the other two lakes. As it has happened, however, it should be regarded as a giant experiment and the opportunity should be taken to study the effect of the introduction on the overall ecology and fish production from Lake Victoria when the *Lates* population has stabilized.

6

The Neotropical fish fauna

We have looked at riverine and lacustrine communities with African faunas subject to African conditions. How do studies in other parts of the tropics corroborate, refute, or enlarge these findings?

The Neotropical fish fauna, examples of which are shown in Fig. 6.1, is the most diversified and richest freshwater fish fauna in the world, with more than 2400 species already described. It differs from the African fauna in being derived from fewer basic stocks of fish. It lacks the rather primitive families endemic to Africa and cyprinoids and is richest in characoids and siluroids, developed by spectacular adaptive radiations initiated during South America's long isolation during the Tertiary. In the Amazon, 85% of the fishes are otophysans (compared with 54% in the Zaire), 43% are characoids, 39% are siluroids and 3% are gymnotoids. Roberts (1972) has suggested that the present Amazonian fish fauna with its large number of species might be the product of a million years of evolution from an original stock of two or three hundred founder species, but we do not know phyletic time scales. The numbers of species in families endemic to South America shown in the Appendix are only very provisional, as whole river systems have yet to be explored.

The Neotropical fauna now consists of: (1) a few representatives of widely distributed groups, such as the lungfish *Lepidosiren*, the osteoglossids *Arapaima gigas* and *Osteoglossum* (2 species) and two small nandid species; also, numerous cyprinodontiform killifishes (see Parenti, 1981, 1984*a,b,c*) which include the endemic families Poeciliidae, Anablepidae and a group (Orestines) which has speciated in Lake Titicaca and other high lakes of the Andes; (2) about 50 representatives of predominantly marine groups, such as stingrays (*Potamotrygon*), clupeids, engraulids, sciaenids, achirine flatfish, belonid needlefish and endemic freshwater species of some six other families (the general low level of the land and wide river mouths with greater areas of euryhalinity have allowed easier penetration of freshwaters by marine fishes here than in Africa); and (3) fishes of 'Gondwanaland' stock – otophysan characoids and siluroids (catfishes), and percomorph cichlids.

Cichlid genera are more numerous, *ca* 20 genera with 100–200 species, and on the whole are better differentiated in South American rivers than they are in African rivers: they are prominent in the lateral lakes of the river systems. There are no cichlid species flocks comparable to those in the African Great Lakes, though the genus *Cichlasoma* has speciated in Central America (*ca* 70 species) and cichlids are well represented in Nicaraguan lakes (Miller, 1966; colour photographs in Mayland, 1984). (Neotropical cichlid nomenclature has been revised by Kullander, 1983.)

The adaptive radiations of the characoids (Fig. 6.1b) are as dramatic as those of the cichlids in the African lakes, but differentiation has proceeded to a higher phyletic level. Over 1100 Neotropical characoid species are known. There is as yet no complete agreement about their family groupings. Greenwood (1975) recognized 13 Neotropical families, all but one (Characidae) endemic; Gery (1984) considered Amazon species to fall into 11 families. The family Characidae represents more than half of the group, with 10–12 subgroups mostly regarded as subfamilies, though their relationships are not yet clear. Characoids include families of small fishes, many of which are well-known to aquarists, such as brilliantly coloured nannostomine pencilfishes (Lebiasinidae), hatchetfish (Gasteropelecidae) which skitter along the surface of forest rivers like freshwater flyingfishes, and darter-like characinids, bottom-dwellers in small streams. But the majority of the small species belong to the huge complex of tetragonopterines (Characidae *s.s.*), the 'tetras' of aquarists. These are mostly less than 12 cm SL, elongate or deep-bodied, dwellers in shallow water at the sides of rivers; many are very similar in appearance and hard to identify (including some examples of mimicry, see Gery (1969)). *Triportheus* species, keeled with large pectoral fins and living in surface waters, are fished for food when larger species are not very abundant. Larger characids include the omnivorous *Brycon* species, valued food fishes.

Most ecological studies have been concerned with the larger species: (1) the 30+ cm deep-bodied serrasalmines, members of an apparently monophyletic group with two main feeding adaptations, teeth specialized for cracking tough fruits (as in *Colossoma*) or for shearing flesh (as in *Serrasalmus* and other piranhas); (2) at least four independent lines of piscivorous fish which swallow their prey whole – (i) pike-like *Acestrorhynchus* (Characidae), (ii) the widely distributed shallow-water

Fig. 6.1(a). South American freshwater fishes I. (A) *Potamotrygon* stingray (30 cm, Elasmobranch); (B) *Lepidosiren* lungfish (50 cm, Dipnoi); (C) *Anableps* foureyefish (15 cm, Anablepidae); (D) *Osteoglossum* (50 cm, Osteoglossidae); (E) *Electrophorus* electric eel (60 cm, Electrophoridae); (F) *Colomesus* pufferfish (Tetraodontidae); (G) *Cichlasoma* (10 cm, Cichlidae); (H) *Cichla* (40 cm, Cichlidae); (I) *Crenicichla* (25 cm, Cichlidae); (J) *Hoplosternum* (15 cm, Callichthyidae; (K) *Ancistrus* (15 cm, Loricariidae), (L) *Megalodoras* (70 cm, Doradidae); (M) *Hypostomus* (15 cm, Loricariidae); (N) *Pimelodus* (30 cm, Pimelodidae); (O) *Arapaima* (150 cm, Osteoglossidae); (P) *Pseudoplatystoma* (120 cm, Pimelodidae); (Q) *Vandellia* (4 cm, Trichomycteridae).

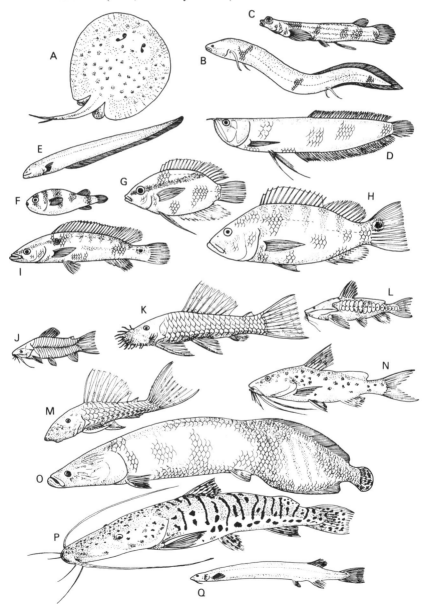

Fig. 6.1(b). South American freshwater fishes II, illustrating the adaptive radiations in characoid fishes. (A) *Gasteropelecus* (6 cm); (B) *Tetragonopterus* (12 cm); (C) *Brycon* (50 cm); (D) *Leporinus* (30 cm); (E) *Anostomus* (12 cm); (F) *Characidium* (4 cm); (G) *Poecilobrycon* (4 cm); (H) *Metynnis* (12 cm); (I) *Colossoma* (50 cm); (J) *Serrasalmus* (30 cm); (K) *Prochilodus* (40 cm); (L) *Boulengerella* (45 cm); (M) *Acestrorhynchus* (20 cm); (N) *Hoplias* (30 cm); (O) *Hoplerythrinus* (25 cm); (P) *Hydrolycus* (60 cm); (Q) *Salminus* (50 cm).

dwellers *Hoplias* and *Hoplerythrinus* (Erythrinidae), considered to be very primitive, (iii) *Boulengerella* (Ctenolucidae) living near the water surface, (iv) *Hydrolycus* (Cynodontidae) with greatly elongated canine teeth; (3) a group of detritivorous fishes with reduced dentition, the Prochilodontidae of *ca* 30 species of large migratory mud-feeders, and the Curimatidae of smaller species schooling near the bottom in open water; (4) hemiodids, streamlined schooling fishes, good swimmers, the toothless lower jaw probably used to select food from detritus or bottom sediments; and (5) anostomids, including medium-sized fish with forceps teeth such as *Leporinus, Schizodon* and *Rhytiodus*, powerful swimmers caught in open waters (where they leap over nets), and small aquarium-sized *Anostomus*, fishes with a vertical mouth and complex teeth used to strip pieces off macrophytes. Much systematic work is needed to unravel the relationships of these complex groups.

The gymnotoids, electric fishes showing remarkable convergences with the unrelated mormyrids in Africa (see Fig. 11.2), are now considered to be an offshoot of siluroid stock (Fink & Fink, 1981). Nocturnal fishes, like the mormyrids, hide away in crevices or other sheltered places by day.

The siluroids range in size from the giant *Brachyplatystoma* species, growing to over 2 m long, which move long distances up- and downriver, to minute trichomyterids living in the gill cavities of other catfishes. *Hypophthalmus* is a midwater plankton-feeder with long sieve-like gill rakers. Many small catfishes such as auchenipterids live in crevices. Aspredinids ('banjo cats' which strum audibly) resemble bits of gnarled wood. Three families are armoured: doradids, with hooked dermal plates along their sides, which live mainly in rivers and lagoons; callichthyids, which live in forest streams where they hunt insect prey such as ephemeropteran nymphs amongst the leaf debris; and loricariids, which live mainly in rocky or stony places, often in streams which become torrential seasonally, where they graze algae off rock surfaces. Bottom-dwellers, all of these, their armour may assist them to keep close to the bottom undamaged by abrasion in fast water. It may also assist them to withstand desiccation when pools dry up. Callichthyids can move through damp leaf litter from pool to pool, using their pectoral spines and body flexures, and are able to withstand great changes of temperature. The armour may also deter predators. Loricariids with the body covered in bony scutes are beautifully adapted for bottom living, with a suctorial ventral mouth and the long coiled intestine of an algae- or detritus-feeder. They live among stones or roots, mainly in small streams, but some penetrate into the torrents of the Andes. Unlike most catfishes, which have sombre colours for nocturnal life, hypostomine ('plecostomine') loricariids are often distinctively marked. Male *Ancistrus* carry branched tentacles on the snout, and male *Pseudancistrus* bristles on the side of the head when in breeding condition. Algae-grazing *Hypostomus* feed by night but hide under rocks by day in clear streams where the light is very bright; an iris lobe, a peg-shaped process, expands over the pupil in the light. Secondarily naked catfishes of the related Astroblepidae (*ca* 40 species) live in Andean torrents. Fishes of the largest family of naked catfishes, Pimelodidae, superficially resemble those of the Bagridae in Africa and Asia, but both families are now thought to be polyphyletic (Howes, 1983). Many pimelodids are very common omnivorous bottom-scavengers, belonging to large genera such as *Pimelodus*; the family includes some very large species of *Brachyplatystoma* and *Pseudoplatystoma*. (See Reid (1983, 1986) for biology of latter.)

Good general accounts of the South American freshwater fish fauna are given by Gery (1969, 1984) and by Fink & Fink (1979). Keys to the different groups of fishes and for different regions are listed in Lowe-McConnell & Howes (1981). Food fish species are listed in Lowe-McConnell (1984), cultured species in Bonetto & Castello (1985).

The distribution of fishes in South America

Gery (1969) recognized eight faunal regions in South America: (1) the Guyanan–Amazonian region with interconnections to (2) the Orinoco–Venezuelan region to the north, and (3) the Paranean to the south; (4) Magdalenean and (5) Trans-Andean in the northwest; (6) Andean and (7) Patagonian south of this, with (8) the East Brazilian in rivers flowing to the Atlantic coast (see Fig. 6.2). The main pattern of fish distribution is of central richness, greatest in the Amazon, with somewhat less in drainages north and south, less in the isolated Magdalena and a moderately poor fauna in the western drainages of Columbia and Ecuador. The Andes have a poor fauna, mainly of specialized torrent fishes and a radiation of cyprinodontoid orestine fishes in the high lakes (Parenti, 1984*a,b*); in the far south there is a transition to a completely different fauna of Antarctic peripheral fishes. In this book we are concerned mainly with studies in the central regions, in the equatorial Amazon system, comparable with the Zaire system in Africa, and with more seasonal rivers in Guyana and Venezuela north of the Amazon and the Paraná–Paraguai system of the Paranean region to the south.

The world's richest freshwater fish fauna, probably more than 1300 species, 85% of them otophysan, is in the Amazon basin, which drains 6.5

million km^2 in the centre of the continent. To the north this connects to the Orinoco system through the Casiquaire canal, which alternates its direction of flow seasonally, and to the Essequibo system in Guyana across the seasonally flooded Rupununi savannas. To the south there were former connections with the vast Paraná–Paraguai system. Many Amazonian fishes penetrate south to the Paranean region which comprises the La Plata–Uruguay–Paraná–Paraguai, the second largest drainage system in South America (3.2 million km^2). Many of the larger fish species are now widely distributed from the Orinoco to the Paraná–Paraguai.

The main Amazon river presents an ecological barrier to many of the smaller fish species; ecological conditions in the peripheral streams are much alike and these share many of the same species. This suggests that a primitive fauna spread around the Amazon basin (then possibly a marine gulf and later large lakes), and that the rich fauna of the Amazon proper diversified later. How much distributions are accounted for by ecological conditions and how much by the geological history of the area is not yet clear. The Amazon basin drained westwards to the Pacific before the rise of the Andes reversed the drainage, probably in Miocene times. The waters were then evidently dammed to form a huge lake, or lakes, before they broke through to the Atlantic in the Pleistocene (Sioli, 1964). During Pleistocene glacial phases, when the sea level was lowered, the rivers would have run in deep valleys. Conditions as the fishes evolved may therefore have been very different from those prevailing today (as in the Zaire system which also had lakes). Today, with the exception of the high Andean Lake Titicaca (lying at 3800 m altitude), South America lacks large deep lakes; the most lake-like waters are the drowned valley mouths of Tapajos and Xingu tributaries of the Amazon ('river-lakes' described by Sioli, 1964).

Sources of ecological information

Many South American rivers have yet to be explored ichthyologically, and fish ecology has only been studied at the relatively few places where facilities for such work exist (Fig. 6.2). To study the ecology of fishes in such a vast, remote and little-explored area is a daunting task, and it is further complicated by the presence of so many fish species, many of them greatly resembling one another, making their correct identification very difficult. Many studies have been made around Manaus, where the blackwaters of the Rio Negro join the whitewaters of the Solimões to form the Amazon proper, by Sioli, Marlier, Fittkau, Knöppel, Junk,

Geisler, Goulding, Bayley, Petrere and a whole new generation of
Brazilian limnologists, many of whom start their field work by autecological studies of a species or group of species in this area (theses from the
University of Amazonas later published in Amazonian journals); see for
example Marlier (1967, 1968), Knöppel (1970), U. Santos (1973), Kramer

Fig. 6.2. Sites of South American studies: 1. Central Amazon;
2. Madeira; 3. Itacoatiara; 4. Negro; 5. Aripuaña; 6. Ecuador;
7. Peruvian Amazon; 8. Rupununi; 9. Mato Grosso; 10. Orinoco;
11. Magdalena; 12. Parnaíba; 13. Paraná; 14. Pilcomayo; 15. Mogi
Guassu; 16. Gran Chaco; 17. L. Titicaca; 18. Panama streams. Cc
Casiquaire canal; M. Manaus; MI Marajo Is.; Tn Tocantins;
Tp Tapajos; X Xingu. (See p. 123 for numbered regions on small map).

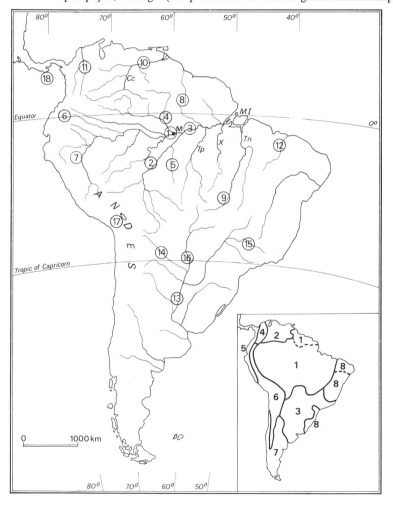

et al. (1978), Petrere (1978*a,b*, 1982, 1983*a,b*, 1985), Fink & Fink (1979), Soares (1979), Almeida (1980), F. M. Carvalho (1980), Goulding (1980, 1981), Bayley (1982), Goulding & Carvalho (1982, 1984), G. M. Santos (1982), Goulding & Ferreira (1984) and other papers listed by Junk (1984). Amazon studies have now been collated in a monograph on The Amazon edited by Sioli (1984). Many of the same fish species have also been studied in other rivers: in Peru, Guyana and the Mato Grosso, also in the Paraná, but these more seasonal waters are considered later. *Arapaima, Cichla, Astronotus, Prochilodus* and *Colossoma* are cultured as food fish in the dry northeast of Brazil where their breeding biology and growth has been studied in ponds (references in FAO, 1978).

Our knowledge of the Amazon fishes and their ecology comes from only a few places scattered in this huge complex of waters: (1) from the central basin – (i) fishes are brought to Manaus fish markets from a wide area, stretching from the Brazilian border 1700 km up the Solimões to 600 km down the main Amazon from Manaus, and from long distances up many tributaries (sites listed and mapped by Petrere (1978*a,b*); (ii) from ecological studies in the Manaus area; (iii) from Itacoatiara, 250 km downriver from Manaus (Smith, 1981); and (iv) from the tributary Madeira (Goulding, 1980, 1981); (2) from peripheral regions – (i) Andean headwater streams at 600 m in Peru (Patrick, 1964), at 340 m in Ecuador (Saul, 1975), and at 200 m in the Peruvian Amazon and its tributaries (Lüling, 1962, 1963, 1971*a,b*, 1975), (ii) far to the north of the central basin in Guyana in waters draining to the Branco/Negro (Lowe-McConnell, 1964, 1975); and (iii) far to the south in Brazilian Mato Grosso waters draining to the Araguaia and Xingu tributaries (Gery, 1964; Lowe-McConnell, personal observations.)

Relatively little information is available on fishes in high-gradient Andean streams. At the other end of the system, 5000 km to the east, little has been published about fishes in estuarine stretches (but see Barthem, 1985). The main body of information comes, therefore, from the very low-lying central basin with its mosaic of biotopes – rivers, streams, swamps, lateral lakes and flooded forest, into and out of which the fishes move with changing water levels.

Central Amazon fish market statistics, available from Manaus (Petrere, 1978*b*), Porto Velho on the Madeira (Goulding, 1981) and Itacoatiara (Smith, 1981), indicate seasonal changes in catches, but these are influenced by the gear used (described by Goulding (1980) and Smith (1981, see also p. 317)); and where and when this is set – whether in the main river, side channels, flooded forest or lakes. For example, in Manaus markets seasonality is most closely related to the low-water migrations in the whitewater rivers and to spawning runs; the total catch hides this as *Colossoma macropomum* is caught throughout the year by heavy fishing for it with gillnets in flooded forest where other migratory species are barely exploited (Goulding, personal communication). Over 100 species appear in Manaus markets (grouped under some 30 local names in market statistics), and nearly 90 species at Itacoatiara (Table 6.1). At Manaus in 1976, *Colossoma* were of prime importance, nearly half the catch handled, followed by prochilodontids. In Porto Velho on the Rio Madeira in 1977–79, prochilodontids, *Brycon* and *Mylossoma* were again important, but here *Brachyplatystoma* and some 17 other large pimelodid and doradid catfish species (Goulding, 1981) were gaffed while migrating up the nearby Teotônio rapids; these were exported to parts of Brazil lacking local fish supplies. Large catfishes were not handled at Manaus as preferred types of fish were abundant. In Itacoatiara market in 1977, migrating prochilodontids were again the main species.

The Amazon system: comparison with the Zaire

Both the Amazon and the Zaire are equatorial forest rivers straddling the equator, with tributaries stretching far to the north and south into more seasonally flooding areas. Compared with the Zaire the Amazon drainage system is much larger (6.5 million km^2) and deeper (90 m in some places), its water mass four times that of the Zaire. Some tributaries rise high in the Andes but in the central basin 2.5 million km^2 are less than 200 m above sea level, mostly covered by dense rainforest. The slope in the central basin is very gentle (contrast a highly variable slope in the Zaire) and tidal influences are felt far upriver. The Amazon water level fluctuations are much greater than those of the Zaire (15 m at Manaus, rarely more than 3 m in the Zaire). The Andean tributaries bring down a very heavy load of silt; this is deposited as a floodplain (*várzea*) of allochthonous material along the main river, averaging 48 km wide and up to 100 km wide at some points. The complex channels and lateral lakes on this floodplain change their form with rising and falling water levels, accompanied by chemical changes and associated fluctuations in availability of oxygen, presenting extremely variable environments to the fishes.

Thus in the Amazon there are great chemical differences between the relatively nutrient-rich silt-laden whitewaters and the black and clear

Table 6.1. *Relative abundances of the most important fishes in Amazonian fish markets*

MARKET	MANAUS	ITACOATIARA	PORTO VELHO (MADEIRA)
Period	1976	1977	1977–79
Total	30 929 tonnes	4701 kg	2675 tonnes
No. species recorded	31 sp. groups	86+	18+
Data source	Petrere (1978)	Smith (1979)	Goulding (1981)
	% weight of total catch		
Characoids			
Colossoma macropomum (tambaqui)	44.0	1.9	0.4
Colossoma bidens (pirapitinga)	4.3	–	0.2
Mylossoma spp. (pacu)	5.5	–	10.0
Semaprochilodus spp. (jaraqui)	16.4 (2)	61.7 (1)	10.0
Prochilodus nigricans (curimatá)	12.0 (3)	5.3	13.0 (3)
Brycon sp. (matrinchão)	3.3	–	–
Brycon cephalus (jatuarana)	–	–	16.0 (2)
Anostomids (aracu)	3.4	–	–

Triportheus spp. (sardinha)	2.3	1.2	0.4
Curimatids (branquinha)	1.8	–	0.3
Siluroids			
Brachyplatystoma flavicans (dourado)	–	2.5	21.0 (1)
Pseudoplatystoma fasciatum (sorubim)	–	1.2	–
Pimelodus spp. (mandi)	–	1.1	–
Loricariids (acari)	0.6	3.7 (3)	–
Other			
Osteoglossum (aruana)	1.5	5.7 (2)	–
Arapaima gigas (pirarucu)	–	1.6	–
Cichla ocellaris (tucunaré)	3.0	1.9	0.3
Other cichlids (acara)	0.6	1.5	–
Plagioscion sp. (pescada)	0.9	–	–
Other species	present	present	present

waters from other tributaries. How much do the various tributaries of white-, black- or clearwater have their own distinctive fish faunas? How much do the fish faunas of the district tributaries interact with one another and with local terrestrial ecosystems?

Ecological conditions in the central Amazon basin

The key to Amazon community structure lies in the mobility of the fishes, as many species move from rivers in and out of lateral lakes, flooded forest and tributary streams with changing water levels and associated oxygen availability. Many species make separate movements in relation to oxygen levels and trophic, breeding and dispersal migrations. So communities are dynamic, their species composition and the relative abundances of different species continually changing, in great contrast with the structured communities in the African Great Lakes.

In the central Amazon basin many large tributaries (17 of them more than 1600 km long) join the main river, known as the Solimões above the Rio Negro confluence and as the Amazon below it. These tributaries fall into the following groups:

1. Andean headstreams, such as Ucayali, Huallaga, Marañon, all tributaries of the Solimões (Upper Amazon) which drain the Cordilleras between 3°N and 20°S. These rivers, swollen by rains from September to June and by snow meltwater, carry vast loads of sediment from mountains to plain to be deposited alongside the main river, forming the *várzea* floodplain of relatively fertile soil allochthonous to the central Amazon region. These 'whitewater' rivers have numerous lateral lakes on the floodplain. The Zaire lacks such a whitewater source as the Andes.

2. Left-bank tributaries, such as the Rio Negro, stained black with humic acids, which rise in the highlands of Guyana and Colombia where they have a northern hemisphere rainfall regime with high water in June.

3. Right-bank tributaries: (i) 'clearwater' rivers that drain much of the Brazilian plateau, such as the Tapajos and Xingu; with headwaters south of 10°S, high water is in March or April; (ii) rivers such as the Purus and Jurua which receive whitewater from Andean foothills then meander in large loops across the floor of the basin, and the Madeira which is a whitewater river with clearwater tributaries.

The whitewaters with their high content of suspended materials throughout the year (transparency only 10–50 cm, pH 6.2–7.2) form an effective barrier to some biota, such as submerged flora and lamelli-branchs. Fishes living here feed mainly on foods of terrestrial origin, or around the floating meadows which develop in whitewater *várzea* lakes, or are piscivores. These lakes have somewhat different faunas from the clear or blackwater *terra firme* lakes. A comparative study throughout the year of two such waters near Manaus, the whitewater *várzea* Lake Redondo and the blackwater *terra firme* Rio Prêto da Eva, less than 100 km apart, illustrated the influence of water chemistry on the ecology of the fishes (Marlier, 1967, 1968). Of 47 species taken from Lake Redondo (mainly by gillnets) and 49 species from R. Prêto da Eva, only six were common to both places, species widely distributed in South America.

The blackwater rivers, with very acid (pH 3.8–4.9) dark waters, extremely poor in inorganic ions (almost distilled water), lack primary production as plankton and floating meadows only develop rarely (Goulding, personal communication). They also lack aquatic insects. Only certain fish species can live here and these have to depend largely on allochthonous forest foods collected from the floodforest (*igapó*). Decomposing vegetation reduces the oxygen content of the water. In the Rio Negro fish kills occur when cold winds blow in May, causing an upwelling of water with little or no oxygen in slow-flowing stretches (Geisler, 1969). These blackwaters are the least productive in the Amazon basin (but see Goulding, Carvalho & Ferreira, in preparation).

Clearwater rivers, such as the Xingu and Tapajos, carry only small amounts of suspended matter, the transparency of their greenish water ranging from *ca* 0.6 to 4 m in rainy and dry seasons, and the pH from 4.5 to 7.8. Drowned river mouths form enormous trumpet-shaped openwater surfaces (110 km long in the Tapajos) extending to the main river. Limnologically these resemble lakes rather than rivers, with clear bright sandy beaches and some deposits of very soft mud. Such river-lakes are characteristic of the Lower and Middle Amazon. Phytoplankton blooms develop in some of them (Xingu and Tapajos) despite the apparent paucity of nutrients in the clearwaters, and shoals of the plankton-feeding catfish *Hypophthalmus* are caught here; these make vertical migrations to the surface and inshore at night.

Fish stocks are poor in the main rivers, except during migration runs, but the density increases near to banks and beaches. Large pimelodid catfishes such as *Brachyplatystoma* and *Pseudoplatystoma* are fished from the main riverbed on lines. Accumulations of fishes occur where rivers meet and at the entrances to side channels, and populations are larger in the numerous side arms where there is less current. The slower the current, the more the fish communities acquire the characters of lacustrine ones, with a high proportion of characoids and cichlids, and fewer large catfishes.

These inland waters thus include a huge system of anastomosing channels, creeks and streams, side arms and mouthlakes, lagoons on the floodplain (*várzea* lakes) and lakes above the floodplain on *terra firme*, flooded forest both on the floodplain and islands and in the inundation zone along the water courses (*várzea* forest on the *várzea* plain, *igapó* forest alongside black and clear rivers), swampy valleys, and in places periodically flooded grassland (*campo*). The relationships of these various types of water are shown in Fig. 6.3.

In the main river the water temperature fluctuates no more than one degree on either side of 29°C. In small shaded forest streams temperatures may be as low as 23.4°C, in wide stagnant pools as high as 33.9°C (Sioli, 1964). Throughout most of the central basin the rainfall (*ca* 3000 mm) is well distributed around the year. In the Upper and Middle Amazon the huge rainfall causes a large annual variation in water flow, leading to intense lateral flooding of forest and savanna. The most spectacular anastomoses of channels (*furos*) occur between the Solimões and Rio Japurá; these facilitate the dispersal of the fishes into the forest at high water. The seasonal flood rises about 6 m at Iquitos, 15 m at Manaus, 12 m at Obidos and 4 m at Belem. Below the Rio Negro confluence the current at low water flows at 2–3 knots (*ca* 1.7 m s^{-1}) increasing to 5 knots

Fig. 6.3. Cross-section through the Amazon valley (after Sioli, 1964).

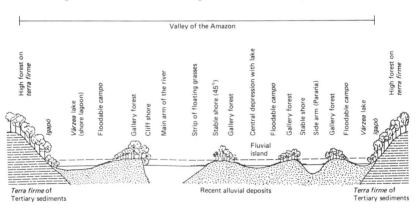

(2.5 m s^{-1}) in some places at high water. The river here averages 25–30 m deep, but is much deeper (nearly 70 m) where it contracts into a narrow channel below Obidos.

The **estuary** has many large islands, and the northern shore has a notable bore (the *prororoca*). The effects of tides backing up the freshwater are felt as far upstream as Obidos; the mixed zone of saltwater and freshwater may move 200 km up- and downriver according to the river flow (Egler & Schwassmann, 1964). Muddy freshwater discharged into the Atlantic is detectable *ca* 100 km offshore. In the estuary much of Marajó Island is covered by 2 m of water from January to May; when the rains subside in May, the island slowly dries up, mostly by evaporation, and the biota become restricted to a few shallow lakes and interconnecting rivers. The largest lake becomes a highly concentrated brine solution. This island lies only just south of the equator, yet a seasonality is imposed on its ecology by events outside the area, by floods which come to it from regions of higher latitude, a seasonality which is reflected in the fish breeding seasons (Schwassmann, 1978).

Floating meadows, formed mainly of aquatic grasses such as *Paspalum* and *Echinochloa* covering many square kilometres in the Middle Amazon, support enormous numbers of individuals and species of animals which use them partly as substratum and partly as food (Junk, 1973). Fishes characteristic of the floating meadow biotopes include many species widely distributed in South America: small *Hemigrammus* and *Hyphessobrycon* characoids, the cichlids *Cichlasoma severum* and *Mesonauta festivum*, also *Gymnotus carapo*, *Synbranchus marmoratus* and a small eleotrid. These fishes remain mainly in the peripheral areas where there is more available oxygen, except for the eel-like *Synbranchus*, well adapted to wind its way among the dense plant masses and able to use atmospheric air. Species also found here by Junk included the predatory characoids *Hoplias malabaricus* and *Serrasalmus*, also *Mylossoma* species, *Leporinus fasciatus*, small *Cheirodon* and *Pyrrulina*, the gymnotoids *Eigenmannia* and *Sternopygus*, the siluroids *Hoplosternum* and *Andoras*, the cichlids *Cichlasoma bimaculatum*, *Acaronia nassa*, *Apistogramma*, and the cyprinodont *Rivulus*. Attachment to floating islands may have helped to distribute some of these species.

Deoxygenation

Decomposition of the abundant vegetable matter at the prevailing high temperatures leads to widespread deoxygenation (hypoxia) in these tropical waters. The seasonal movements of fishes in and out of a *várzea* lake in relation to the oxygen conditions were studied over 20 months by Junk, Soares & Carvalho (1983). Lago Cameleão, a narrow 6.5 km-long lake in a *várzea* island in the Amazon about 15 km above its confluence with the Rio Negro, floods from April to October and almost dries out at low water. This lake was found to be strongly deficient in oxygen for long periods. Of 25 000 fish captured here (132 species belonging to 31 families), about 40 species were able to live at extremely low oxygen concentrations. Twelve of these were air-breathers, another eight species were known to possess adaptations to cope with oxygen deficiency. The authors suggested that the colonization of oxygen-deficient habitats reduces interspecific competition for food and diminishes predation pressure on the fish community, as the total number of predatory fish species and individuals was much lower at low oxygen concentrations.

Respiratory strategies, their evolution and costs (Kramer, 1973), and their relationships with other aspects of biology – feeding niche, habitat type, antipredator strategies and capability of overland movements, were discussed by Kramer *et al.* (1978). Adaptations to cope with low oxygen in these South American fishes include using the oxygenated surface layers for branchial respiration in many species (as described by Carter & Beadle (1931), Carter (1935), Kramer *et al.* (1978) and Dorn (1983) amongst others). This is helped by vascular extensions of the lips which develop in *Colossoma* and *Brycon* (Braum, 1983), also in *Osteoglossum* and many small species. Air-breathing organs of many species are derived from the alimentary tract: the mouth cavity in *Electrophorus* (an obligate air-breather) and *Synbranchus* (a facultative air-breather), the air-filled stomach in some loricariid catfishes (*Hypostomus* and *Pterygoplichthys*), the intestine in some callichthyids (*Hoplosternum, Brochis*). The highly vascularized swimbladder is used in *Hoplerythrinus, Arapaima* (see Lüling, 1964) and *Lepidosiren*. The use of stomach or intestine seems to be a South American speciality, perhaps related to the reduced feeding at low water when respiratory adaptations are most needed. Elsewhere, swimbladder modifications assist respiration in the African lungfish and *Polypterus* and the branchial chamber epithelium or diverticula do so in African *Clarias* and in Asian *Anabas, Heteropneustes, Osphronemus, Macropodes, Betta* and *Amphipnous*; in the sea the pharyngeal epithelium is used by *Periophthalmus*.

Fish movements

The Amazon fishes are very mobile. Many species move into *várzea* lakes soon after the rising water level. rich in nutrients. leads to plankton blooms, but leave these lakes temporarily to spawn in better-oxygenated river water, then return to the lakes to feed while the water is high (Santos, 1973). Decaying vegetation leads to deoxygenation of the lake bottom waters until wind-mixing occurs; this mixing sometimes causes fish kills. Even small 'resident' characids (*Moenkhausia dichroura, Ctenobrycon spilurus*) move from lake to flooded forest when oxygen falls in the lake. These species also have highly synchronized movements from marginal vegetation to open water to feed on zooplankton for only a few hours each day (Zaret, 1984*b*). Such 'resident' fishes really live an ephemeral existence in the shallow floodplain lakes which may dry up every year, or every ten or hundred years depending on water levels.

Among the larger species, Goulding recognized two basic types of migration: spawning migrations and low-water upriver movements (Goulding, 1981; Goulding & Carvalho, 1982). Ribeiro (1983), in his studies of *Semaprochilodus* migrations in the Rio Negro, distinguished dispersal movements (mainly of young fishes) and trophic movements, in addition to spawning migrations. Dispersal migrations of small species have been observed personally on the Rio das Mortes (Mato Grosso). Junk (1984) distinguished migrations due to changing water levels from spawning and feeding migrations.

On the Madeira, spawning migrations occur early in the annual flood when fishes move from the floodplain to the river to spawn, the exact timing depending on the species. Those in the nutrient-poor tributaries where they have been feeding in the flooded forest descend to the turbid water of the main river to spawn (Fig. 6.4). Spawning places are not yet known but local fishermen believe them to be near to the river conflu-ences. After spawning the spent characoids return to floodplains or tributaries to feed (a trophic movement), though not necessarily the ones from which they came. Eggs and alevins are probably swept downstream.

Low-water upriver movements occur after several months of feeding in floodplain or flooded forest. The first schooling characoids begin to move out or descend the nutrient-poor tributaries as the water level starts to fall. They enter the main river, then move upriver intermittently through-out the low-water season. This is not a steady migration, rather a movement from one tributary or floodplain to another further upriver. The factors which determine how far the fish move, and which tributary or floodplain they enter, are still obscure. This upriver movement of well-fed fish is known in the central Amazon as the *'piracema'* (= time of fish), a term used for the upriver spawning migration in southern Brazilian rivers such as the Mogi Guassu (p. 155). Dispersal movements of the young *Semaprochilodus* are also upriver movements, into tributaries probably other than that from which the parent fish came. Such upriver movements will help to counterbalance eggs and alevins washed downriver.

These studies have shown clearly that whitewater and clear- or blackwater rivers are not completely distinct entities with their own distinctive faunas, since individual fish of many species move from one type of water to another seasonally and during the course of their life history. We know little as yet about the spawning grounds of many of the main food fishes, such as the large catfishes. Nor is anything known of the movements of juvenile catfishes. No large catfishes were caught moving downstream (Goulding, 1980), though large pimelodid catfishes (of at least 14 species)

Fig. 6.4. Seasonal changes in water level in Amazon tributaries and associated fish movements (based on data for *Colossoma macropomum* in the Madeira system (Goulding, 1980; Goulding & Carvalho, 1982)). On FALLING FLOOD: adult fish move from floodforest (A), down nutrient-poor tributaries and upstream in main whitewater river, sheltering along wooded shores (B) at low water; juveniles move from shrinking floodplain lakes and upriver. On RISING FLOOD: adults move to spawning areas (D) in whitewater river and after spawning move up tributaries into floodforest to feed (E); newborn young find nursery areas in floodplain lakes (F).

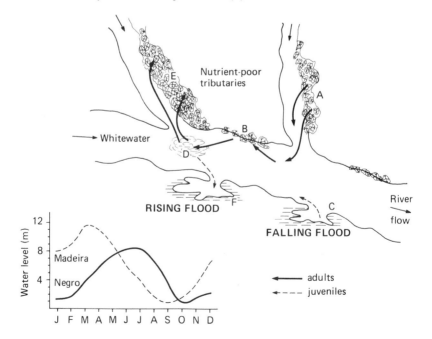

and doradid catfishes (4 species) are caught moving up through cataracts on the Rio Madeira, mainly at low water (Goulding, 1981). Nearly all the large characoids form schools which migrate in the rivers at some time of year. In the Madeira these include the characoid *Brycon, Colossoma* and *Mylossoma* species, *Triportheus* species, the prochilodontids *Prochilodus* and *Semaprochilodus*, curimatids, the anostomids *Leporinus, Schizodon* and *Rhytiodus*, also *Hemiodus* and *Anodus* species. Major seasonal fisheries are based on these movements.

Goulding (1981) considered that the need to escape predation has been very influential in the evolution of migrations to spawn in headwater streams or in turbid waters. Species producing pelagic or semipelagic eggs spawn on rising floods which carry eggs and young to marginal areas where they can then develop without risk of a sudden rise and fall in water level stranding them. The young stages generally live in different biotopes from the adults, in marginal lakes or on floodplains, places where zooplankton is abundant. The young stages of numerous species here feed on zooplankton.

Trophic studies in the Central Amazon basin

Marlier (1967, 1968) initiated studies into trophic conditions and foods eaten when making comparative studies on the *várzea* Lake Redondo on the Solimões 25 km southwest of Manaus and some neighbouring blackwaters. The basic study on Lake Redondo showed that food webs are very complex even in such a small lake (35 ha $= 0.35$ km^2, 3.5 m deep). Here the littoral floating meadows provided the main source of substances at the base of the food webs. Marlier commented that before the banks were cleared of forest, floating meadows were probably less extensive and allochthonous fruits and leaves from the trees would have been a main food. Marlier distinguished between fishes which took only plant **or** animal food (his 'stenophages') and those which combined these in their diet (his 'euryphages'). Among the euryphages he distinguished predominantly carnivorous species and a rather larger number of vegetarian ones. Of those with restricted diets, one group was carnivorous, another vegetarian (Table 6.2). The carnivores included a number of non-specialists taking any kind of animal food according to its availability, and specialists using either other fishes, or insects, or zooplankton. Stenophagic plant-eaters included mud-feeders which get their nourishment from bacteria and other small organisms in the mud, feeders on higher plant material, mainly allochthonous fruits and leaves, and non-specialists having a mixed diet of higher plants and algae. More recent

Table 6.2. *Trophic categories of Amazon fishes from Lake Redondo (data from Marlier, 1968)*

I. Stenophages
Non-specialists

Carnivores
Geophagus surinamensis
Plagioscion squamosissimus
Eigenmannia virescens
Colomesus asellus
Serrasalmus nattereri
Serrasalmus elongatus
Pimelodella cristata
Apistogramma taeniatum

Herbivores
Cichlasoma bimaculatum (filt. algae + plants)
Cichlasoma festivum (epiphytic diatoms, seeds)
Anodus laticeps (phytoplankton, epiphytic diatoms)

Specialists

Piscivores
Cichla ocellaris
Arapaima gigas
Boulengerella cuvieri
Ageneiosus ucayalensis
?*Synbranchus marmoratus*

Colossoma bidens (fruit)
Ctenobrycon hauxwellianus (grass seeds)
Metynnis hypsauchen (plants)
Leporinus maculatus (plants)
Poecilobrycon trifasciatus ⎫ (algae and
Poecilobrycon unifasciatus ⎭ aufwuchs)

II. Euryphages

Insectivores
Triportheus elongatus
Oxydoras niger (burrows)

Zooplankton-feeders
Astyanax fasciatus
Hypophthalmus edentatus

Predominantly carnivorous
Osteoglossum bicirrhosum
Serrasalmus rhombeus

Mud-eaters (bacteria in bottom mud)
Curimatus sp.
Potamorhina pristigaster
Prochilodus sp.
Pterygoplichthys multiradiatus

Predominantly vegetarian
Acarichthys heckelii (seeds, molluscs)
Pterophyllum scalare (littoral zooplankton and plants)
Metynnis lippincottianus (algae, plants, and *Cladocera*)
Corydoras sp. (aufwuchs)
Pyrrhulina brevis
Cheirodon piaba
Hyphessobrycon 3 spp.
Anchoviella brevirostris

Table 6.3. *Principal habitats and trophic groups of the adults of the main food fishes in the Central Amazon basin (Data from Smith, 1981; Goulding, 1980)*

	Habitats				
Species	River	Lateral channel	Lake	Várzea forest	Trophic groups
Characoids					
Colossoma macropomum	−	+	+	+	forest feeders[a]
Colossoma bidens	+	−	+	+	forest feeders[a]
Mylossoma spp.	+	−	+	−	forest feeders[a]
Brycon spp.	+	+	−	−	forest feeders[a]
Prochilodus spp.	+	+	+	+	mud-feeders
Semaprochilodus spp.	+	−	−	−	mud-feeders
curimatids	+	−	+	−	mud-feeders
Leporinus spp.	+	−	+	−	forest/omnivore[a]
Schizodon fasciatum	+	−	−	−	forest/omnivore[a]
Triportheus spp.	+	−	+	+	surface/omnivores[a]
hemiodids	+	−	−	−	omnivores?

Hydrolycus spp.	+	−	−	piscivores
erythrinids	+	−	−	piscivores
Siluroids				
Brachyplatystoma spp.	+	−	−	piscivores
Pseudoplatystoma spp.	+	+	+	piscivores
Pimelodus spp.	+	−	+	omnivores
Hypophthalmus spp.	+	+	+	planktivores
loricariids	−	+	+	algal-grazers
Others				
Cichla ocellaris	−	+	+	piscivore
other cichlids	−	+	+	invertebrates/fish
Arapaima gigas	−	+	+	piscivore
Osteoglossum sp.	−	+	+	surface omnivore
Plagioscion sp.	+	−	−	piscivore
Pellona sp.	+	−	−	piscivore

[a] Diet has large allochthonous component (fruits/seeds/leaves, arthropods).

studies include those of Goulding (1980) on fishes feeding in flooded forests, those on foods used by cichlids (Ferreira, 1981), those on feeding by adults and young during autecological studies of the main food fishes. On the tributary Aripuaña River, Soares (1979) looked at the diets of 20 species, distinguishing between diurnal and nocturnal feeders.

Principal habitats and trophic groups of the adults of the main food fishes are indicated in Table 6.3. Allochthonous foods (fruits, seeds, arthropods) are most readily available in flooded forests, and it is during high water that fishes feeding on them most clearly display their feeding adaptations. Goulding (1980) found that the fishes divide the available resources along lines of their own specialized, but not necessarily restricted, feeding adaptations. Nearly all of the taxa studied had adaptations for getting certain types of food that most of the others did not utilize to the same extent. The predatory species also displayed a wide variety of adaptations for capturing their prey, but no piscivore was found to specialize on one prey species. Most predatory genera were represented by more than one species living sympatrically. Fish predators are numerous, not only piscivorous fishes, but also mammals, birds and reptiles. The Amazon system has well over 40 species of primarily or exclusively piscivorous characoids, as well as piscivorous cichlids (*Cichla* and *Crenicichla*), large siluroids, large gymnotoids, sciaenids, *Arapaima* and *Plagioscion*.

The large fruit-eating characids, such as *Colossoma* and *Brycon*, were nearly 90% full when feeding in flooded forests at high water; mean fullness dropped considerably during the low-water period when these fishes were restricted to channels or lakes. The development of fat reserves during the highwater season is an important adaptation in these large vegetarian fishes for survival and ova development during the low-water season. Other species of fish here included allochthonous invertebrates (insects and spiders) in their diets at high water and detritus at low water. Carnivorous species, including piscivores, showed less seasonal difference in diet. Microphagous species (prochilodontids and curimatids) also migrate into flooded forests at high water, where they build up fat stores; this indicates that the detritus they eat is more abundant in flooded forest, and perhaps more nutritious, than in open rivers and lakes; they can obtain detritus from submerged tree debris above deoxygenated bottom water.

Marlier's comparative studies in various lakes and creeks indicated that biomass and production are higher in whitewaters than in the black- or clearwaters where food cycles depend very much on allochthonous forest products and most of the nutrients are locked up in the bodies of the fauna and flora. These *várzea* lakes are very important feeding grounds for juvenile fishes of many species such as *Colossoma* (see for example Goulding & Carvalho, 1982), many of which consume zooplankton. Marlier pointed out that the number of animals present is important for production, since they store nutrients which would otherwise be lost to the effluents of the lakes (a density-dependent process) and also increase the speed of mineralization of littoral plant material which falls into the water (a density- and diversity-dependent process). Herbivores such as the Sirenian manatee or seacow (*Trichechus inunguis*), which used to be fairly common in the Amazon, have a beneficial role on production as they feed on water meadows and excrete into the water; herbivorous fishes have a comparable role (compare the role of herbivores in African lakes, p. 302). Predatory species may also make important contributions to the nutrient salts. Fittkau (1973) noted that when caiman disappeared from the mouthlakes of Amazon tributaries, following intense hunting for their skins, the populations of all kinds of fish diminished markedly. Bayley (1982) has examined the efficiencies of transfers from prey to predatory fish in these waters (see p. 280).

The whitewater rivers, in which nutrients are replaced each flood, are the most productive, but in waters where most of the nutrients are bound up in the fauna, the mass removal of fishes or other creatures is likely to lead to rapid impoverishment of the whole ecosystem. This has to be kept in mind when planning fishery development in these waters. As Marlier (1967) stressed, the myth of the economic richness of Brazilian equatorial waters must be seriously reconsidered.

Stream studies: central Amazon streams

The small streams in the rainforest of central Amazonia present extreme biotopes for the fishes as the trees closing over the stream prevent light from reaching the water surface; also nutrient salts are scarce. Aquatic plant life is virtually non-existent and food webs are dependent on allochthonous forest debris raining onto the stream: pollen, flowers, fruits, leaves, insects and spiders (compare streams in Zaire, p. 44, also streams in Borneo studied by Inger (1955)). Small species of fish, are, however, often surprisingly abundant; in the central Amazon, 30–50 species may be taken from one stream, mainly Characidae, with siluroids, gymnotoids, cichlids and, in muddy sections, cyprinodontids.

These streams generally have three sections: an upper source and erosion zone where the gradient is steep, fast-flowing stretches over rock alternating with sandy-bottomed pools; a middle sedimentation zone, with meanders over a sandy bed; and the *igapó* zone, where seasonal variations in the water level of the main river back up the stream so that it ceases to flow for much of the year (Fittkau, 1967). In the upper and middle courses the fauna changes with the current velocity, current-loving species living where the flow exceeds 20 cm s^{-1}. At lower speeds organic materials sediment. Piles of drifted leaves provide cover on the bottoms, alternating with bare sandy or rocky stretches. Tree and epiphyte roots and the debris of fallen trees lying in the water provide cover for the invertebrates on which the fishes feed, as well as daytime hiding places for nocturnal fishes.

   Stomach contents of fishes from *terra firme* streams near Manaus were examined in the 'dry' season when the water level was falling (October–November) and again in May and July (Knöppel, 1970). One site was a 2–3 m wide blackwater stream in high forest, above 15–20 m falls; this yielded six species (2 characoid, 1 siluroid, 1 gymnotoid, 1 cichlid, 1 cyprinodont). Another site was a 1 m wide clearwater sandy-bottomed stream which had 17 species (5 characoids, 4 siluroids, 5 gymnotoids, 3 cichlids): the third site included a clearwater rivulet and a sunlit *várzea* lake with Solimões water, where 41 species were taken (18 characoid, 4 siluroid, 3 gymnotoid and 16 cichlid species). As elsewhere, cichlids were numerous in the lake. Fishes were abundant, despite the low primary production in all but the lake open to the sun. Five collections at these three localities produced over 3000 fishes of 53 species. Of 1296 stomachs examined, 94% were at least half full of food, and contents in a species collected at different times of year were generally similar. The main items of food were: (1) nymphs and larvae of aquatic insects (in 27 species, main food in 9), though no species showed a clear preference of any particular kind of insect prey; (2) terrestrial insects, especially ants (Formicoidea) (in 13 species) together with some termites (Isoptera), Coleoptera and Diptera; (3) plant remains, coarse litter (in 19 species), certain small species making it their main food; (4) fine sandy mud, which filled the intestines of *Prochilodus, Chilodus* and *Curimatus*; (5) Crustacea, including Copepoda, Ostracoda, Cladocera and palaemonid prawns (in 21 species); and (6) fishes, found in 20 species, all of which contained other food as well.

Mato Grosso headwater streams

   In Mato Grosso streams far to the south of the main Amazon (*ca* 12°S), and in the Rupununi, Guyana, far north of the main river (see p. 150), many of the same fish species or their ecological equivalents were encountered. These Mato Grosso streams flowed to the main Amazon either via the Rio das Mortes, Araguaia and Tocantins, or via the Suia-Missu to the Xingu. Stomach contents again stressed the importance of allochthonous foods (vegetable debris and aerial insects, such as flying ants) for fishes living in these forest streams, also of aquatic insects, particularly Ephemeroptera and Odonata nymphs (personal observation). The rainy season is here from November to April. Gonad states and the abundances and length frequencies of young fishes in March–May showed that breeding is here very seasonal, the main spawning season for most species being early in the rains (possibly just before them in some species), as in Guyana. Many of the fishes had well-defined growth checks on their scales; by the end of the rains most of the larger species had well-developed fat ribbons along the intestine, indicating that the rains are the main feeding time here too.

The fish fauna sampled here, at only one time of year and with inadequate gear, included about 40 species of which a surprisingly high proportion of the small characids were undescribed species. This illustrates the difficulties of ecological work when identifications of the species present such problems. Apart from this high rate of endemicity, the characoid fauna was here more like that of Guyana than that of central Amazonia, with many of the same species as in Guyana. About 60% of the species were characoids, about 21% siluroids of eight families, 8% cichlids, 4% gymnotoids, together with individuals of eight other families (clupeids, garfish, cyprinodonts, poeciliids). The numbers of species diminished towards the headwaters of the tributary streams. Here only about a dozen species were found, all hardy and ubiquitous ones. In temporarily flooded areas young-of-the-year of the fast-growing *Hoplerythrinus* lived in small schools of up to a dozen fishes, feeding largely on allochthonous matter, moving downstream as the water level fell; the adults had already left the area. Further downstream and in pools open to the sunlight, the numbers of species were much greater.

˙Andean streams

The composition of the fish communities and foods eaten at 340 m altitude in the Rio Napo in Ecuador were examined by Saul (1975) from (1) the 100 m wide whitewater Rio Aguarico, (2) the 6 m wide Rio Conejo (local drainage, its waters joining those of the Aguarico 1088 km away to the east), (3) rainwater pools and swamps, (4) small (2 m wide) creeks, and (5) two eutrophic lakes. The communities were all dominated by characoids except in the lakes, where cichlids were more numerous. In the Aguarico, 40 of its 53 species were characoids, which made up 90% of the fauna, with loricariids (3%), cichlids (5%). In the Conejo (42 species), characoids made up 64% of the fauna with lorcariids (15%), cichlids (21%). The small creeks had diverse faunas (48 species) of characoids (64%), pimelodids (8%) and cichlids (12%). Swamps and lakes both had only 17 species: swamps with characoids (58%), callichthyids (4%), gymnotoids (26%), cichlids (11%), and the lakes cichlids (62%) and characoids (25%). Of the 101 species (21 families) collected here, 46 were common, 55 only occasionally seen (42% of them represented by only one or two individuals); 38 species occurred in both Aguarico and Conejo drainages. The widespread species were generally abundant in at least one biotope. Widespread species appeared to be tolerant of a wide range of ecological conditions.

In these Ecuadorian waters Saul found the following items in fish stomachs: detritus (in 16 species, main contents in none); plant material (in 34 species, main contents in none); aquatic insects (in 57 species, main contents in 9); terrestrial insects (in 41 species, main contents in 21); molluscs (in 17 species, main contents in 2); crustaceans (in 16 species, main contents in none); and fish (in 20 species, main contents in 2).

Primary consumers (taking mainly plant material and bottom debris) included loricariids, aspredinids, anostomids, curimatids, prochilodontids and a cyprinodont; secondary consumers (feeding on insects, crustacea, molluscs) included stingrays, small characids, lebiasinids, nannostomids, small gymnotoids, auchenipterid, pimelodid and callichthyid catfishes and cichlids; tertiary consumers (eating mainly fish, with some crustaceans and insects) were the larger characoids and gymnotoids.

Most fish species were carnivorous, feeding mainly on insects (Plecoptera, Odonata, Trichoptera, Ephemeroptera, as adults, larvae or nymphs), but many also took some crustaceans, molluscs, fish or vegetable debris. The main bulk of food eaten was of terrestrial origin, with ants the single most abundant item.

Again, the apparent lack of specialization was striking, many species including the same kinds of food items in a rather varied diet. This was particularly true of the ichthyofauna of small streams. Stream fishes appear to be able to cope with many kinds of foods; they have to take whatever is available seasonally and the terrestrial component is high. Saul concluded that the great contribution of allochthonous forest products accounted for the large numbers of fish and fish species present despite the lack of primary production. Many kinds of fish ingest these forest products directly; the role of fishes in recycling them needs investigation. These findings agreed with those from streams near Manaus and in the Mato Grosso.

Andean streams above the forest

The fauna of high-gradient Andean streams is mainly of small catfishes: astroblepids (one genus which has speciated extensively), loricariids, trichomycterids and small pimelodids, with various small to medium-sized characoids. The paucity of this fauna and other evidence suggests that it is a relatively recent one (Roberts, 1972). In shallow headwater streams near Tingo Maria in Peru (600 m altitude) and in a section of the Rio Nanay near Iquitos, characoids dominated the fauna (47–77% in four streams, 38% in the Rio Nanay and 58% in the nearby Amazon), followed by catfishes with relatively few other species (Patrick, 1964). The total numbers of species in streams (17–31) were comparable with those in temperate waters, but the Nanay and the Amazon had much richer faunas (124 and 96 species, respectively). The more diverse niches in these larger rivers with floodplains, their greater age and larger drainages all help to account for their ability to carry higher numbers of species.

The fish communities of meadow ditches, brooks and streams flowing into the Rio Huallaga, an Andean tributary in eastern Peru near Tingo Maria, were described by Lüling (1963, 1971*b*). Some of the same species as in Mato Grosso streams were present, including the very aggressive *Hoplias malabaricus* and the hardy catfish *Callichthys callichthys*, while alternative species of *Aequidens* and *Crenicichla* lived in the quieter waters with numerous small charaoids, such as *Astyanax* species, in the current. The fauna in these Peruvian headwater streams is scanty compared with that in the whitewaters in the rainforest, the huge richness in species and numbers beginning below the 300 m contour.

Lüling (1971*a*) also studied the behaviour of the annual cyprinodont *Rivulus beniensis* in the meadow ditches. These fishes die after spawning and hard-shelled eggs remain viable through the dry period from June to late December. As the pools diminish in size in late May, the *Rivulus* jump repeatedly over the damp soil until they reach the deepest pools, behaviour which concentrates the breeding populations. Changes in water chemistry, particularly conductivity and increases in pH and alkalinity, were thought by Lüling to stimulate spawning. These temporary meadow ditches provide a habitat free from predatory fishes such as *Hoplias malabaricus*.

The river continuum concept (see Vannote *et al.*, 1980) suggests the importance of allochthonous food sources in headwater streams, giving way to autochthonous benthic and planktonic foods as a river widens and deepens, and accumulations of mud as feeding grounds in lower reaches (as indeed had already been noted for many tropical rivers by Lowe-McConnell (1975)). But in the Amazon system the high-gradient headwater streams are above the forest, the predominant fishes here algae-grazers. In the Amazon allochthonous forest foods are main food sources in the nutrient-poor rivers throughout the central basin, enabling these rivers to carry much higher fish populations than would be expected from their lack of autochthonous primary production. Zooplankton is important for a few specialized species, and for young stages of others, in the open waters of *várzea* lakes and in the mouthlakes of some tributaries (Tapajos and Xingu), but the fish fauna lacks phytoplankton-feeding specialists. Detritus is of great importance in flooded forests, where migratory microphagous species move to feed seasonally. We have little information about foods used in estuarine reaches, but mud-feeding grey mullets (Mugilidae) are known to be common (see also Barthem, 1985). The Amazon system is too vast to fit neatly into this continuum concept. Studies so far made have, however, brought out clearly the role of fishes in contributing to the interdependence of the various types of water (both of river and lateral lake and of nutrient-rich turbid waters with nutrient-poor clear- and blackwaters), and the close ties that exist between aquatic and forest ecosystems.

The diversity of the Amazon fish fauna
The high diversity of the Amazon fish fauna has been attributed to many factors: (1) the age and size of the drainage system; (2) the succession of habitats offered by the meandering rivers, with certain ones such as high-gradient streams being separated by long distances; (3) the diverse niches in the lowland rivers and their adjacent lakes; (4) the high proportion of the basin at low level with comparatively stable conditions capable of supporting large numbers of individuals; and (5) river captures and faunal exchanges. The large total species list might result from each section of the Amazon system having its own faunal elements, but how much is this richness due to 'alpha diversity', namely the numbers of species sharing a community, utilizing different niches, and how much to 'beta diversity' of habitat-orientated species added along environmental gradients? (See also p. 287.)

It is clear that alpha diversity is very high, up to 50 species being found together in many of the water bodies. Generally only a handful are common at one place and many of them are represented by only one or two individuals. This is perhaps not surprising as most of the fishes are so mobile that communities change very much through the year and from year to year. A distinction should be drawn between true feeding communities and aggregations of species concentrated from a wide area in dry-season pools, where chance may have a particularly large role in what species happen to be present. Within the feeding communities all the niches appear to be exploited, but with some overlaps between species, and often with very closely related species sharing the same resources (for example *Colossoma macropomum* and *C. bidens* both prefer the same fruits (Goulding, 1980)). There are even species which appear to mimic others with which they live (Gery, 1969). So alpha diversity is here very complex. The time-sharing of resources, seasonally by migrations and diurnally for feeding, is marked: cichlids and characoids active by day, siluroids and gymnotoids by night.

We know less about replacements along environmental gradients. The high-gradient streams have impoverished and specialized faunas, mainly of algae-grazing catfishes. Fish faunas in the long estuarine reaches include many marine species. Between these two extremes, over 4000 km apart, there is virtually no fall in river level, and in this central basin biotopes have a greater influence on species distributions than gradients. However, many fish species change their biotope as they grow and with seasonal changes in water level.

Headwater streams within the forest generally harbour a dozen or so ubiquitous species, euryphagous hardy fishes able to withstand low oxygen and lowered temperatures. Many of those found in Mato Grosso streams were the same species, and feeding on the same kinds of food, as Saul found in comparable habitats in Ecuador and Knöppel found in *terra firme* streams near Manaus. This emphasizes the role of biotope.

Among the smaller and less mobile species, however, new ones are found in each previously unworked area. Gery (personal communication) expected *ca* 15% of the small characoids to be new, and many new ones were found in the Upper Xingu headwaters for example. Populations of many species in the remoter areas show slight meristic differences and whether they are given specific status or not often depends on the taxonomist concerned (see discussion on microgeographical diversity in *Eigenmannia macrops* (Lundberg & Stager, 1985)). We cannot, as yet, therefore assess linear replacements of species over this vast system, but we can say that some small and readily recognizable species are very widely distributed within it wherever suitable biotopes exist. Among the more sedentary species, such as loricariids, which from their distinctive colours and behaviour appear to be territorial, numerous (over 600) species are known, in contrast with the large migratory pimelodid catfish genera which have very few species widely distributed over much of South America.

Relationships between stability and diversity are discussed in later chapters. Unlike the 'stable' tropical forest, Amazonian waters present extremely variable environments. Water depth varies over 10 m annually in many places; shoreline and shallows can change by many hundreds of metres. Some habitats dry out seasonally. Annual changes in sizes of water bodies will affect the physical and chemical conditions. Productivity varies with interacting cycles of black- and whitewaters. Sudden drops in temperature affect oxygen levels. On the whole the small *terra firme* streams and the open waters of the larger rivers appear to be more stable over the annual cycle than are *várzea* waters.

Floodplain rivers of South America

The more seasonal events in savanna floodplain rivers and their effects on the fish faunas have been studied north of the Amazon in the Rupununi District of Guyana (Lowe-McConnell, 1964), in neighbouring Venezuela (Mago, 1970) and on the Magdalena floodplain in Colombia (Kapetsky, 1976), and south of the Amazon basin in the Paraná–Paraguai system.

The seasonal cycle in northern floodplain rivers

The Rupununi savannas stretching between the Amazon and Essequibo drainages at 2–4°N have one well-defined annual rainy season, May to August, during which rivers flood the extensive savannas to a depth of 1–2 m and the growth of water plants is exuberant. After the rains the country dries up rapidly. The fish fauna was found to include numerous small (less than 5 cm) aquarium-sized fishes and about 150 species of larger fishes (characoids of 61 species, siluroids 35, cichlids 19, gymnotoids 9, with a few representatives of about nine other families). In the dry season savanna ponds contained 60–70 of the larger species, many of them different from the species in the main riverbed and river pools: for example, *Serrasalmus nattereri* was common in savanna ponds, *S. rhombeus* was only caught in the river. Relative numbers fluctuated greatly from year to year; this appeared to be the result of spawning success or failure in the previous year or two, largely determined by the way in which the savannas flooded. Inadequate rains failed to release fishes trapped in some of the ponds, and fluctuating water levels stranded fish spawn. Larger pools carried larger-growing species, the small pools only small species; the size of the dry-season refuge pools appeared to be an important factor governing fish distribution. Savanna pond species were in almost all cases found in both the Amazon and Essequibo drainage systems; species which can withstand pond conditions are well equipped to cross the drainage divide. Among the riverine species on the other hand, some were found only on the Amazon drainage side.

These Rupununi waters are mostly very clear natural aquaria, where fish behaviour can be observed. The cichlids and most of the characoids were diurnally active, hiding away motionless against the bank or amongst the tree litter at night, when the nocturnal catfishes and electrogenic gymnotoids emerged to feed from the crevices where they hide by day. Such crevices, in rocks and tree litter which provided shelter above the often deoxygenated bottom water, appeared to be of greater significance for the nocturnal fauna. Up to 17 fish species (15 of them catfishes of six different families) were taken from tree-litter crevices at one dry-season site. Other small fishes hide in the leaf litter carpeting the pools; the gymnotoid *Gymnorhamphthichthys* burrows in bottom sand.

When rivers began to rise at the start of the rains, large characoids (*Myleus pacu, Hydrolycus scomberoides, Boulengerella cuvieri*) moved upriver and along sidecreeks onto the flooded savanna. *Prochilodus* spawned as they moved out; they could be heard on their spawning runs (see p. 266). At the end of the rains there was a general return to the main rivers, but many fishes became trapped in the savanna ponds. Trap catches showed that species moved downriver in a definite order, and that different species travelled at different times of day or night (Lowe-McConnell, 1964).

The main breeding season for most of the species is here in the main rains. This is marked in the otophysan fishes, in which the eggs are shed more or less at one time, and less marked in species which brood their eggs and produce fewer eggs at a time, at frequent intervals in cichlids, *Osteoglossum* and loricariid catfishes; Chichocki (1976) found some cichlids spawning throughout the year. The main spawning season for *Arapaima gigas*, in which the ova do not all ripen at once, is also in the rains. Sudden changes in water level present great dangers to eggs and larval fishes, putting a premium on rapid development. This well-defined spawning season enabled growth rates of young fishes to be determined from length frequency data. Scales of many species showed growth checks in the 'physiological winter' of the dry season. These indicated that many species here could reach maturity in 1 or 2 years' growth.

Foods were extremely limited in the dry season. The many species crowded together in small drying pools shared whatever was available, mainly bottom debris, so there was much overlap in foods eaten by different species. Competition was, however, reduced as feeding levels are lowered at this time, many of the larger species living off their fat stores. In the wet season many more foods became available and the fishes developed fat ribbons along the intestine, showing that this is a rich feeding time; little information is available on how much fishes specialize on different foods, as they become widely scattered in the enlarged flooded environment and are then very difficult to catch.

In neighbouring Venezuela, the ecology of fishes in lagoons and channels near the River Apure tributary of the Orinoco (7–9°N) is very like that in the Rupununi (Mago, 1970). Here, too, the annual rains control the lives of the fishes. As in Guyana, the start of the rains (May–June) marks the start of the spawning seasons for many species: the characoids *Serrasalmus notatus, Salminus hilarii, Hydrolycus scomberoides* and *Hoplias malabaricus*, the catfishes *Pseudoplatystoma fasciatum, Sorubim lima, Pinirampus pirinampu, Ancistrus, Hypostomus* and *Pterygoplichthys*, the cichlid *Astronotus ocellatus* and the sciaenid *Plagioscion squamosissimus*. However, Mago thought that specializations to get food were probably needed when food was short in the dry season, rather than in the rains (in contrast with conclusions from Rupununi studies).

Lagoons formed alongside highways where earth had been removed to build the roads, became naturally stocked with fish. Chemofishing one such lagoon (1225 m^2) produced 25 species of fish, 2500 individuals totalling 120 kg, equivalent to 1000 kg ha^{-1}. The biomass of predators was very high (*ca* 75%); *Hoplias malabaricus* formed 47% of the catch (compare Argentinian lagoons, p. 155). Fishes adapted to special biotopes in Apure waters included: (1) species which bury themselves in sandy bottoms (stingrays, the gymnotoid *Gymnorhamphthichthys* and characoid *Xenagoniates*); (2) crevice-dwellers, which included the many nocturnal electric eels and other gymnotoids and catfishes; (3) fishes associated with carpets of floating vegetation, including many cichlids, small characoids and catfishes, with juveniles of many species; and (4) annual fishes, the cyprinodonts *Pterolebias* and *Austrofundulus* which withstand dry periods as resistant eggs. Special habits discussed by Mago included mimicry (for example, young *Colossoma* resembling predatory *Serrasalmus*), the ability to move overland through wet vegetation (as in *Hoplias, Hoplerythrinus* and callichthyid catfish), sound production by doradid catfishes and others, and colour blotches which appear important for sexual recognition and schooling.

Fish communities in the Paraná–Paraguai system

The ecology of fishes in the immense Paraná–Paraguai system has been investigated in the Mogi Guassu and other tributaries of the Upper Paraná in Brazil and in the Middle Paraná river which, although south of the tropics, carries many of the same species as the Amazon. Fish migrations up the Pilcomayo, a 1500 km tributary of the Rio Paraguai which rises in the Bolivian Andes, have been studied. The special problems of the Chaco swamps were observed by Carter & Beadle (1931). From these and other studies we have a picture of fish spawning migrations and of how the dry-season fish communities vary from pool to pool and from year to year with hydrological conditions. (See Bonetto in Davies & Walker, 1986.)

Middle Paraná pool communities
The Paraná river near Santa Fé reaches its maximum level in summer (February–March) and its minimum in August–September. During the flood the river reaches numerous ponds and lakes on the many islands on the floodplain; fishes then migrate between river and ponds. The shallow pools dry up at the end of the low-water season. The large numbers of stranded fish which then perish have been estimated to have a biomass equal to that cropped by the fisheries (Bonetto *et al.*, 1969). These riverside pools and channels contain abundant diverse small characoids (such as *Astyanax, Acestrorhynchus*) and the young of larger-growing species, such as 7–12 cm long *Prochilodus*, along with some 40–50 cm adults of this species. The oxbow lakes, riverine in origin but lentic in conditions, lack rooted vegetation but have extensive covers of floating hydrophytes.

The 19 ha Don Felipe lagoon contained 52 species of fish (779 128 individuals, a mean of 6.2 per m^{-2} of water (Ringuelet *et al.*, 1967)), but only five large species were relatively abundant (*Prochilodus platensis, Leporinus obtusidens, Schizodon fasciatus* and two pimelodid catfish species); the remaining 47 species were small species or juveniles of large species. The proportions of different trophic types were: mud-feeders, 7 species (14% of the total number of individuals); detritus-feeders, 5 species (8%); herbivores, 12 species (12%); omnivores, 12 species (3%); invertebrate-feeders, 12 species (62%); piscivores, 11 species (0.5% of the individuals).

Six temporary oxbow lakes and another two island lakes in a Paraná channel near Santa Fé studied by Bonetto, Dioni & Pignalberi (1969) produced 42 fish species (representing 13 families, 7 orders). Characiform fishes predominated in all of them, but the proportions of different species varied stochastically from pool to pool (Fig. 6.5) dictated by what fish happened to be trapped in them as the water level fell. The numbers of individuals were very high, though numbers and biomasses varied very much from pool to pool (175 to 6500 kg ha^{-1}). A large proportion of the

Fig. 6.5. The proportions of different types of fish taken by total fishing temporary oxbow lakes in islands in a tributary of the Paraná river near Sante Fé. Plant cover varied from complete cover (+ +) in pool A to lack of cover (−) in pool B (adapted from Bonetto et al., 1969a).

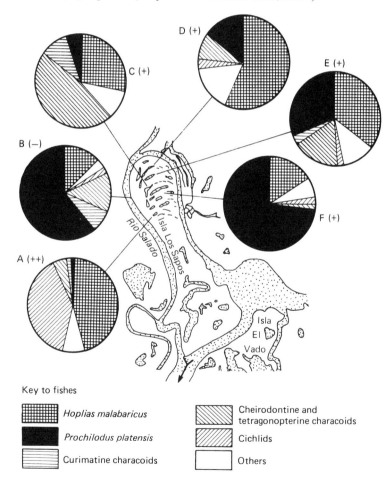

Key to fishes

Hoplias malabaricus	Cheirodontine and tetragonopterine characoids
Prochilodus platensis	Cichlids
Curimatine characoids	Others

biomass was made up of relatively few species – over 30% of mud-feeding *Prochilodus platensis*, 25% of piscivorous *Hoplias malabaricus*. Cichlids (mainly *Aequidens*) were abundant in oxbows with plenty of plant cover. Most other fish were juveniles of lotic species, *Leporinus, Schizodon, Pimelodus, Synbranchus*, hypostomine and loricariid catfish, *Serrasalmus* and other small carnivores.

Fish populations of four permanent pools were examined using chemofishing and mark–recapture methods (Bonetto *et al.*, 1969). Here too *Prochilodus platensis* were predominant (61% of the biomass), with loricariids (6%), *Pimelodus clarias* (5%), *Hoplias malabaricus* (4%). Juveniles of some large species (*Salminus maxillosus, Pseudoplatystoma coruscans, Colossoma mitrei*) occurred in permanent pools but not in temporary ones, and certain species found in temporary pools were absent from the permanent pools. The biomass was again high, equivalent to 1100 kg ha^{-1}. Some species, such as *Hoplias malabaricus*, were the same as in Venezuelan or Guyanan waters; others, such as the *Prochilodus*, were ecologically equivalent species.

Lakes in this Middle Paraná were fished again in 1968–72, the results showing notable variations in population structure between ponds and from year to year in relation to the hydrological regime (Cordiviola de Yuan & Pignalberi, 1981).

Fish tagging experiments

Large-scale tagging experiments have been carried out in the Mogi Guassu, a 473 km long river with rapids flowing westwards into the Upper Paraná in Brazil (Godoy, 1959, 1967, 1975) and in the Middle Paraná in Argentina (Bonetto *et al.*, 1971). In the Mogi Guassu, 27 000 fishes (16 861 (62%) of them *Prochilodus scrofa*) were tagged between 1954 and 1963. Between 1954 and 1971, 2734 tagged fish (*ca* 10%) were returned. Most of these were tagged on their upriver reproductive migration (here called the '*piracema*' = fish swarms) near the fish ladder in a hydroelectric barrier at the Cachoeira de Emas rapids. Some tagged fish were transplanted to other parts of the river system and associated rivers, though most were returned to the river at the tagging site. Returns showed clearly that the rivers Mogi Guassu and Pardo plus the Grande into which they flow (Fig. 6.2, site 15) form one ecosystem for these fishes (an estimated 39 611 km^2). These migratory fishes feed in the lower reaches of the Pardo and Grande Rivers, then ascend each year in vast numbers to spawn in the Mogi Guassu between the C. de Emas and the Salto de Pinhal falls another 160 km or so upriver. An estimated 100 000–

140 000 fishes collected in the shallows below the C. de Emas barrage. After spawning, these fish return some 600 km downriver to their feeding grounds. This entails a round trip of *ca* 1200 km a year.

Tagged *Prochilodus scrofa* have been returned up to 5 yr after tagging, *Leporinus fasciatus* up to 4 yr 5 months, *Salminus maxillosus* over 2 yr, and *Pimelodus clarias* over 1 yr 7 months. Returns showed that *Prochilodus scrofa* move upriver at a mean speed of 5–8 km day^{-1} (though exceptional fish moved 10, 15, 20 and 43 km day^{-1}). *S. maxillosus* averaged about 20 km day^{-1}, *Leporinus copelandii* 3 km day^{-1} and *L. elongatus* 3.5 km day^{-1}. Downstream movement was more leisurely at *ca* 3.5 km day^{-1}. One remarkable 58 cm long *P. scrofa* tagged at C. de Emas on 6 November 1962 at 1430 h was recaught here and tagged a second time the following year at almost the same time (5 November, 1505 h); this fish was caught yet again the following year, on 16 October at 1500 h a few kilometres downstream of C. de Emas, providing evidence of appearance here in three consecutive years at very regular times.

In the Middle Paraná, 40 000 fishes of 25 species (70% *Prochilodus platensis*, 10% *Hoplias malabaricus*, 10% *Pimelodus clarias* and smaller numbers of *S. maxillosus*, *Leporinus obtusidens* and others) were tagged starting in 1961 (Bonetto & Pignalberi, 1964). From the estuary near Buenos Aires *Prochilodus* moved up both Paraná and Uruguay rivers; from Santa Fé they moved down into the estuary; those tagged near the junction of the Paraná and Paraguai moved up both these rivers, while *Salminus* tagged here were only recaptured downstream (one travelled down 1000 km in 60 days, 16.7 km day^{-1}). These recaptures indicated that there were reproductive migrations upriver and trophic movements downriver, but patterns were complex, suggesting that there might be separate upriver and downriver populations of *P. platensis*. The maximum distance travelled by *P. platensis* was 650 km downstream at 3.3 km day^{-1}, though the mean speed was 7 km day^{-1}. A considerable number of the fish stayed near the release site. In a second series of experiments a further 11 288 fish (70% of them *P. platensis*, 18% *S. maxillosus*) were tagged and returned at six centres in the Paraná and La Plata system and tributary rivers (Bonetto *et al.*, 1971). Returns gave more details of migration distances; the most spectacular movements were of *P. platensis* 700 km upriver at 8.7 km day^{-1} and *S. maxillosus* upriver 237 km in 11 days (21 km day^{-1}), downriver 610 km at 5 km day^{-1}. These indicated that upstream migrations undertaken under powerful physiological stimuli associated with breeding are likely to be generally faster than the more leisurely return, with deviations for resting and feeding on the way.

In the Pilcomayo tributary of the Paraguai, the main feeding grounds for most of the large fishes appeared to be the swamps in the Lower Chaco. These fish then migrate over 400 km into Andean reaches of the river to spawn (Bayley, 1973). The main ascent of the most important food fishes here (*P. platensis, Leporinus obtusidens, Schizodon fasciatus, Colossoma mitrei, Salminus maxillosus* and *Pseudoplatystoma coruscans*) occurs in the winter months, bringing them to the Andean foothills in May or June. The gonads mature during the migration. *Prochilodus platensis* spawn by the day after the first floods of the rainy season (as described and photographed by Bayley (1973)). Most of the spent fish descend when the large floods come in November–December. There are no lateral flood-plains near these spawning grounds and Bayley supposed that the young are swept downstream.

In the Mogi Guassu wheeling circles of spawning fish occur on a rising flood between November and February; the first fertilized eggs are found in December. Some species spawn in inundated grassland, others in the main river, often in special places out of the main current. Shoals of small fishes gorge themselves on the eggs, and eggs in shallow water perish if the level falls. Development is extremely rapid, eggs hatching 2 or 3 days after fertilization. In the river, floating eggs are carried down to quiet stretches where they develop fast in the warm shallows.

Prochilodus scrofa males here move upriver in the centre of the channel, emitting sounds as they do so. The females, slightly larger fish, move up near the bank and out into the centre of the river to spawn; this species spawns only at the end of the afternoon and at night. *P. scrofa* migrate in schools: Godoy reported aggregations estimated at 60 000–80 000 in the Mogi Guassu at the Emas rapids at the end of August. During upriver migration they move both by day and night, in heterogeneous groups of several age classes, including first-year-class fish (*ca* 20 cm) not yet mature. Growth data from scale rings showed that males mature when 2 yr old (*ca* 25 cm) and are most abundant from 3 to 7 yr old in the upstream runs; they reach 58 cm total length (2720 g) and 9 yr of age; females mature when 3 yr old (*ca* 31 cm) and are most abundant from 4 to 8 yr; they grow to 68 cm (5800 g) and live to 13 yr old (Godoy, 1959).

Prochilodus are all mud-feeders, for which they have many adaptations including a suctorial mouth and very long intestine. They are preyed on by large *Pseudoplatystoma* pimelodid catfish and the characoid dorado (*Salminus maxillosus*), a rather salmon-like food and sport fish. In the Rio de la Plata great accumulations of *Prochilodus platensis* (an estimated 200 million) occur in summer, but break up in February–March as the water gets colder (Ringuelet *et al.*, 1967).

Migrations in other South American rivers are included in Petrere's (1985*b*) survey; their effects on fish production in rivers and reservoirs and suitable species for fish culture are discussed by Bonetto & Castello (1985).

Fishes of the Chaco swamps

Fishes living in the vast swampy region of the Paraguayan Chaco faced two problems: deoxygenation and desiccation. The fish fauna of these swamps was considerable and varied but never rich. The swamps examined were mostly 1–1.5 m deep, over black mud and choked with floating rooted plants; the oxygen content of the water was very low, especially where floating plants protected the water surface from wind disturbance. Many areas dried up completely seasonally or during droughts. Of the 20 species found in these swamps, many of them the same species found in deoxygenated water elsewhere in South America, 18 of them lived here all year round; eight species had special adaptations for aerial respiration (Carter & Beadle, 1930).

In the Chaco swamps the lungfish *Lepidosiren paradoxa* makes a burrow 1 m vertically downwards into the bottom mud in which it aestivates, sometimes for many months while the whole country is dry. *Synbranchus* also aestivates in similar burrows (Carter & Beadle, 1930; Lüling, 1958, who studied this species in Peru). Both species leave their burrows as soon as the ground above them is flooded. The dry season here is usually the winter. *Lepidosiren* breeds early in the summer after rain, thus shortly after leaving its burrow. The nest is an almost horizontal burrow in the bottom mud into which dead leaves and grass are taken, and in which the eggs are laid and guarded by the male; he remains coiled around them and may help to keep them oxygenated.

7

Far Eastern freshwater fish faunas and their distributions

Far Eastern freshwaters considered here include those of tropical Asia (Oriental Region) and of Australia/New Guinea. The Asian tropics comprise the subcontinent of India, mainland Southeast Asia, and the islands lying on the continental (Sunda) shelf, Borneo, Sumatra, Java and certain others, which were connected with the mainland at intervals when the sea level was lowered during the Pleistocene glaciations. An extinct river system flowed into the sea between the present coastlines of Borneo and the Asian mainland. The rivers of eastern Sumatra and western Borneo, and even some from India, Thailand and Indochina, were tributaries of this vast system, which accounts for many similarities in fish fauna between the islands and mainland. North Borneo rivers did not join this system, and the fauna here is somewhat impoverished. Islands lying off the Sunda shelf have completely different freshwater faunas.

Wallace's Line down the Makassa Straits between Borneo and Sulawesi (Celebes) is the world's most spectacular zoogeographical boundary affecting freshwater faunas. To the west, Borneo teems with 300+ species of primary freshwater fishes (17 families); only 140 km to the east, Sulawesi has but two primary freshwater fish species, *Anabas testudineus* and *Channa striatus*, both probably introduced by man. Sulawesi, like Borneo, does have a few secondary freshwater fishes (cyprinodonts, a synbranchid and vicarious atherinids). Wallace's Line, separating Oriental and Australian zoogeographical regions, continues south between Java and Lombok; here differences are not quite so spectacular, as Java's freshwater fauna is much less rich than that of Borneo, but Java still has nearly 100 species of primary freshwater fishes (12 families) compared with 5 species (3 of them probably carried by man as food fish) in Lombok.

Freshwater fish evidence suggests that even in the Mesozoic there was never any dry land connection between Asia and Australia. Only three relicts of the Mesozoic fish fauna remain in this area today, an osteoglossid

(*Scleropages formosus*) in Sundaland, another (*S. leichardti*) in Queensland and in New Guinea (which connects with the Australian continental shelf), and the lungfish *Neoceratodus forsteri* in Queensland. The rest of the New Guinea and Australian freshwater fish fauna is of marine origin.

Indian rivers fall into two major groups (Mani, 1974): (1) peninsular rivers, many of which dry up seasonally, over 600 short coastal streams on the west coast, and eastflowing rivers with wide fan-shaped catchments and often large deltas; and (2) extrapeninsular rivers, fed by melting snow from the Himalayas, which never dry up completely and have extensive deltas. Those draining the northern slopes of the Himalayas, such as the Brahmaputra, are antecedent rivers older than the Himalayas, which kept their channels open by faster erosion as the mountains rose. Rivers draining the southern slopes, such as the Ganges, assumed their present form in the Pliocene or Pleistocene and now wander as slow streams over the plains of northern India.

The tropical Asian fish fauna (Fig. 7.1) is dominated by cyprinids. Siluroids are also very abundant, and there are very many species of marine origin. The freshwater fish fauna of Thailand, for example, in its total of 549 species (48 families) has 214 cyprinid species (39%), 99 siluroids (18%), 80 gobioids (15%), 71 species of other marine families (13%), 18 anabantoids (3%), and 67 species of other freshwater families (12%); the gobioids and other species of marine families together make a marine component of 28% (Smith, 1945). Asian siluroid families include Ariidae, Bagridae, Schilbeidae, Clariidae (all found in Africa) and nine endemic families. They include the giant (2 m long) *Pangasianodon gigas* of the Mekong, a strict vegetarian which lacks teeth in the adult, also *Heteropneustes fossilis*, an Indian catfish with very long airsacs used as

Fig. 7.1. Representative Asian freshwater fishes cultured or used for food. (A) *Chanos chanos* milkfish (150 cm, Chanidae); (B) *Notopterus* featherback (75 cm, Notopteridae); (C) *Trichopterus pectoralis* (25 cm, Anabantidae); (D) *Osphronemus goramy* (65 cm, Osphronemidae); (E) *Helostoma temminckii* (30 cm, Helostomidae); (F) *Toxotes* archer fish (20 cm, Toxotidae); (G) *Mystus* (50 cm, Bagridae); (H) *Heteropneustes* (30 cm, Heteropneustidae); (I) *Mastacembelus* (40 cm, Mastacembelidae); (J) *Channa* (90 cm, Channidae); (K) *Barbus* ('*Puntius*') (50 cm, Cyprinidae); (L) *Catla catla* (180 cm, Cyprinidae); (M) *Etroplus* (30 cm Cichlidae); (N, O, P) Chinese carps (Cyprinidae): (N) *Hypophthalmus nobilis* bighead (60 cm); (O) *Hypophthalmus molitrix* silver carp (60 cm); (P) *Ctenopharyngdon idella* grass carp (120 cm).

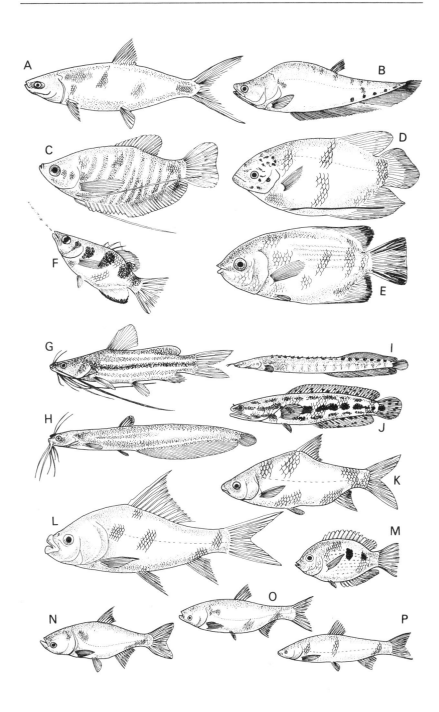

accessory respiratory structures, and *Sisor rhapdophorus* with dermal armour. The cyprinids are all non-piscivorous fishes which forage for plant and invertebrate food. As in most tropical communities the numbers of predators are very high; in Asian waters these include various siluroids (such as *Wallago*), also species of *Channa* and *Notopterus*, many of them large fishes. Archer fishes, *Toxotes* (Toxotidae), are common in sea-level swamps, where they obtain insect prey from the overhanging vegetation by squirting them down with jets of water; there are several species, of which *T. jaculatrix*, growing to about 23 cm long, has been most studied.

Fish culture is very important throughout tropical Asia. Its history, going back 4000 yr, stems from two separate traditions: the Indian tradition, mainly in India; and the Chinese tradition which spread with immigrants to Thailand, Malaysia and Indonesia. In Indonesia cultured fishes include anabantids, planktivorous or vegetarian species which live in very weedy places, and which, like many of these Asian fishes, have accessory respiratory organs enabling them to live in deoxygenated water. The milkfish *Chanos chanos* (Chanidae) is an important cultivated fish in brackish ponds in the Philippines. In Java fish have been introduced and cultured for so long that it is difficult to know how much of the present fauna is indigenous. Man-made lakes in Java produced very high yields for large water bodies, for example 500–600 kg ha^{-1} yr^{-1} from a 25 km^2 (2500 ha) lake in West Java, mostly of the carps *Cyprinus carpio, Puntius javanicus* and the anabantoid *Helostoma temmincki* (Vaas & Sachlan, 1952).

The extensive freshwater fish research in India centres around the biology of species used to stock dams and lakes, rather than on natural fish communities, but life history studies have been made of most of the main commercial species (Jayaram, 1974; Jhingran, 1975). In India the 'major carps' *Catla catla, Labeo rohita, L. fimbriatus* and *Cirrhina mrigala* are the main cultured fishes (Hora & Pillay, 1962). Carp spawning places in rivers have been mapped, as carp fry are much in demand for stocking tanks and other water bodies. Most of these carps spawn in the wet season, July–September; should the rains fail, the fish may not spawn that year. Methods have now been developed to induce riverine fishes to spawn by means of pituitary hormone injections. Chinese carp, the grass carp *Ctenopharyngodon idella*, silver carp *Hypophthalmichthys molitrix* and bighead *H.* ('*Aristichthys*') *nobilis*, are also now much used for fish culture in tropical Asia, where they grow much faster and mature earlier than they do in their home territory north of the tropics (Amur River). In India, life history studies include those of *Mystus* catfishes found in all principal rivers (Saigal, 1964). The male *M. seenghala* guards its nest until the young are 45 mm long.

Tidal influences are felt 50 km up some Indian rivers and the fish communities have many species of marine origin. Anadromous clupeids such as *Hilsa ilisha* support important fisheries. In the Hooghly River flowing past Calcutta these fishes ascend twice a year, the main run being in the monsoon season (Pillay, 1958). The construction of dams has affected some of these fisheries (Ganapati in Ackermann, White & Worthington, 1973).

Sri Lanka, once part of the Indian mainland, has no natural lakes but over 10 000 have been created by man over 1500 years, mostly small (less than 300 ha) in the low country and dry zone, plus over 50 larger lakes created recently for hydroelectric power. The 54 indigenous fish species are essentially stream- and marsh-living forms; there are about 17 kinds of some economic importance, but the main economic production is of introduced tilapia. The 65 ha Beira Lake enriched by heavy pollution yielded 2230 kg ha^{-1} yr^{-1} of *Oreochromis mossambicus* (Fernando in Ackermann *et al.*, 1973). *Osphronemus goramy*, *Trichogaster pectoralis* and *Cyprinus carpio* are among other introduced species. Unlike other tropical Asian countries, Sri Lanka has no history of fish culture. Competition for food and spawning grounds among cohabiting cyprinids in Sri Lanka streams has been examined by Schut, de Silva & Kortmulder (1984).

From the freshwaters of Peninsular Malaysia, Mohsin & Ambak (1983) have listed 384 species representing 56 fish families and indicated the distributions of these species in neighbouring countries. Over half (55%) are otophysan fishes (142 cypriniform species, plus 72 siluriform fishes of 10 families); the fauna also includes 20 anabantid species, 9 channids, 8 mastacembelids, 19 gobies and numerous freshwater representatives of marine families. These authors include ecological data and give faunal lists for upper, middle and lower zones of a number of streams, and for riverine fishes occupying surface, bottom and middle strata. Malaysia has few lacustrine species but many species can adjust to conditions in man-made lakes. They conclude that the majority of Malaysian fishes are omnivores, eating plant matter and invertebrates, or carnivores eating mainly insects and other invertebrates; higher-order predators are few. Terrestrial insects and plants contribute significantly to the fish food.

In Peninsular Malaysia, Johnson (1967) had distinguished associations in blackwaters, in 'tree-country' streams and in ricelands. In tree-country streams fish faunas are very diversified, in blackwaters and ricelands they are restricted by low pH and low oxygen, respectively. Faunas contain both specialists and generally tolerant species, the latter more important in the more extreme habitats. Cyprinids, especially *Rasbora* species, are the predominant species in the acid and unproductive forest streams and blackwaters, a contrast with European waters where cyprinids generally live in the more productive waters. The riceland fauna in Malaysia is impoverished compared with that in neighbouring countries; of the 14 more frequent species, nine are air-breathers. *Anabas testudineus* and *Trichogaster trichopterus* are common riceland species. An intensive study of the Malayan blackwater swamp Tasek Bera (Furtado & Mori, 1982) included studies on the composition of the fish communities (based on 75 recorded species which included 45 cyprinoid, 19 siluroid, and 11 anabantoid species), fish associations in relation to habitat types (such as open water, *Utricularia* and *Lepironia* stands, forest streams), the diurnal rhythm of activity, food habits and seasonal growth rates of several fish species (Mizuno & Furtado, 1982).

In the Philippines certain islands on the Sunda Shelf were colonized with primary freshwater fishes from Borneo. The fishes of Lake Lanao on Mindanao were of major evolutionary interest as this lake, thought to be only 10 000 yr old, had about 18 endemic cyprinid species, some with very specialized features. Since 1962 introductions of exotic species have led to the extinction of most of these endemic species. The lake now contains a mixture of species from many geographical areas. Hardly anything was known about the ecology of the endemic species before this occurred; their taxonomic validity, possible evolutionary rates and speciation scenarios have been examined by Kornfield & Carpenter (1984).

Southeast Asian forest rivers and their fishes

Fish communities have been studied in forested rivers on the large equatorial island of Borneo (7°N–4°S) – in North Borneo by Inger (1955) and in the westflowing Kapuas River and adjacent waters by Vaas (1952) and Roberts (unpublished). Vaas, Sachlan & Wiraatmadja (1953) also examined fishes in Southeast Sumatra where much of the forest has been cleared. These large islands have much the same fauna as the mainland.

In North Borneo heavy rainstorms are often very localized and there are wide fluctuations in heavy rainfall from year to year at any one place. The mountainous west coast has swift clear streams over gravel and rock; rivers running to the mangrove swamps on the flatter east coast are slower and turbid over silt bottoms. During Inger's survey, streams 1–10 m wide flowing largely through untouched forest were fished with rotenone and the larger rivers fished with lines, traps and nets. Among the 77 species collected (21 families), cyprinoids predominated, with siluroids of eight families, five anabantids and three mastacembelid species; most of the rest were freshwater representatives of marine families (especially gobies). Insect larvae and nymphs were abundant in every stream; Plecoptera, Trichoptera, Neuroptera and Diptera were the most numerous in fish stomachs. Many fishes contained terrestrial invertebrates, especially ants, spiders and Orthoptera, washed into the water by the heavy rain. As many fish species fed on terrestrial insects as fed on aquatic insects. Decapod Crustacea were also important fish food.

Studies on the vertical zonation of these fishes showed that the clear forest streams had one or two surface-dwelling species, one or two species living just below the surface, three to five midwater species, two to four species living just above the bottom, and three to ten living on or in the bottom. Within a drainage system the nature of the bottom and the current speed dictated the numbers of fish species present; the number was smaller when the bottom was of silt or sand than when it was of large rocks. Close to the sources of small streams the fish communities became impoverished in a way that affected the strata differently; the layers occupied by surface and benthic species remained, but the relative importance of midwater species decreased. The stratification of species in the major rivers is virtually unknown.

The food habits of fishes from the River Kapuas and its adjacent lakes and side arms in Borneo, and from rivers in Southeast Sumatra and nearby *lebaks*, areas inundated at high water and gradually drying out, studied by Vaas (1952) and Vaas *et al.* (1953) showed many species to be omnivorous, sharing a wide variety of animal and vegetable foods. In the forested waters of Borneo, inundated vegetation was an important source of food, more so than in neighbouring Sumatra where much of the land had been cleared for rice cultivation. In Bornean waters predatory fishes were very abundant. Although vegetarian cyprinids and the anabantoid *Helostoma* were caught in great numbers, the bulk of the catch was of piscivorous siluroids, *Channa* and *Notopterus*, particularly in pits retaining fishes in the dry season.

The longitudinal zonation of fishes in the Gombak River near Kuala Lumpur in Malaysia was studied by Bishop (1973) in an attempt to define limnological conditions in a small unaltered river not subject to seasonal change. Physical, chemical, floral and faunal surveys were made at five stations over two years; quantitative studies included measurements of the input of allochthonous materials. Fish populations were estimated by electrofishing, and the fishery yield by riparian fishermen assessed by questionnaires backed by sampling. In the upper zone the river flowed through forest, in the lower zones through cleared land where there was some pollution. Light and temperature conditions were governed mainly by presence or absence of forest. River temperatures varied by only one or two degrees seasonally, and daily ranges were also small. Considerable rain fell in all months; river discharges were highly variable with rapidly rising and falling flood peaks after each intense storm.

The Gombak River had 28 fish species (12 families), predominantly cyprinids (30%), channids and anabantids. Species of *Channa* (4), *Clarias* (2) and *Mastacembelus* (2) were segregated according to altitude and stream size, but apart from these geographical replacements, the longitudinal zonation was characterized by the addition of species downstream, where physical and chemical conditions were more stable and feeding niches more complex. The species diversity increased with stream order, generally where tributaries joined, and with associated changes in stream size, gradient, substrate, and sometimes temperature. Some niches were also lost or reduced at these points (such as fast water or riffles from lower stretches).

A large part of this ichthyofauna was indiscriminately euryphagous, deriving much of its food from allochthonous sources (fruits, leaves, flowers, terrestrial insects), but also taking benthos as available. Specialized feeders formed only a minor part of the community. Few of the fishes were strictly carnivorous, none cited as piscivorous. The catfish *Macrones* was present in all streams of sufficient size to support a mixed predator, often together with a predatory *Channa* or *Clarias* species. The only dependent herbivore, the detritus-feeding deep-bodied cyprinid *Osteochilus*, was only common in the lower zone where adequate decomposing material was available. In the upper zone, terrestrial insects, especially ants, were important food. In the very small channels not all of the allochthonous food was consumed, possibly because the water surface was too disturbed here to allow any species to specialize on using this food in this zone, but it was gradually swept downstream and utilized in lower reaches.

Electrofishing produced 35, 11 and 9 fish per hour in the upper, middle and lower zones, respectively, but the biomass was the same, *ca* 250 g ha^{-1}, in all zones. The mean catch of 0.2 fish m^{-2} was equivalent to *ca* 3.6 g m^{-2} of fish flesh. This was probably a considerable underestimate for the standing stock of fish, many fish escaping the electric shocker. The fishing questionnaire suggested that an annual crop of 30 tonnes was taken, representing a yield of 87 g m^{-2} (very much more than the standing stock estimates from electrofishing). The ability of the fish population to support this fishing pressure was debated, but such a high yield was considered possible with the high temperatures, continuous recruitment and surfeit of most foods.

Later studies (Geisler, Schmidt & Sookvibul, 1979) of diversity and biomass of fishes in three streams in Thailand, using electrofishing, found a biomass equivalent to 81–186 kg ha^{-1}; there were considerable differences in diversity, numbers and biomass between the three streams. Of the 33 fish species found, 18, 17 and 11 were collected, respectively, at the three stations. There was no relation between stock of fishes and total mineral content of the water.

Thus, as in South American and African streams, studies in Southeast Asia stressed the role of allochthonous forest products, tree debris and arthropods, in headwaters, whereas detritus becomes increasingly important as fish food lower down the river. Fishes in headwater streams are euryphagous, taking whatever food drops into the water and feeding at different levels, rather than just taking it from the surface. Lower down the river, bottom omnivores become more abundant. Apart from a few ecological replacement species, successional changes are mainly additional, diversity increasing with stream order and size, reflecting the greater complexity of niches in lower zones.

The Mekong: a seasonal Asian river

The Mekong, one of the ten great rivers of the world, arises in the snow-covered mountains of the Tibetan plateau and flows southwards through six countries on its way to the sea through a complex delta at 10°N in Vietnam. In Kampuchea it floods back through the Tonle Sap into the Grand Lac, a shallow seasonally fluctuating lake (some 11 000 km^2 at high water, shrinking to 2500 km^2 at low water and from 8–10 m to less than 1 m deep). In the wet season, June–October, this lake used to flood back into forest to support one of the world's largest freshwater fisheries. The lakeside villages are on barges which move with the lakeshore as it advances and retreats. Much of the forest is now cleared, the lake is shallower and fish catches have declined (Dussart, 1974). The importance of the Mekong fisheries has stimulated research on how the planned Mekong control operations will affect them; as part of this overall plan, several large man-made lakes have been created (Lagler, 1976; Beeckman & de Bont, 1985; Pantulu in Davies & Walker, 1986).

The Mekong system has a fish fauna of about 500 species. Their distributions within the system and in neighbouring basins were discussed by Taki (1978), who also examined the movements, feeding habits, spawning seasons and production of different groups of fish. He showed that the mountain streams are populated by predominantly carnivorous or omnivorous cypriniform fishes, the brooks by carnivorous cypriniform species, stagnant water by non-ostariophysans (carnivores or less frequently omnivores or herbivores), and the rivers by non-migratory fishes in all three trophic categories (carnivores predominating) and by numerous species of migratory cypriniform fishes (mostly herbivores or omnivores).

Fish movements are controlled by water levels. Many fishes move into the Tonle Sap Grand Lac as it fills, then out into the flooded zone (formerly forest) returning to the lake and migrating downriver as the water falls between October and February. Fishes are confined to the river bed in the dry season. Dry season conditions check fish growth, leaving scale rings which can be used for age determination (Chevey & Le Poulain, 1940). These provided evidence that cyprinid growth was faster in the lake than in the river, that growth was faster in the flooded forest than in the open lake, and that fishes in the delta grew faster when debris was brought down with the floods (this marked their scales annually, in contrast with the unmarked scales of fishes in the Indo-China Sea away from the river mouth).

The high production from the Grand Lac was attributed by Chevey & Le Poulain to the fish food in the flooded forest; they predicted that should the forest be cleared the fishery would decline. The feeding potentialities of flooded forest have direct relevance for forest clearance programmes in man-made lakes. Some ripe fishes were found at all times of year except for a short period in April–May when the water was very low, the time of cessation of growth when rings formed on scales. All reproductive stages were found in the populations at one time. It seemed that each fish had its own spawning rhythm and these were not synchronous. There was, however, a peak in spawning activity about June–July as the water was rising.

The downstream movements between October and February, as the water level falls, were studied at barrages on the river below the Tonle Sap where over 100 fish species were caught (Blache & Goossens, 1954). The fishes moved downstream in a definite order and were caught during only part of each lunar month, mainly between the first quarter and full moon from October to February. The largest fishes migrated down in the first lunar period (October), the siluroids travelling by night, the large cyprinids by day. The migration built up to a maximum of species in the third lunar period (December). Many Mekong fishes, including the giant catfish *Pangasianodon gigas*, are believed to migrate many hundreds of kilometres up- and downriver.

The Mekong Delta also has very important fisheries. Many catadromous fishes move into the estauries and freshwaters to feed in the wet season, returning to the sea to breed, and anadromous clupeids breed in the lower reaches. Local impoundments are made by the villagers to retain diadromous fishes and culture them in the delta region. Fishes living in the lower Mekong are classified broadly as 'white fishes', fluvial fishes feeding on plankton and small fish, and 'black fishes', bottom-dwellers feeding on benthic organisms; most white fishes, including cyprinids, migrate down by day, and the black fishes, including many siluroids and *Channa*, by night.

Southeast Asian reservoirs

The considerable literature on freshwater fisheries in Southeast Asia consisted until recently, largely of fish culture studies. Jhingran (1975) listed the larger Indian reservoirs and their fisheries and Fernando (1984) has reviewed the literature on limnology and fisheries in Southeast Asian waters from India through to the islands of the Sunda Shelf. Reservoirs form the major component of standing waters in Southeast Asia, where reservoir building commenced about 4000 yr ago as part of an extensive system of irrigation for ricefields, and there are now about half a million reservoirs here (sites shown in Fernando (1984)), most of them of modest size, the largest less than 1000 km^2. The various types of standing water in Southeast Asian countries total an estimated 3 million ha (30 000 km^2) of reservoirs (to be augmented to *ca* 20 million ha by AD 2000) and 85 million ha of ricefields (much of this useable for fish culture), compared with only 3.5 million ha of natural and floodplain lakes and 12.5 million ha of freshwater swamps.

Fish yields from Southeast Asian reservoirs vary greatly from area to area. This is due partly to levels of fishing efficiency but also to lack of indigenous fishes which thrive in lakes, as most of the indigenous fishes are riverine forms. Jhingran (1975) gives a mean range of only 6–7 kg ha^{-1} yr^{-1} for Indian reservoirs stocked with indigenous species, whereas in Sri Lanka with introduced African tilapia the mean yield is now 100 kg ha^{-1} yr^{-1}, and yields of up to 400–500 kg ha^{-1} yr^{-1} are being obtained in some man-made lakes in Sri Lanka and Indonesia. Fernando attributed the huge impact that tilapia have had on fish yields in Southeast Asian reservoirs to be partly due to their ability to digest blue-green algae, bacteria and non-protein amino acids in detritus, as many of these reservoirs are green with algae from human use and pollution. A comparable lack of lacustrine species to colonize reservoirs has been noted in Brazil and in Papua New Guinea, in both of which places tilapia have now been introduced.

In Lake Parakrama, a 2262 ha 1600-yr-old reservoir in Sri Lanka, the fish yield before the introduction of tilapia was very low, though the lake had 34 of Sri Lanka's 60 species (Fernando, 1984). *Oreochromis mossambicus* was introduced in 1952; by 1955 it was appearing in sufficient numbers to be recorded, and by 1967 the fish yield had risen from 10 kg ha^{-1} yr^{-1} to 180 kg ha^{-1} yr^{-1}. Complementary species of tilapia were introduced to Sri Lanka in 1969 and 1974. Parakrama's 1978 yield was *ca* 450–500 kg ha^{-1} yr^{-1}; the species composition, consisting of 80% *O. mossambicus* in 1967, changed as *Tilapia rendalli* became the dominant species. Indigenous carps still predominated in daytime catches, tilapia in gillnets set overnight. Beach seines took more diverse catches and the diversity of the indigenous fishes was not reduced by the introduced tilapia.

Introductions of exotic species are, however, only permissable under very exceptional circumstances, and after thorough studies of the conditions and search for suitable indigenous species. Once an exotic is introduced it is generally impossible to eradicate it, and escapes occur to other parts of the river system. The well-documented effect of the South American piscivorous cichlid *Cichla ocellaris* on the fauna of Gatun Lake in Panama Canal may be cited as an example (Zaret & Paine, 1973; Zaret, 1984a; see pp. 301–2). Introductions often have unforeseen side effects and they have led to extinctions of indigenous species in numerous cases. Introductions into lakes and reservoirs of Indo-Pacific islands are listed by Petr (1984) and data on distributions, biology and management of exotic species are collated in Courtenay & Stauffer (1984).

The Australia/New Guinea region
The Australia/New Guinea freshwater faunas are interesting as they lack the otophysan primary freshwater fishes dominating other tropical faunas and are mostly of representatives of marine families with many catadromous species. New Guinea is divided into two faunal regions by the central highlands; northern rivers have about 84 known species, 36 of them only known north of the mountains. About half the

Table 7.1. *Fish families represented in New Guinea's freshwaters, all except Osteoglossidae with marine affinities. Number of species present in Fly River (Roberts, 1978) and Purari River (Haines, 1983)*

	Fly R.	Purari R.		Fly R.	Purari R.		Fly R.	Purari R.
Carcharinidae	–	1	Melanotaeniidae	4	1	Toxotidae	1	1
Pristidae	1	1	Atherinidae	5	–	Scatophagidae	1	1
Osteoglossidae	1	1	Ambassidae	5	–	Mugilidae	3	3
Megalopidae	1	–	Centropomidae	1	–	Blenniidae	1	–
Moringuidae	1	–	Latidae	1	1	Gobiidae	8	3
Muraenidae	1	1	Chandidae	1	3	Eleotridae	10	6
Anguillidae	–	–	Lobotidae	1	2	Periophthalmidae	2	1
Clupeidae	4	1	Teraponidae	5	1	Gobioididae	1	1
Engraulidae	8	–	Apogonidae	4	1	Kurtidae	1	1
Ariidae	13	12	Lutjanidae	1	2	Soleidae	2	1
Plotosidae	7	1	Sparidae	1	1	Cynoglossidae	1	–
Belonidae	2	1	Sciaenidae	2	1	Tetraodontidae	2	1
Hemirhamphidae	2	1						

fauna of the southward-draining rivers is endemic, half shared with northern Australia (Berra, Moore & Reynolds, 1975). At least 12 families have endemic species in New Guinea rivers. The catfishes Ariidae and Plotosidae are best represented. Almost all the ariids are restricted to freshwater where they have radiated to use different foods (16 ariid species in the southward-draining Fly River, 12 species in the Purari). But there are no extensive radiations comparable to those undergone by primary freshwater fishes in other regions (Roberts, 1978). Another feature of the New Guinea fishes is their large size, nearly half the species growing to over 30 cm long (in contrast with a majority of small species in the Amazon and Zaire). The 103+ species known from the southward-draining Fly River represent 33 families (Table 7.1). North of the Central Highlands, the eastward-flowing Sepik has a much less diverse fauna and the main fishery on its vast floodplain is for the exotic tilapia *Oreochromis mossambicus* (Glucksman, West & Berra, 1976; Coates, 1985).

Ecological studies in the southward-flowing Purari, stimulated by proposed engineering projects (Petr, 1983; Haines, 1983) showed detritus, derived almost entirely from allochthonous sources, to be the main basis of the food webs (Table 7.2). The role of mud-feeding fishes and prawn-eaters was significant, prawns becoming more important foods in lower reaches. Foods used varied with the zone: the archer fish *Toxotes chatareus* for example, feeds on insects and fruit in the riverine

Table 7.2. *Numbers of fish species in various trophic categories in the main zones of the Purari River, Papua New Guinea. Data simplified from Haines (1983). (Same species entered in more than one category where applicable)*

	Small creeks	River[a]	Delta I[b]	Delta II[c]
Detritivores	4	8	4	9+
Fructivores	0	3	2	3
Other plant-eaters	0	3	0	2
Insectivores	4	5	3	4+
Molluscivores	0	0	4	4
Prawn-eaters	0	12	14	20
Crab-eaters	0	0	0	11
Piscivores	1	6	7	15+
Omnivores	0	3	2	2

[a] River = gorge to delta
[b] Delta I with flowing freshwater
[c] Delta II = lower mangrove zone

environment and on crabs and fruit in estuarine reaches. The larger fish species migrate to estuaries or the sea to breed or exhibit parental care (as in oral-brooding ariid catfishes). Some ripe fish may be found at any time, but the main breeding season appears to be October–December at the end of the wet season. The few small species (melanotaeniid rainbowfish, ambassids, apogonids, garfish, gobies, eleotrids) tend to be restricted to small side streams.

The important commercial species the barramundi (*Lates calcarifer*) spawns off the southern coast between October and February; larvae move into coastal nursery swamps when 5 mm long, returning to coastal waters when 20–30 cm (or when swamps dry out), when 6 months to 1 yr old. Tagging experiments have shown that during their second or third year most move into inland waters, moving far up the Fly River and into Lake Murray, then migrating back when 3 or 4 yr old to the coastal waters from which they came (Moore, 1982; Moore & Reynolds, 1982; Reynolds & Moore, 1982). Barramundi are protandrous hermaphrodites, males changing sex after spawning when 85–90 cm long, though not all females are males first (Moore, 1979; Davis, 1982). Maturation sizes vary in different river systems draining into the Gulf of Carpentaria, Northern Australia, though the reason for this is not known (Davis, 1984). Opportunistic predators, the barramundi change their diet from invertebrates to fish, and from benthic to pelagic fishes as they increase in size; in salt water, 87% of the food is fish, including ariids, dorosomatids, engraulids, *Lates*, mugilids, polynemids (Davis, 1985).

PART III

Marine fish studies

8

Underwater observations: coral reef fishes

Coral reef environments

Coral reefs harbour the most colourful and diverse of any fish communities, with the most complex relationships between species. The clear shallow warm waters provide natural aquaria, ideal conditions for direct observations of fish behaviour. The most important families of reef fishes are indicated in Table 8.1. Some species are not obligate reef fishes, but have a much wider distribution in association with hard substrates, and some reef fishes are particularly associated with marginal habitats, such as sand patches, lagoons, mangroves, or eel-grass beds (studied by Weinstein & Heck, 1979).

Coral reefs are of three main types: (1) barrier reefs along continents, most extensive along their eastern coasts; (2) fringing reefs, around islands; and (3) atolls, broken rings of reefs and islands around a central lagoon. The Indo-Pacific is particularly rich in corals and reefs, especially in the Indo-Australian archipelago, Malaysia, Sri Lanka, Madagascar, and around Indian Ocean and western Pacific islands. The Australian Great Barrier Reef is an intermittent series of reefs and cays stretching over 1900 km along the coast of Queensland north to New Guinea. Other major coastal reefs lie off the coast of East Africa and in the Red Sea. In the western Atlantic, reefs and cays extend 200 km southward from Yucatan in Central America, and many Caribbean islands are fringed with coral reefs.

Reef-building corals grow only where the mean sea temperature is at least 20°C; they need clear, saline water and so are unable to grow where rivers dilute the sea or deposit sediments. The coral polyps extend their tentacles at night to feed on zooplankton; their tissues house symbiotic green algae (zooxanthellae) and these and other algae living on their calareous skeletons photosynthesize in the sunlight. Reef-building corals reach their optimum development in water less than 30 m deep, where

Table 8.1. *Faunal composition of Virgin Island (western Atlantic) coral reefs (from data for 13 pavement and five Porites reefs* (Gladfelter & Gladfelter, 1978))

Trophic group	No. of families	No. of species	Families (species)
Herbivores	5	18	Scaridae (10); Acanthuridae (3); Pomacentridae (3); Blenniidae (1); Kyphosidae (1)
Omnivores	13	36	Labridae (7); Pomadasyidae (6); Pomacentridae (4); Ostraciontidae (4); Chaetodontidae (3); Monacanthidae (2); Mullidae (2); Gobiidae (2); Diodontidae (2); Sparidae (1); Carangidae (1); Gerridae (1); Canthigasteridae (1)
Planktivores	7	13	Apogonidae (6); Pomacentridae (2); Holocentridae (1); Grammidae (1); Priacanthidae (1); Sciaenidae (1); Pempheridae (1)
Crustacean-plus fish-feeders	9	23	Serranidae (8); Holocentridae (5); Lutjanidae (2); Scorpaenidae (2); Sciaenidae (2); Pomadasyidae (1); Muraenidae (1); Ophichthidae (1); Grammistidae (1)
Piscivores	9	15	Serranidae (3); Lutjanidae (2); Carangidae (2); Sphyraenidae (2); Muraenidae (2); Synodontidae (1); Fistulariidae (1); Aulostomidae (1); Bothidae (1)
Others	4	5	Pomacanthidae (2); Balistidae (1); Pomadasyidae (1); Gobiidae (1)

photosynthesis can occur; formation of new reef stops below 50 m. Dead reefs, relicts of reefs built during the lowered sea levels in Pleistocene times, are, however, found at greater depths, as off Guyana, where they shelter reef-dwelling fishes. Many kinds and species of coral make up a reef, the general form of the coral (glomerate, racemose or branching) being related in part to the degree of exposure and water movements. For the many reefs in Trade Wind belts the windward side is very exposed to wave action, the leeside very sheltered. In the Central Pacific Hawaiian Islands surges of 2 m wash over the reefs (compared with a tidal difference here of only 1 m). Coralline algae also contribute to tropical reef development.

In equatorial seas physicochemical conditions on the reef may be very constant throughout the year. At higher latitudes, for example at One Tree Island near the southern end of the Great Barrier Reef, conditions are more seasonal. Coral reefs offer an enormous range of ecological niches, perhaps more than any other biotope. The increased surface area of bottom and the innumerable crevices and caverns provide hiding places for diverse invertebrates on which many of the fishes feed. The diversity, abundance and biomass of fishes increases with the complexity of habitat; the effects of habitat structure on species diversity has been explored using artificial reefs. Community structure and zonation is also greatly influenced by actions of territorial damselfishes (Pomacentridae) and roving carnivores. The relations between fish body form and ecology has been studied in detail in damselfishes by Emery (1974) and in some surgeonfishes (Acanthuridae) by Robertson, Polunin and Leighton (1979).

-Coral reef environments are old, dating back some 50 million years or — more (many to Eocene times), so there has been time for very complex interactions between fishes and numerous groups of invertebrates to evolve. But how 'stable' are reefs and their communities? Despite the great age of some of the main reef systems, conditions in small patch reefs may be continually changing. Storms, which make the water turbid, or sudden lowered temperatures, may kill off sections of coral. Coral domes - collapse from the weight of their own growth, and invertebrates such as sponges and bivalves bore into their structure; the fishes, too, affect the reef. The algal gardens of damselfishes kill off patches of coral and these — are then eroded; in deeper areas tetraodont pufferfish bite off pieces of coral, adding an enormous amount of material to the bottom deposits.

Photosynthesis by zooxanthellae and numerous algae encrusting the reef makes reefs into areas of high primary production in what are often rather barren tropical seas. On reefs primary production is associated with the substrate, in contrast with upwelling areas where primary production is planktonic. Reef systems are often bioenergetically more or less self-sustaining, complete ecosystems in themselves. But the snappers (Lutjanidae), grunts (Haemulidae) and other fishes which shelter on the reef by day but move elsewhere to feed at night are bioenergetically not truly part of the reef fauna, though they enter the food webs when they are preyed on by large resident piscivores, such as groupers (Serranidae). Some fish biomass is lost from the reef when residents are preyed on by roving carnivores, large fishes from the outer sea which cruise along the reef wall or visit surge channels and reef flats mainly at night feeding on reef-dwellers. There is also a large exchange to the outer system in the eggs and larvae of the many kinds of reef fishes which have pelagic offspring; attempts are just beginning to estimate this exchange (Lobel & Robinson, 1986).

Why are reef creatures so colourful? There is unlikely to be an all-embracing reason, but the extreme clarity of the water enables distinctive colours and patterns to be used for signalling, signals which are understood by at least some of the other creatures present both intra- and interspecifically (see for example Wickler, 1967). Some are warning colours, for many of the reef creatures have evolved toxic and other defences against the many predators. The reef itself is so brightly coloured that some disruptive colours, aided by countershading, may in fact be concealing, and the selective absorption of light rays in seawater makes red fishes appear dull grey or black at depth. Many of the reef-dwellers other than fish, the coelenterates, worms, molluscs, echinoderms and crustaceans, are brightly coloured too. Hypotheses to account for sexually dimorphic colours and for the distinctive juvenile colours shown by many reef fishes are discussed by Thresher (1984). The reef is also full of sounds, though less is known about their significance: noises of wave surges, of coral-eaters crunching coral, clicks of crustaceans and of many fishes, including triggerfish, and an evening chorus of squirrelfish (see Winn, Marshall & Hazlett, 1964; Schwarz, 1985). Chemical communications are also undoubtedly important too, though as yet little studied, for many nocturnal fishes, such as moray eels, appear to seek their prey by chemical means (Bardach, Winn & Menzel, 1959).

Coral reef fish research in progress
    The very clear warm waters provide ideal conditions for underwater observations and conditions in the shallow waters are also good for

in situ experiments, such as creating artificial reefs or denuding patch reefs to study colonization and recolonization. Since the pleasures of SCUBA diving to observe fishes in their natural habitats became apparent, there has been an explosion in information on coral reef fish ecology and behaviour. Of the 200+ papers consulted for this chapter, space only permits the citation of key papers or recent ones with good bibliographies summarizing earlier literature. The aspects of research considered here were chosen mainly to supplement the observations made in freshwater fish communities in earlier chapters (Fig. 8.1).

Much of the research has centred on the high diversity in reef communities. The Indo-Malaysian archipelago has *ca* 2000 fish species; over 1500 fish species are known from the Great Barrier Reef (*ca* 1200 at the northern end and some 900 near its southern limit); the Caribbean reefs have *ca* 500 species. How are these fishes distributed, zoogeographically and ecologically within the reef? Within-reef (alpha) diversity is known to be high (see Sale, 1978*a*, 1980*b*), for instance 150 species from less than 50 m and some 75 species from a small patch dome. Analogy with studies of terrestrial vertebrates suggests that there must be fine resource partitioning of food and space to permit so many species to coexist, and many studies have suggested that this is so for reef fishes. There is, however, much overlap in resource use, and there are many known cases of reef species with seemingly identical ecological requirements. This led Sale (1977, 1978*b*, 1979, 1980*a*) to suggest that it is a matter of chance which

Fig. 8.1. Some aspects of coral reef fish biology considered here.

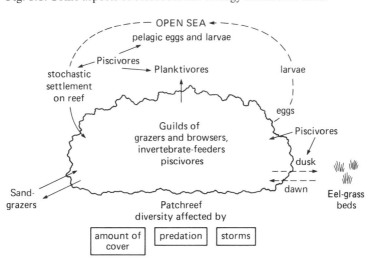

species of a particular guild of fishes with comparable ecological requirements gain access to a particular reef when settling down after their
pelagic larval phase. Sale's 'stochastic recruitment' hypothesis has stimulated an immense amount of research into recruitment to reefs and also
detailed studies of the partitioning of resources.

Useful review papers include those by: Randall (1967) on food habits;
Goldman & Talbot (1976) concerned mainly with zonation, trophic
groups and standing crops; Ehrlich (1975) concerned mainly with population ecology of reef fishes; Sale (1978a, 1980b) on the social organization
and ecology of reef fishes; Anderson et al. (1981) who carried out a
large-scale survey of chaetodontids across 50 miles of the northern Great
Barrier Reef; and Talbot, Russell & Anderson (1978), reporting on their
experiments on recruitment to artificial reefs. In the Red Sea the
sociobiology of feeding behaviour of coral fishes was studied by Fischelson (1977). Data on reproduction in reef fishes have been collated and
assessed by Thresher (1984). During the 'Tektite' experiment in the
Bahamas, biologists lived on the sea floor while observing reef fishes
(Collette & Earle, 1972). Reef studies have been centred: in the western
Atlantic from the Virgin Islands, Panama, Jamaica and Curaçao; in the
Indo-Pacific in the Red Sea, along the East African coast and Aldabra
Atoll; from Indian Ocean islands such as the Maldives and Laccadives;
along the Great Barrier Reef, especially from One Tree Island at the
southern end and Lizard Island at the northern end; and in the Pacific
around island groups such as Hawaii and the Marshall Islands. But huge
areas of reef have yet to be explored. These studies have been particularly
concerned with the following: (1) how important is resource partitioning?; (2) how important is stochastic recruitment from a guild of similar
species, as suggested by Sale?; (3) how much zoogeographical replacement is there of species across and along a reef?; (4) are reef communities
'stable' in the sense of being in equilibrium or are the species combinations
in a constant state of flux?

Use of space: zonation of coral reef fish faunas

A typical section across a coral reef (Fig. 8.2) would often
include:

1. Supratidal intertidal pools (or splash or surge pools where there
   is little tide and continuous surge as in Hawaii): fluctuations of
   temperature and salinity are very great in these pools and they
   are inhabited by euryhaline fishes such as blennies and the young
   of some reef fishes (e.g. surgeonfishes).

2. A reef flat, over which tides rise and fall, and over which some fishes may come to feed.
3. An outer reef wall or slope into deep water: communities here include those in the breaker or surf zone, those of the surge zone, where water movements are felt down to *ca* 10 m, and the infrasurge zone below the range of water movements.
4. Surge channels, breaks in the reef wall which allow access from the outer waters onto the reef flats or into enclosed lagoons in atolls, often tidal with strong currents.

Reef flats and lagoons have communities of bottom-dwelling fishes living round the heads of the various types of coral and on the sandy patches between them, also midwater and surface communities above the corals, and other species moving in and out with the tide. Glomerate corals (e.g. *Porites*), generally devoid of deep crevices, are frequented by polyp-grazers such as triggerfish (Ballistidae) and butterflyfish (Chaetodontidae). Branched corals (such as the stagshorn *Acropora*) provide cover for numerous small fishes (e.g. gobies and damselfish) which may swarm up to feed on zooplankton and dash back into the coral. A coral standing a metre or so high increases the total area of substratum, giving increased cover for the fishes and for the invertebrates on which they may feed. The increased surface area both adds new niches and enlarges niche size, leading to higher numbers of fish species and of individuals. In addition to the highly visible reef fishes, smaller species such as blennies and gobies hide in holes made by invertebrates which bore into the reef structure and in sponges, an invisible fauna of fishes which interacts with the other residents.

Fig. 8.2. Cross-section of a coastal coral reef with typical locations of reef fish groups.

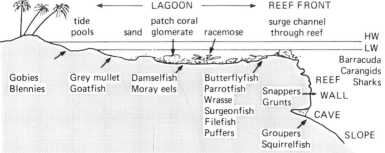

The zonation of reef fish faunas has been studied in many places: for example in the Pacific around Hawaii, where surges affect the fauna down to 6 m, below which the coral grows more freely in the quieter water (Gosline, 1965), and around Fanning Island where Chave & Eckert (1974) made semiquantitative abundance estimates in seven habitats – lagoon shoreline, turbid lagoon, clear lagoon, channel, tide pools, outer reef flats and outer reef slope – concluding that strong surge and tidal currents are here prime factors influencing fish distribution. These habitats were not sharply defined but of the 214 fish species (37 families) found here, 14 ranged widely through most habitats, the lagoon had 19 species, turbid lagoon patch reefs 38 species and clear lagoon patch reefs 76 species (twice as many), tidepools 83 species, the outer reef flat *ca* 64, outer reef slope 91 and reef channel 99 species.

On Virgin Island reefs (western Atlantic) Gladfelter & Gladfelter (1978) showed that diversity and abundance of fishes increased with the complexity of the habitat from coral pavement to racemose stagshorn coral. Abundance and diversity are also affected by whether the reef is adjacent to other kinds of feeding ground, such as eel-grass beds or tidal lagoons and mangroves where some reef fishes go to feed at night. Damselfish zonation has been examined by several workers (see Emery, 1974). Many of these fishes are highly territorial and very aggressive both inter- and intraspecifically so their distribution on the reef has a marked influence on the distributions of other species.

At One Tree Island, southern Great Barrier Reef, quantitative sampling produced 395 fish species from five major habitats: the off reef floor, reef slope or fronts, algal rim or reef surface, reef flat, and lagoon (Goldman & Talbot, 1976). Only 7% (26 species) occurred in all five major habitats and nearly half (188 species, 49%) were restricted to one or other of the major habitats, thus demonstrating a certain degree of habitat selection across the reef.

Site attachment in coral reef fishes

Most adult coral reef fishes are strongly site-attached. A variety of functions has been proposed for this, either singly or in various combinations: defence of space (for feeding, breeding, sleeping sites); defence of food; defence of mates; and defence of spawning sites and offspring.

Tagging experiments on the readily visible species, showing that individuals remain in small areas for long periods and how far they wander

while feeding, include those made by Bardach (1958) and by Springer &
McErlean (1962). Fishes readily visible on the reef by day included: (1) –
diurnal feeders, such as the algivorous pomacentrids which defend their
feeding territories vigorously, and other herbivorous and carnivorous
species (scarids, labrids, acanthurids) which wander through the reef
while feeding, swamping the defences of the territory-owners by moving
in mixed schools; and (2) aggregations of nocturnal feeders, such as
holocentrids, pempherids and apogonids in caves and haemulon grunts
on the open reef, all of which probably range extensively at night. Hobson
(1973) recorded *Haemulon flavilineatum* moving up to 1.6 km from its
resting site to feed. The small cryptic gobies and blennies living in worm
holes and sponges probably have a very restricted range of movements
(see Luckhurst & Luckhurst, 1978*b*).

The many different types of social structures, such as aggregations for
feeding, nesting or spawning, schools, individual territories (mainly for
feeding), pair territories (mainly for spawning), nest site, group and lek
territories, have been the subjects of numerous studies (see discussion in
Sale, 1978*a*, 1980*b*; Thresher, 1984).

Tagging experiments have shown that highly territorial damselfish may
remain round one coral head for at least 4 years. In the Caribbean,
tagging experiments have indicated that many reef fishes remain near
their homes for long periods; white grunts (*Haemulon plumieri*) showed
a tendency to home when transplanted (Springer & McErlean, 1962).
Transplantation experiments in French grunts (*H. flavolineatum*) have
shown that traditional routes are learned, i.e. that there is a non-genetic
transmission of social information across age classes or generations
(Helfman & Schultz, 1984). Parrotfish have been observed to use the
same sleeping caves month after month (and probably year after year),
travelling the same routes daily between feeding and resting sites, probably using sun orientation to do so (Winn, Salmon & Roberts, 1964).
However, two red groupers (*Epinephalus morio*) were recaptured after 3
yr in deeper water 33 and 42 miles (53 and 67 km) away from the tagging
site (Moe, 1969).

Lorenz had suggested that the vivid poster colours of many reef fishes,
such as chaetodontids, advertise their feeding territories, but later field
experiments on 20 chaetodontid species at Lizard Island by Ehrlich *et al.*
(1977), including experiments with models, did not support this view.
These chaetodontids were not territorial by day when feeding, though
they showed both intra- and interspecific aggression when defending
regular resting places. The significance of their poster colours remains

largely unresolved. Kelly & Hourigan (1983) stressed the role of conspicuous patterns for the maintenance of visual contact in following chaetodontids.

Many other reef fishes, including some acanthurids and pomacentrids,
adopt feeding territories when young but abandon then when mature, or
adopt a home range system with other members of the species (sometimes
even with other species) which then defend the home range against
intrusion by conspecifics and others. Constant activity to maintain a
territory consumes a good deal of energy, and whether or not they attack
other species moving through the territory often depends on whether
these compete for food. For example, the algae-grazing '*Pomacentrus*'
flavicauda at Heron Island (Great Barrier Reef) responded aggressively
to 38 fish species (of 13 families) which were also algae-feeders, but not to
16 species which were carnivores: the experimental removal of six *P.
flavicauda* led to an increase of algae-feeding fishes in their territory
(Low, 1971). But in another pomacentrid, *Abudefduf saxatilis*, which
lives in large plankton-feeding schools off Colombia (eastern Pacific),
territories were only established for reproduction and brood care by the
male: here the male only chased away species which nip the substrate and
not fishes which feed in other ways (Albrecht, 1969). Experimental
manipulations of bluehead wrasse (*Thalasomma bifasciatum*) have shown
how local population density can strongly affect the economic defensibility of a mating territory (Warner & Hoffman, 1980).

Use of time: diurnal/nocturnal activity and diel movements

By the day the reef is crowded with fishes. Most of those that feed
on the reef, herbivores and invertebrate-feeders, are diurnal, and there
are also large aggregations of nocturnal snappers and grunts resting there;
other nocturnal species hide there by day, holocentrid squirrelfish in
caves, moray eels in crevices. At night the reef appears deserted as the
resident diurnal fishes hide, jamming themselves into crevices, and the
nocturnal species are away feeding. The diurnal fishes help to protect
themselves from predators by erecting their spines or by inflating the body
in puffer and porcupine fishes. Some parrotfish (*Scarus*) secrete a mucous
sleeping cocoon round the body; preliminary experiments suggested that
this may protect them from predation by nocturnal moray eels which hunt
by chemical means (Winn & Bardach, 1959). Tagging has shown that reef
fishes may use the same sleeping places over long periods.

Most nocturnal fishes move away from the reef to feed at dusk, either into the water column (holocentrids and apogonids studied by Luckhurst & Luckhurst (1978a) and pempherid sweepers (Gladfelter, 1979)), or over considerable distances to sandy flats or sea-grass beds. These migrations are often very regular, both as to time and route taken, distances ranging from just a few metres to several kilometres in different species. Hobson (1965, 1968, 1972, 1973, 1974, 1975), who made a special study of diel activity in marine fishes, concluded that patterns of migration are strongly influenced by the relative threats from predation at different parts of the diel cycle. Most predators that threaten reef fishes are visual feeders whose mode of attack becomes less effective when light falls below a certain level; many of them are crepuscular feeders, i.e. most active at dawn and dusk. By day most of the smaller reef fish remain relatively secure by staying close to shelter or by schooling, mechanisms which are less evident at night when reef fishes range more freely into open water and their schools are loosely defined. The tendency for loose associations and ranging further afield increases on a dark night. On Curaçao reefs, territorial pomacentrids held their territories throughout an 18-month study (Luckhurst & Luckhurst, 1978b); the presence of holocentrids resting here by day affected the space available for other species. The school is especially important for migrating species; submarine topographical features provide reference points for the fish and schools follow well-defined routes (Hobson, 1975). Those followed by grunts between reefs and sea-grass feeding grounds persisted over years; tagging experiments have shown that juvenile grunts can move over 3 km to reach a home reef (Ogden & Ehrlich, 1977).

Stomach contents of 103 species speared on Hawaiian reefs at all times of day and night led Hobson (1972) to conclude that the generalist feeders, including carnivores whose morphology places them close to the main line of teleostean evolution, are predominantly nocturnal or crepuscular feeders (holocentrids, scorpaenids, serranids, apogonids, priacanthids, lutjanids). Of these, the nocturnal species prey mainly on small motile crustaceans which are most readily available when they leave their shelters after dark, while the crepuscular species take mainly small fishes, whose defences are less effective during twilight. Diurnal predators have to ambush or stalk their prey: the ambushers include highly cryptic scorpaenids, synodontid lizardfishes and bothid flatfish; the stalkers include aulostomids, fistulariids (trumpet- and cornetfish), belonid garfish and sphyraenid barracudas, all of them long, attenuated fishes.

Some of these use non-predatory fishes as stalking horses to give cover as they approach their prey (as described by Eibl-Eibesfeldt, 1965; Ormond, 1980). More specialized predators include muraenid eels, which hunt deep in rock crevices by day and night, finding small animals resting there or recovering from injuries or distress, and mullids which drive prey into the open by probing the bottom sand with their long barbels. But the majority of reef fish belong to the acanthopterygian teleosts (see Appendix) which have diverged into a wide variety of specialized carnivores and herbivores. These are mostly diurnal and very colourful (chaetodontids, pomacentrids, labrids, scarids, acanthurids, with blennies and balistids, monacanthids, ostraciontids, tetraodontids, canthigasterid puffers and diodontid porcupinefish). Diurnal feeders have more varied feeding habits and include all the algivores and cleaners. Specialized feeding structures and techniques enable some fishes to consume sponges, coelenterates, large molluscs and tunicates. Diurnal planktivores eat tiny crustaceans (in contrast to nocturnal planktivores, such as holocentrid *Myripristis*, which feed on larger plankton elements, such as crab zoea larvae). The fish have to cope with the varied defences of the food sources, such as toxins, hard bodies and spines. Specialized feeding is only made possible by manoeuvrability and specialized alimentary structures; many of these fishes have a small mouth, and the herbivores have developed very long intestines.

Aquanauts who spent several weeks living under the sea on a Bahamas coral reef investigating the nocturnal/diurnal changeover and space-sharing mechanisms concluded that inter- and intraspecific competition for space is more important than food in limiting the number and kinds of reef fishes (Collette & Earle, 1972). The same crevices may be occupied by different species at different times of day and night.

The importance of schooling to avoid predation was brought out very clearly by Hobson's study of nocturnal/diurnal activity in the Gulf of California (Hobson, 1965), though this was on rough ground, not coral reef. In this 3.5-year study involving 45 species (22 nocturnal, 19 mainly diurnal, 5 crepuscular), all the nocturnal fish were predators, large schools of which moved offshore at night to feed on mobile crustaceans. Most of the large inshore piscivorous fishes were crepuscular, preying on these nocturnal fishes as they moved to and from their nocturnal feeding grounds along well-defined routes at dusk and dawn, waiting for their prey at particular points *en route* where the school tended to become disorientated, such as when rounding a headland. By day the piscivores did not seem to recognize the schooling fish as prey unless stimulated in some way, by disorientation within the school or by damaged fishes behaving in an irregular way – these were picked off at any time of day, as were fishes bolted by diving pelicans.

Trophic interrelationships

Food webs on coral reefs are very complex, but some of the main trophic relationships are indicated in Fig. 8.3. The reef algae support large populations of herbivorous fishes. Unicellular algae and detritus are sifted from sandy patches by grey mullets (Mugilidae) and blennies, but the more characteristic reef fishes are browsers with cutting teeth which bite pieces from algal fronds (some acanthurid surgeonfish, pomacentid damselfish, chaetodontid butterflyfish, balistid triggerfish, siganid rabbitfish, kyphosid pilotfish, tetraodont puffers) and grazers which remove some of the substrate (blennies, gobies) or actually bite pieces off the coral.

Fig. 8.3. Trophic relationships among coral reef fishes. Up to five trophic levels are represented; reef communities also include many omnivores, cropping food from several trophic levels.

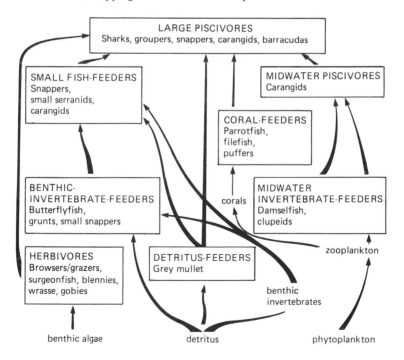

The coral-feeders include browsers on polyps (e.g. chaetodontids, monacanthid filefish), grazers on living coral heads (chaetodontids, monacanthid filefish) and feeders on branching coral tips (balistids, monacanthids, tetraodontids, canthigasterid puffers). Even the stinging *Millepora* coral is used by one kind of filefish (*Alutera scripta*). Innumerable kinds of invertebrates of many phyla live in crevices of the reef or burrow in the sand between the corals. Convergent adaptations for extracting these include the forceps-like snouts of certain chaetodontids and wrasses.

Off Tanzania, Talbot (1965) recorded 46 carnivorous fish species of 19 families feeding on invertebrates other than coral. In the Pacific Marshall Island reefs over 122 fish species of 37 families cropped the benthonic bottom fauna, in addition to the many species taking the fossorial infauna of invertebrates which emerge from burrows in the sand and the coral surface at night (Hiatt & Strasburg, 1960). In the West Indies, Randall (1967) distinguished between (1) fishes taking sessile animals (sponges, anthozoans, polychaetes, gorgonians, tunicates), (2) shelled-invertebrate-feeders, mostly fishes with strong jaws or pharyngeal teeth able to crush prey such as gastropod and bivalve molluscs, echinoderms and crabs, and (3) ectoparasite-feeders (cleaners). Many species are omnivores: some facultative, others primarily herbivorous or primarily carnivorous, others incidental omnivores.

The piscivores of many kinds, some of which also take invertebrates, include both reef residents and transient roving carnivores. The zooplankton above the reef is exploited both by members of the reef community such as damselfish, which live in the shelter of the reef but move up to feed, and by clupeids living in more open water above the reef. These fishes in their turn support the predators of the mid- and surface-water communities, such as carangid jack mackerals, belonid garfish and barracuda (*Sphyraena*). Within these marine fish families the different species tend to specialize on different food organisms, but many species take a variety of foods and the fishes often change their habitats and food as they grow. Attempts to determine the trophic level of a particular species are therefore very complicated.

Comparison of the foods used by reef fishes in the Pacific (Marshall Islands), Indian Ocean (coastal reefs off Tanzania), and West Indies, showed that: (1) the same categories of foods were utilized by the same families of fishes, even though the families were represented by different genera and species in the three areas (only a very few species are circumtropical); (2) zooplankton-feeding fishes were less numerous on the Tanzania reef (a fringing reef where the coral polyps themselves eat zooplankton) than at the other two sites; (3) at all three sites the majority of the fish families were carnivores, using a great variety of animal food – a herbivorous diet involves many changes, both morphological and physiological; and (4) foods eaten by carnivores vary very much according to what is available at the time – for example, stomach contents of the West Indian Nassau grouper (*Epinephalus striatus*) included fishes of 16 families (in 53% of the stomachs), crustaceans of many kinds (in 39%), also cephalopods, gastropods and bivalve molluscs.

Despite their defensive properties (mineralized sclerites, noxious chemical compounds, tough fibres), sponges were found in 21 species of West Indian fishes, making up 70–85% of the food ingested by *Pomacanthus* and *Holocentrus* angelfishes and a filefish. The other species took a little sponge with other food. Herbivorous scarids scrape live coral or gouge algae from rocks, grinding pieces in the pharyngeal mill. Acanthurids feed on calcareous algae, grinding them in the gizzard-like stomach. These two families are among the most abundant reef fishes, one or other dominating the fish biomass on most reefs. The Aldabran benthic algivore community is dominated by acanthurids (12 of 24 species, 51% of individuals, and 402 of 697 kg ha^{-1} standing crop), which Robertson, Polunin & Leighton (1979) attributed to Aldabran habitats lacking adequate types of shelter for use by larger species of territorial damselfish, such as those which affect acanthurid distributions on Caribbean reefs where such shelters are plentiful. Many carnivores also contribute to bottom deposits by crushing hard-shelled invertebrates. Triggerfish and filefish chop up crabs and sea urchins, and puffers and porcupinefish use beak-like jaws to break up molluscs, echinoderms and crustaceans. Wrasses, grunts and porgies seize prey and crush it with pharyngeal teeth. Many sparids have molariform teeth. Eaglerays (Myliobatidae), which travel widely, often in large schools, use their flat pavement-like teeth to crush bivalve molluscs. Fishes thus help to break down reefs and contribute sediments to bottom deposits.

The proportions of producers, herbivores and carnivores on reefs have been of great interest to those concerned with the bioenergetics of reef systems, but the use of many different foods by so many of the fishes, together with the switch from one trophic level to another as a fish grows, makes it almost impossible to assign trophic levels with any accuracy. Furthermore, it is now known that many fish included in the reef community biomass estimates feed away from the reef, which will also have introduced errors into calculations. Proportions of herbivores to carnivores are also likely to be different on the various types of reef. In a classic study at Eniwetok Atoll in the Marshall Islands (Odum & Odum, 1955) the biomass of herbivores (fish plus invertebrates) was estimated to be four to five times that of the carnivores. At Tatia Reef, off Tanzania, herbivorous fishes were found to form less than half of the total biomass (herbivores 39%: carnivores 61%: Talbot, 1965). Tatia Reef is an actively growing reef, deriving much energy from zooplankton, whereas Eniwetok Atoll is an enclosed, almost autotrophic, lagoon.

The composition of western Atlantic reef faunas from 25 natural patch reefs censused in the US Virgin Islands (13 pavement and five *Porites* reefs) by Gladfelter & Gladfelter (1978) is shown in Table 8.1. A later comparison of the numbers of individuals from the different trophic groups on these reefs with the fauna on comparable patch reefs in the Central Pacific, Eniwetok Atoll (Gladfelter, Ogden & Gladfelter, 1980), indicated that the main differences in faunal composition reflected conditions surrounding the reefs. The Virgin Island reefs were near to grass beds, into which many invertebrate-feeding fishes moved to feed at night, and these showed greater segregation in times and place of feeding than did the herbivores and piscivores on the reef. The Eniwetok reefs lacked any nearby grass beds but the reef was washed with plankton-rich water, reflected in the abundance of diurnal planktivores.

The relative proportions of different trophic categories and how they vary across a reef were studied at One Tree Island (Great Barrier Reef) by Goldman & Talbot (1976). For the reef as a whole, the total sample biomass was made up of 54% piscivores, 18% benthic invertebrate-feeders, 18% grazers (on both algae and coral), and 10% planktivores. Carnivorous fishes constituted 3.4 times the biomass of grazers, but these carnivorous fishes were also feeding on second- and third-trophic-level invertebrates. The biomass on the leeward slope contained all four trophic categories in about equal amounts, in the reef lagoon the main biomass was of benthic-invertebrate-feeders and grazers, and on the windward slope and transition to offreef floor the main biomass was of piscivores. These piscivores, however, included many carnivorous species resting here by day but feeding away from the reef at night, so these are not accurate representations of the standing crops supported by the zones.

Studies of carnivorous feeding habits include those by Randall & Brock (1960), by Thompson & Munro (1978) and those of groupers and snappers considered below (p. 215). A number of species may be found containing the same food items (see also p. 291), but the view that reef fishes are generalists with broadly overlapping resource utilization is countered by more detailed studies, for example that of Gladfelter & Johnson (1983) on holocentrids (7 species) around St Croix (West Indies). In these nocturnally active crustacean-feeders, food was partitioned by taxon between four species that consumed mainly shrimps and three species that ate mainly crabs, and by body size of prey, while differences in foraging microhabitat (position in the reef zone) were as important as food in separating the species. Again, on Madagascar reefs, stomach contents of eight serranid species and 17 scorpaenid species indicated to Harmelin-Vivien & Bouchon (1976) that in each species the foods consumed depended on time of feeding (day or night), size of individual, biotope and position on the reef. Among the serranids one species was strictly piscivorous, two species contained crustaceans as well as fish (mainly brachyurans in one, shrimps in the other) and six had more varied diets, one including cephalopods. Among the scorpaenid species, only one was strictly piscivorous, others had diets of fish and crustaceans, and one took mainly polychaetes. The hunting behaviour of the large roving carangid *Caranyx melampygus* and the reactions of its fish prey have been studied in Aldabran channels (Potts, 1980).

The partitioning of resources in these holocentrids suggested that competitive interactions have played a role in the evolution of this assemblage but, as Gladfelter & Johnson stressed, this is not to say that competition is occurring at present, or that competitive interactions are structuring the community. The differences in food utilization observed appeared to permit competition-free existence over a spectrum of holocentrid and prey densities, the limits of which would be set by a lower limit of prey availability and an upper limit of holocentrid density; these holocentrid populations are limited by factors other than competition for food, at least under some conditions.

Feeding guilds

Several families of marine fishes have guilds of species utilizing similar food resources. Amongst the chaetodontid butterflyfishes of the Great Barrier Reef, Anderson *et al.* (1981) recorded guilds of hard-coral-feeders (8 species – 6 small, 2 large), soft-coral-feeders (4 species – 3 large, 1 small), four species utilizing non-coralline invertebrates (one with a very deep body found inshore in calm water) and four generalists of varying sizes. Every community included a generalist species, but this was never the most abundant species present.

In acanthurid surgeonfishes, a pantropical family of *ca* 75 species in six well-defined genera, feeding guilds include: (1) detritus-feeders, small to medium-sized species tied to the hard reef which feed busily most of the time; (2) reef grazers (the largest group), mostly *Acanthurus* species, which also feed almost continuously during the day, with a broad overlap of foods consumed but some division of resources according to where on the reef the species characteristically feeds; (3) sand grazers, all *Acanthurus* species, large mobile species living in schools of mixed species at selected places on the reef or wandering widely among patch reefs removing diatoms and algae from sand grains – there is no competition for feeding areas on open sand as there is on the reef; (4) browsers, which feed on patchy leafy algae, food which is easy to crop and digest, leaving more time for other activities, large mobile species (*Naso*) in which competition between species seems minimal; and (5) planktivores of two species which feed in schools in deeper (10–20 m) water off cliff faces and pinnacles.

Associated with acanthurid guild membership are size and mobility, time free from feeding and extent of competition (Barlow, 1974*a*). Detritus-feeders and reef grazers are small busy feeders with time for little else; reef grazers also compete with many other species. These all tend to remain in a relatively small area (though the reef flat invaders roam considerably). Browsers, on the other hand, are large mobile species which feed quickly and have much spare time; competition between them seems minimal. Sand grazers are intermediate in size and free time, often travelling in mixed species groups and joining with other species when venturing away from the protection of the reef. The Hawaiian *Acanthurus triostegus* has two forms of school: dense column-forming schools are used as a tactic in getting to the bottom to feed when the bottom is held by territorial competitors, but in atolls where habitat partitioning exists schooling is used as an antipredator tactic among the *A. triostegus* feeding in the intertidal area (Barlow, 1974*b*).

The interrelatedness of various aspects of a fish's ecology is well shown by studies of morphology, feeding strategies and social and mating systems in acanthurids. Of three surgeonfish species at Aldabra studied by Robertson *et al.* (1979), adults of which defend feeding territories intra- and interspecifically, the largest species, *Acanthurus lineatus*, because of its morphological limitations relies on food that has to be defended against many species; this species forms large colonies in which single fishes defend small territories containing high standing-crop algal mats and territory defence is aided by colony formation. Morphological adaptations in the smallest species, *Zebrasoma scopas*, enable this to rely on food that the other species cannot exploit efficiently. This species forms pairs that defend territories containing a thin algal mat and it is restricted to the poorest quality habitat by the aggressive tendencies of the more dominant species. *A. leucosternon*, which also forms pairs, has an intermediate feeding strategy. The local coexistence of these three and ten other acanthurid species at Aldabra is enabled by partitioning of both habitat and food resources, aided by the populations of two of the most dominant species being below the carrying capacity. In acanthurids, territoriality is associated with pair formation; permanently territorial fishes usually form pairs, but colonial species do not do so; presumably the colonial habit would facilitate interference of males in each other's spawning activities.

The roles of herbivorous fishes on the reef

Herbivores and the reef algae on which they feed represent a co-evolved system of defence and counterdefence (see Ogden & Lobel, 1978). Many reef fishes are highly selective of the algae they consume. Some may reduce the abundance of competing algae, allowing corals and cementing coralline algae to survive. Herbivorous fish are conspicuous and active by day, often moving relatively long distances at twilight to hide within the reef. By day they have one of three main foraging strategies: (1) territorial defence, which in some damselfishes includes tending algal 'gardens' in their territories; (2) feeding within a home range, as do some pomacanthid angelfishes and acanthurid surgeonfish; or (3) feeding in foraging groups, often of mixed species, as in some surgeonfishes, parrotfishes, angelfishes.

The herbivorous damselfishes are very aggressive; they dominate other species and affect coral reef communities in three basic ways (Lobel, 1980): (1) the green algal mat in their territories affects the corals and coralline algae essential for reef structure, so the territories tend to erode to loose rubble; (2) the damselfish feed on the blue-green algae with their included diatoms, bacteria and detritus – these blue-green algae are a site for nitrogen fixation on the reef; and (3) the fish leave a basal layer of red algae which forms a refuge for juvenile motile invertebrates (crabs, ophiuroids) which are more abundant than outside the territory. Damselfishes are very aggressive, particularly to other herbivores. Their territorial defences are, however, overcome by the arrival of schools of mixed species of feeding fishes (scarids and their associates: Robertson *et al*. 1976; see also Ehrlich & Ehrlich, 1973; Alevison, 1976; Itzkowitz, 1977). In the omnivorous territorial *Eupomacentrus planifrons* defences are overcome by parties of *Scarus 'croensis'* and associated benthic herbivores *Acanthurus* and *Sparisoma*, with browsing carnivorous *Pomacanthus paru*, predatory wrasses *Halichoeres* and goatfish *Mulloidichthys*. Removal experiments have tested ideas on coexistence (Waldner & Robertson, 1980); aggressive *Pomacentrus* were removed by Low (1971). Robertson & Lassig (1980) found a marked degree of habitat partitioning between damselfish species. Exclusion of fish led to increased algal growth (Stephenson & Searles, 1960).

Acanthurid faeces may be an important component of the detritus in shallow waters of windward reefs at Eniwetok Atoll; Chartock (1983) estimated that acanthurids apparently required 59% of the net primary production along transects of the reef flat, but because of their low assimilation efficiency, *ca* 0.83 g C m^{-2} day^{-1} was defaecated onto the reef flat by the fish population. This cycling of algae may provide one important allochthonous detritus source to the food webs.

Symbiotic relationships in coral reef fishes

Under the conditions of close proximity and high density on coral reefs a whole range of interspecific interactions occur, many of which have led to the evolution of stable symbiotic relationships. Fish–fish relationships include: (1) cleaner fishes and their hosts; (2) symbiotic sharing of feeding territories and algal food by damselfish and surgeonfishes; and (3) feeding commensalism between the bottom-probing red mullets (Mullidae) and the many other fishes (15 species in the Gulf of Aqaba) feeding in the sand clouds stirred up by them (see Fricke, 1975). Relationships between fishes and various invertebrates include: (1) that between pomacentrid clownfish (*Amphiprion*) and sea anemones, the fish protected from the host anemone's stinging cells (much studied – see Allen, 1972); (2) the pearlfish *Carapus* (Carapidae) living in holothurian sea cucumbers (see Smith, Tyler & Feinberg, 1981); (3) that between gobioid fishes and *Alpheus* shrimps (see Fricke, 1975); and (4) fishes, often with shrimps, living amongst the forests of sea urchin spines.

Ormond (1980) described interspecific feeding associations in 24 species of Red Sea coral reef predators, involving following and scavenging (for example labrids following mullids), interspecific joint hunting by 'riding' (stalking prey under cover of a non-predatory species, as described for the trumpetfish *Aulostomus* which uses parrotfish as cover when approaching prey), and hunting by aggressive mimicry (as blennies do to gain access to prey by mimicking cleaner wrasse). Some species make use of these associations very frequently during feeding and may associate with a selection of species according to their availability. Symbiotic relationships occur more often among coral fish than in the open sea or freshwaters. In the sea there has been longer for such relationships to evolve than in freshwater; furthermore, many of the invertebrate groups involved (sea anemonies, echinoderms) do not occur in freshwater. Numerous cases of protective resemblance and mimicry among these reef fishes have also been described (see Randall & Randall, 1960; Russell, Allen & Lubbock, 1976); in some cases the mimic–model relationship is loose and facultative, in others it appears to be highly co-evolved and obligatory.

Cleaner fish behaviour

Cleaning appears to be essential for the welfare of reef fishes. Ectoparasites, such as copepods and larval isopods, and dead tissue are removed. In the Bahamas the experimental removal of all known cleaner organisms from two small isolated but well-populated reefs led to fish numbers falling drastically within a few days (Limbaugh, 1961), though other attempts at this type of experiment have not been so conclusive (Ehrlich, 1975). Cleaning appears to be more highly developed in tropical than in temperate seas; the cleaning habit has now been described from tropical freshwaters, both in Sri Lanka (Wyman & Ward, 1972) and in the African Great Lakes (Witte & Witte-Maas, 1981; Ribbink, 1984*a*). In tropical seas many fishes have specialized as full-time cleaners; they are very conspicuous in colour and have adopted special displays to make them more so; they generally live in pairs or in small groups (harems, see p. 243) at special 'cleaning stations' which lead to social aggregations of fishes. More than 50 species of marine fishes are now known to clean other fishes at some stage of their life history. They range from species in which only some individuals clean, through others which only clean when young, to the obligate cleaners of the Indo-Pacific only very rarely found with food of free-living origin in the stomach.

Cleaning is a diurnal activity and appears to depend on the clarity of the water. Nocturnal fishes such as moray eels emerge from holes to be cleaned by day. Even large manta rays visit cleaning stations. Convergent evolution has led to the development in different families of analogous structures for cleaning, such as pointed snouts and tweezer-like teeth. The most specialized cleaner, the wrasse *Labroides dimidiatus* in the Indo-Pacific and the neon goby *Elactinus oceanops* in the Caribbean, present a striking example of convergence in colour and markings, blue with a longitudinal black stripe, a cleaner uniform which is emphasized by see-saw dancing movements in front of potential clients which advertises the cleaning station. In the Indo-Pacific this cleaner uniform is also mimicked by the parasitic sabre-toothed blenny *Aspridonotus taeniatus*, enabling it to get near to its host unmolested, where it then nips pieces from the host's skin or fins.

In the Indo-Pacific the cleaner wrasses *Labroides* (the very widely distributed *L. dimidiatus* and three more localized species) are firmly in possession of the cleaner niche. All are non-specific in the fishes they serve. The Caribbean lacks *Labroides* and a whole range of fishes appear to be taking over the cleaning role, of which the neon goby *E. oceanops* is the most specialized. This goby so closely resembles *L. dimidiatus* that it has been suggested that *Labroides* was once present in the Caribbean but died out during the Ice Age cooling, and that knowledge of the cleaner uniform survived amongst descendants of fish once cleaned by it, giving the fish most like the original cleaner a selective advantage. This suggestion is supported by aquarium observations: Caribbean fishes demand *Labroides* to clean them even though they have never previously met one and, conversely, Pacific fishes demand Caribbean cleaner gobies to do the same (Eibl-Eibesfeldt, 1965). Such convergence presents an excellent example of the moulding effects of biotic pressures on the evolution of appearance, behaviour and responses of another species. Species affect one another in many more ways than just through predation or competition for food or space.

Ethological studies have been made of the reactions of cleaners to hosts and of hosts to cleaners (e.g. Potts, 1973). The see-saw dance of the cleaner inviting a potential host is thought to have arisen from conflict between desire to flee and to dive into the safety of the reef, a dance that developed social significance as visiting fishes were generally cleansed following it. Host species are cleaned in a highly specific manner and hosts play a very active role, adopting a special attitude (which varies from species to species) inviting cleaning. Cleaners are only rarely found in the stomachs of host fish. A host such as a grouper warns the cleaner to leave its mouth by ritualized intention movements before it shuts its mouth and moves away. Many studies have been made of the immunity of the cleaner to predation by its host.

Breeding strategies in coral reef fishes

Thresher (1984) who has collated data on reproduction in reef fish, lists pelagic eggs in 36 families, demersal eggs in 13 (mostly smaller species), egg-scattering in two (Siganidae, Tetraodontidae), benthic broadcast eggs in one (Muraenidae) and live-bearing in two (Brotulidae, Clinidae), and he examines costs and benefits of these different systems. Most large species of coral reef fishes produce large numbers of young which are dispersed by a pelagic larval stage. These are produced by repeated spawnings over a lengthy breeding season, by adults which are themselves relatively sedentary. Some examples of the seasonality and frequency of spawning in a representative sample of reef fish observed by various authors are given in Table 8.2 (see also Sale, 1977, 1978a). Mortality of the young is probably very high relative to that of the adults. In many species the eggs are also pelagic – in some mugilids, serranids, lutjanids, mullids, anthiids, chaetodontids, labrids, scarids, acanthurids, scorpaenids, diodontids and pomacanthids.

Among families producing demersal eggs (mostly smaller species), such as pomacentrids, the newly hatched larvae may have a short pelagic phase, but some may be demersal throughout life. Siganid rabbitfish scatter adhesive eggs, which hatch in about a day into larvae which are pelagic for about a month. Blenny and goby parents (one or both) generally care for their eggs, but some species may have a pelagic larval phase. Apogonid cardinalfishes practice oral incubation of eggs, and larvae are restricted to shelf waters (Springer, 1982).

In the Caribbean, seasonality of spawning has been studied in 35 species on Jamaican reefs (18°N) by Munro *et al.* (1973), and the spawning

Table 8.2. *Patterns of reproduction in some representative reef fishes (further data and references given in Sale, 1977, 1978a; Thresher, 1984)*

Species	Duration of season (months)	Frequency of spawning by individuals	Pelagic stages	
			Eggs	Larvae
Acanthuridae				
Acanthurus triostegus	8–12	multiple	+	+
Scaridae				
Sparisoma rubripinne	all year	?	+	+
Labridae				
24 species	6	multiple	+	+
Pomacentridae				
Abudefduf saxatilis	5	multiple	–	+
Chromis caeruleus	8	up to twice a week	–	+
Amphiprion chrysopterus	9	monthly	–	+
Pomacanthidae				
Centropyge bicolor	all year	daily	+	+

patterns and seasonal number of juveniles in 350 species around Puerto Rico (18°N) by Erdman (1977). Reef fishes here comprise: small species, with demersal eggs or oral incubation whose larval young never enter the plankton; and larger species which produce eggs hatching into pelagic larvae with either a short pelagic life, rapidly metamorphosing into the adult form and never found in oceanic plankton, or a long oceanic pelagic larval or postlarval life in which the majority of larvae may drift far from the spawning site. In the western Pacific (Paulau), Johannes (1978) distinguished between: (1) migrating spawners (serranids, mugilids and possibly some carangids, lutjanids, scarids and leiognathids); (2) non-migrating spawners with pelagic eggs, comprising smaller species which remain close to their usual habitats to spawn, probably because small fish are more vulnerable to predation (e.g. some scarids, labrids, mullids, acanthurids); and (3) non-migrating spawners with demersal eggs, found among pomacentrids, siganids, balistids, tetraodontids, gobies, blennies, clinids and apogonids, in most of which the eggs are guarded against predators.

In the Caribbean many of the small species with demersal eggs appear to spawn continuously throughout the year (with slight equinoxial peaks in the pomacentrid *Abudefduf* on Jamaican reefs), as do the pomacentrid anemone commensals at Eniwetok Atoll in the Pacific (Allen, 1972). In the Red Sea the numerous pomacentrid species, with their very varied types of social spawning behaviour (p. 206), spawn over an extended season from March–April to September (Fischelson, Popper & Avidor, 1974).

On Jamaican reefs, among the larger species with a short pelagic larval stage, some lutjanid snappers and pomadasyid grunts spawn more or less throughout the year but with a peak from February to April, the months when water temperatures are minimal (26.5°C in March); the *Haemulon* grunts show some staggering of peak times. In scarids, spawning appears to be confined to January–June (with a February–March peak). Among those with a long pelagic larval stage, carangids spawn throughout the year but, again, most spawn in the cooler-water months, and acanthurids spawn from January to June (peak February–March). The serranid groupers here have an even more restricted season, from January to May (peak February–March). Off Florida, the red grouper *Epinephalus morio*, a protogynous hermaphrodite (like many serranids), is a total spawner with breeding confined to May (Moe, 1969). The Nassau grouper *E. striatus* spawns in late January off the Bahamas; Smith (1972) described a spawning aggregation here of well over 30 000 individuals, many of which must have travelled long distances (25+ km) to the spawning site. Captain Cousteau has filmed a similar aggregation of large groupers off Honduras. Data on seasonality of spawning of other groupers are given by Randall & Brock (1960) and Thompson & Munro (1978).

It was somewhat surprising to find that spawning peaks coincided with lowest water temperatures on Jamaican reefs since maximum spawning occurs at times of highest water temperatures at the southern end of the Great Barrier Reef. Johannes (1978), who collated data on seasonality of spawning in tropical coastal fishes, considered that there are selection pressures for spawning at the calmest times of year, when pelagic larvae will be least endangered by turbulent surface waters and also when inshore gyres have the best chance of bringing the larval young back to the reef. He pointed out that nearly 50 species of taxonomically widely distributed tropical marine fish have lunar spawning rhythms, the majority of them spawning around new or full moon, and suggested that there may be selective advantages associated with spawning on spring tides. Randall & Randall (1963), when describing year-round spawning in the sparid *Sparisoma rubripinne* in the Virgin Isles and the lunar spawning cycle of the labrid *Thalassoma bifasciatum*, stressed that the reproductive potential of a fish is greatly enhanced when large groups assemble to spawn. Lunar cycles should help to synchronize spawning; Thresher (1984) cites lunar cycles in lutjanids, most epinethalines, migratory acanthurids, also some apogonids, pomacentrids and balistids and members of about nine other families; semilunar cycles are common in pomacentrids, some apogonids and balistids and about four other families, while pomacanthids, serranids, scarids, probably most labrids and others of some seven families do not appear to have lunar cycles.

In most species individuals spawn several times during the spawning season, some on a daily, others on a weekly or monthly basis, with perhaps several spawnings over 2 or 3 days each month. Sale (1978*a*) thought such behaviour should ensure wide dispersal of offspring. According to Thresher (1984) most pelagic-egg-producers spawn at dusk, whereas demersal-egg-producers spawn predawn and through the day.

In all the American pomacanthid angelfishes (*ca* 9 species) spawning occurs daily throughout all or most of the lunar cycle, around sunset. Pair spawning is usual, though highly variable social organization occurs in this family – from apparent monogamy (in *Pomacanthus paru*), through male-dominated harems in several species, to apparently promiscuous explosive breeding assemblages (*Pomacanthus arcuatus*) and a lek-like system (*Holocanthus passer*). Some species are protogynous hermaphrodites; males are larger than females in many genera, and both permanent sexual dichromatism and temporary dichromatism during courtship and spawning occur in the family (Moyer, Thresher & Colin, 1983).

The 'continuous spawning strategy' of the pygmy angelfish *Centropyge* (Pomacanthidae) consists of a well-sustained daily gamete production, resulting in larval dispersal which is maximal and continuous (Bauer & Bauer, 1981). In this species a crepuscular spawning ritual is a regular daily activity of the harem group, each female spawning only once a day with the harem male, producing 50–150 small pelagic eggs every day continuously throughout the year (an estimated 20 000–55 000 eggs a year) or during a season depending on the latitude, in a spawning ritual that provides defence from predation by the swift release of a few eggs.

Spawning behaviour in reef fishes with planktonic eggs and larvae is described for pomacanthids and chaetodontids by Lobel (1978) and Neudecker & Lobel (1981). It often resembles that described for the Pacific surgeonfish *Acanthurus triostegus sandvicensis* by Randall (1961). These *Acanthurus* spawn where strong currents set to the open sea. Ripe fishes congregate near the reef edge, in water 2–8 m deep, where they mill around actively. At intervals, three to five fish make a sudden upward rush, release eggs and milt, then swim down to rejoin the main group. This same type of behaviour has also been observed in five species of West Indian parrotfish and a wrasse, which all lay pelagic eggs. The scarid *Sparisoma rubripinne* spawns in about 20 m of water along fringing reefs in all months of the year, and spawns only in the afternoon.

Data on egg counts in females of various species have been collated by Sale (1980b). Large numbers of eggs are produced; for example, four serranid species examined by Thompson & Munro (1978) produced about 160 000 eggs a female, and Randall (1961) estimated 40 000 mature eggs in a single female *A. triostegus*. The available data all lead to the conclusion that even species showing some parental care of eggs produce considerable numbers of offspring each year, since they produce batches of eggs at frequent intervals. Comparisons of data on fecundity and larval recruitment indicate that nearly all these offspring must die before returning to the reef. Recruitment occurs in groups in some fish (some apogonids, and in *Diodon* on St Croix reefs, Ogden, personal communication) which may help their survival by swamping the predators. There are few hard data on survivorship patterns in reef fishes as it is so difficult to determine their ages. Among the smaller reef fish, 4–5 years is probably a common life span (some may even be annual): large species probably live considerably longer. Pygmy angelfish *Centropyge* live 8+ yr live 8+ yr in aquaria (Bauer & Bauer, 1981). *Epinephalus morio* in subtropical waters off Florida may live up to 25 yr (Moe, 1969), changing sex to become males when between 5 and 11 yr old.

Little is yet known of the pelagic life of most species. In *A. t. sandvicensis*, the late postlarval stage (acronurus) is estimated to be 2.5 months old when it enters tide pools as a discoid transparent silvery-abdomened fish, armed with poisonous fin spines and with a short digestive tract (Randall, 1961). During the 4–5-day transformation to the adult shape and coloration the digestive tube lengthens three times, reflecting the change in feeding habits from a zooplanktivore to an algal-browser. In this species spawning, as well as settling on the reef, shows a lunar periodicity (12 days before to 2 days after full moon). Artificially fertilized eggs hatch in 26 h (at 24°C) and the young start to feed in 5.5 days. Lobel & Robinson (1986) examine larval transport.

What selection pressures have led to these pelagic larval phases in reef fishes? Plankton is known to be very good food for larval fish and there is intense competition for it over the reef (from coral polyps as well as from fish) yet strong predation pressure makes moving above the reef to snatch it very perilous. Johannes (1978) considered predation to be the main selection pressure for an offshore pelagic larval existence, to escape high predation on the reef, rather than to effect the dispersal of young (as Sale believed and see Doherty, Williams & Sale, 1985) or to crop food from the open ocean, for plankton in tropical seas is sparse and varies little seasonally. Johannes also thought that there would be selection to spawn at dusk to avoid predation and at special phases of the moon and seasons when gyres (circular currents) in the open sea are most likely to bring some larvae back to the reef when they are ready to metamorphose. Barlow (1981) considered that the major selection pressure was for dispersal of the propagules of reef fish and was caused by the patchiness of reef systems. He noted that it is the larger fishes (more than 100 mm SL) which have a pelagic phase and maintained that smaller fishes could not produce enough propagules to overcome the hazards of planktonic life (estimated to be high); these small species have invested in the reverse strategy of putting resources into fewer propagules with parental care of eggs (as in pomacentrids) or oral brooding (as in apogonids).

The implications of the pelagic larval phase in these marine fishes are: (1) there is no clear information concerning the sizes and areal extent of reef fish populations; (2) the reef is an open system, divorced from patterns of reproduction at that site; and (3) neighbouring individuals of a species on the reef are unlikely to be related to one another. Young are not related to nearby adults; in this they differ greatly from the cichlid populations in the African Great Lakes. The evolutionary consequences of these differences need examination.

Sex changes in reef fishes

Hermaphroditism is now known in over 100 species of 15 families of fish, and it is thought to have evolved independently on at least ten occasions. Many reef fishes of the families Serranidae, Labridae, Scaridae and some Sparidae are protogynous hermaphrodites, individual fish changing from female to male as they grow and under socially induced circumstances. The reverse change, protandry, from male to female, is more usual in species which occur in large aggregations and not on reefs, such as in polynemids on the West African shelf (Longhurst, 1963), some sparids and platycephalids, and in the catadromous *Lates calcarifer* in Australia/New Guinea (p. 173). But protandry does occur in the anemone commensal *Amphiprion* clownfish on reefs, in which the (larger) female of a pair resident in an anemone is replaced by sex change in the formerly active, and smaller, male if she disappears. In freshwaters the cyprinodont *Rivulus marmoratus* is a simultaneous hermaphrodite, as are some small serranids in the sea, and some synbranchids, e.g. *Monopterus albus*, are protogynous (see reviews by Reinboth (1975) and Warner (1975, 1978)).

Hermaphroditism tends to evolve under conditions where it is hard to find a mate, or where there are small genetically isolated populations, or where one sex benefits from being larger or smaller than the other. Protogynous sex reversal occurs in fish species as diverse as coral-dwelling gobies (see Robertson & Justin, 1982), herbivorous parrotfish, piscivores, cleaner wrasses, zooplanktivores (Shapiro, 1984) and pygmy angelfish (Aldenhoven, 1983). The protogynous serranids, labrids and scarids have been most studied (see Robertson & Warner, 1978; Warner & Robertson, 1978). In these, sex change is associated with social systems in which large dominant males have a high degree of reproductive success (as calculated by Warner, Robertson & Leigh, 1975). The large terminal-phase males generally have different colour patterns from females and initial-phase males, the colours changing with changes of sexual state (see, for example, Robertson, Reinboth & Bruce, 1982; Thresher, 1984). Sex change, at least in some species, has been shown to be socially controlled, the females changing sex when dominant males are removed, as in the cleaner wrasse *Labroides dimidiatus* studied by Robertson (1973). With such mating systems sex ratios may be highly skewed towards females. In some communities where population densities are high, non-hermaphroditic initial-phase males (with the colouration of small adult individuals) exist that exploit the abundance of females by sneaking into the dominant (terminal-phase) male's territory to steal spawnings, or streak in to join a male and female at the climax of spawning, or group together to pursue a female till she spawns. In some species the initial-phase male may later assume the behaviour and colours of a dominant male. Population size seems to have a strong effect on whether initial males occur or not; species of labroids which have dense populations have the strongest representation of initial-male phases (Warner & Robertson, 1978). But the social system is probably the most critical factor in determining initial-male success. In pygmy angelfish (*Centropyge bicolor*) on the Great Barrier Reef, which live in harems of 1–10 females within a territory defended by a dominant male and sometimes a bachelor male, the pattern of sex change varies according to the 'social climate' of the reef. Experimental manipulations have shown that sex change can occur when the dominant male is present if the harem number is large, the new male splitting part of the original territory and taking some females to form a new harem, suggesting that change can be triggered by a sex-ratio threshold. Alternatively a new male may become a bachelor male in the original or neighbouring territory (Aldenhoven, 1983). Sex reversal in the small zooplankton-feeding serranid *Anthias squamipinnis*, in which all males are sex-reversed females, has been much studied experimentally in the laboratory by Shapiro (1981 and references listed in 1984), but he points out that the physiological changes underlying sex reversal could be different in the various groups of fishes in which it occurs.

Hoffman (1983) has used experimentally induced sex change in the herbivorous parrotfish *Sparisoma aurofrenatum* to compare time and energy budgets in female and male phases. After changing from female to male, individuals decrease foraging time and allocate more time to reproductive territoriality and mate attraction. The proportion of time spent foraging by males is inversely related to male size. A male's daily mating success is positively related to size of territory and his size relative to that of the males in neighbouring territories – large males obtain most mates. Experimental additions of food led to small males increasing growth rates while large males increased time spent in social and mating activities; both size classes decreased foraging time, increased territory size and increased daily mating success, though large males obtained a proportionately greater increase in mating success than did smaller males.

The social behaviour of pomacentrid damselfishes

Pomacentrids are conspicuous fishes in reef faunas in all warm seas in habitats down to about 45 m (as discussed by, among others, Fricke (1977); see also Anon. (1980)). Small fishes, they are mostly grazers or browsers on a variety of foods; some species feed on plankton over the reef. Observations in the Red Sea over 20 years by Fischelson *et al.* (1974) revealed a series of social relationships. Where the habitat was most complex, 24 pomacentrid species shared resources; food appeared to be more or less unlimited, and space, particularly refuges against predation, appeared to govern the types and numbers of species present.

These pomacentrids were found to fall into two main social groups: solitary species and gregarious species. The few intermediates included the anemone commensal clownfish *Amphiprion bicinctus* in which juveniles live in groups of 5 to 20 but adults live solitarily or in monogamous pairs in their host anemone. Solitary species are persistently territorial, mostly living near the substrate, very aggressive, protecting their shelter sites against both conspecifics and other fishes. Nest sites are cryptic and the home range is generally small. Populations are limited to small numbers of persistent numerical stability in a specific habitat. Gregarious species live in 'shoals' or 'schools'. The shoals are groups of individuals which swim in any direction, only partly in accord with other members of the group, wandering in and out of caves and lagoons, juveniles behaving like adults. The schools are groups which remain stationary in a home range on a solid object such as a coral head, hovering over it with synchronous ascents to feed and descents to shelter, like a flock of birds; the young fish form subschools. In these schooling species (which included *Chromis caeruleus* and *Dascyllus* species) aggression is much reduced within the group, and the group as a whole defends the home range against conspecifics. This high compatibility of individuals within the group allows far greater concentrations per unit area than in solitary or shoaling species (e.g. 100–200 *C. caeruleus* round a single 1 m wide *Acropora* coral head).

Territoriality for reproduction is common to all pomacentrids and is often accompanied by specific courtship behavior with colour changes and sound production. After spawning the male guards the eggs. Most species are polygynous. Small hole- and crevice-dwellers are solitary breeders. Others, including *C. caeruleus* and *Abudefduf saxatilis*, are colonial breeders, laying eggs on a coral head; a single coral head may support a breeding colony of several tens of breeding fishes, all showing reduced aggressive behaviour to one another while retaining it against outside intruders. In some species two or three males may fertilize eggs from several females and each male then guards part of the communal group of eggs.

Fischelson *et al.* (1974) concluded that the general evolutionary trend has been towards restricted movements and the use of constant shelters, and from solitary to schooling behaviour which allows many more individuals per unit area. In *Dascyllus marginatus*, which live in pairs, harems or multimale groups, group size and number of sexually active males are restricted by the number of available hiding places in the coral (Fricke, 1980). Further levels of social complexity are shown by interspecific and intergeneric schools of fishes (e.g. pomacentrids with the serranoid *Anthias squamipinnis*), the several species combining to defend the coral head. In these schools single fish, or even single species, no longer act individually, only as members of the whole group. Such levels of social organization can only develop during long-term shared ecology and evolution, as found in coral reef systems.

Reef communities: the maintenance of diversity

As discussed later (p. 287), most tropical communities are very diverse, but are these very diverse communities in equilibrium? How 'stable' is the species composition (for types of stability see Orians (1975)), or is it that relatively stable conditions over long periods of time have allowed their diversity to develop?

Connell (1978) stressed the role of local instabilities caused by unpredictable events, such as storms, on the successions of reef animals. Sale (1977) pointed out the role of chance in the combinations of fish species gaining access to patch reefs as vacancies occurred, from a pool (guild) of ecologically similar species present in the area; this idea has stimulated a great deal of research on recruitment and habitat partitioning.

Recruitment to coral reefs

In small patch reefs chance factors affecting recruitment appear more important than for fish communities on large reefs. Recruitment patterns have been studied by Luckhurst & Luckhurst (1977), Sale, Doherty & Douglas (1980). Williams & Sale (1981) and after local disasters. Cyclonic storm damage on Lizard Island, Great Barrier Reef, caused high juvenile mortality and redistribution of subadults, but had little effect on adult fish (Lassig, 1983). A hypothermal fish kill on the Florida Keys (Bohnsack, 1983) was followed by increased recruitment success and increased species richness (though individuals were smaller); this was attributed to decreased competition and increased predation from prior residents, supporting the view that some reef fish communities can be resistant to some regional disturbance.

Smith & Tyler (1975) censused a West Indian dome-shaped patch reef in 1970 and again in 1973; they found that when the dome collapsed, decreasing the amount of shelter available for apogonids, holocentrids and haemulids, there was a dramatic increase in populations of gobies and blennies, but little change in the total numbers of species and individuals inhabiting the reef. They concluded that competitive interaction related to the size of the individual fishes regulates the composition of the community.

Experiments with colonization of **artificial reefs**, initiated by Randall (1963) in the Caribbean, have been continued by many others (e.g. Sale & Dybdahl, 1975; Talbot, Russell & Anderson, 1978; Kock, 1982). Talbot *et al.*, in the artificial reefs set up at One Tree Island near the southern end of the Great Barrier Reef, tested the effect of habitat structure on species diversity with replicate pairs of different reef types: plain (no holes), small holes, medium-sized holes and large holes; they looked at seasonal differences in colonization by establishing one set in summer (October, 1971) and another in winter (July, 1972). Visual censuses were made at 32 and 23 months, respectively. Settlement was by juvenile fishes of which 205 resident species were recorded. Species composition differed little from that in natural patch reefs nearby. Between-reef variability was high in the artificial reefs and largely unrelated to habitat structure. Recruitment was, however, markedly seasonal here, occurring mainly in the summer (September–May), and showed little year-to-year constancy. Competitive interactions appeared unimportant in explaining distribution between the reefs. Species turnover was high, ranging from *ca* 17% per species month in winter to *ca* 39% in summer. Most species persisted for less than 12 months; individual survivorship was generally only a few months. Much of the high turnover rate appeared to be due to losses through predation.

These results suggested that communities of small patch reefs are highly unstable and that a persistent species equilibrium will never be attained. Talbot *et al.* envisaged patch reefs as spatiotemporal mosaics, their components in a continual state of flux (compare Zaire fishes, p. 48). They concluded that reef fish populations are maintained in this nonequilibrial state principally by predation and the seasonally varying uncertain nature of recruitment, and that by preventing the monopolization of resources by any one species these factors contribute to high within-habitat diversity of coral reef communities. It was therefore a surprise to these investigators when their survey of chaetodontids across 50 miles of the Great Barrier Reef from Lizard Island did not altogether support this view.

Habitat partitioning

In this Lizard Island survey, Anderson *et al.* (1981) found: (1) that among these chaetodontids there are conspicuous differences in the niches of many of the species which coexist locally; and (2) that geographical replacement does occur between species using the same niche (even on this 50 mile scale – a small one in relation to the vast distances in the Pacific Ocean). This led them to conclude that these chaetodontid communities are structured in ways similar to terrestrial vertebrate communities, and that the data did not support the need for alternative hypotheses based on larval habitat preferences and stochastic recruitment. The difference in view thus seems to be one of scale, stochastic elements being more noticeable in small patch reefs, while structured communities develop on the large-scale long-established reefs (as Smith (1978) also suggested).

Analogy with terrestrial communities had suggested that the high diversity on tropical reefs would probably be due to fine partitioning of resources: specializations to use different foods; territorial behaviour and other devices for the use of space while feeding, breeding, or as a refuge from predators; and temporal partitioning, with well-defined diurnal or nocturnal activity.

All these types of partitioning are indeed present on most reefs (for example among the holocentrids mentioned, p. 193). Territorial damselfishes have been much studied in this respect (see Fischelson *et al.*, 1974; Helfman, 1978). There is often much overlap in resource use (food and habitat), as summarized by Sale (1977) and as, for example, Shpigel (1982) reported for two damselfish species throughout their life histories for all resources examined. Partitioning has, however, been found to be very fine in many other cases; for example, of two small blennies living in holes bored by invertebrates into the coral, one species lives mainly on horizontal surfaces, the other on vertical ones, which brings them to predominate in different parts of the reef, utilizing slightly different foods (Greenfield & Greenfield, 1982).

There is no doubt that reef communities are highly structured, with fine partitioning of food, space and time, but there are also considerable overlaps in resource use (in some cases aided by dominant species being below the carrying capacity). There is a continual flux of species coming and going, and there is now plenty of evidence that alternative members of guilds of fishes with similar feeding habits and other ecological require-ments ('ecological replacement species') may gain access to the reef. On small reefs the success of a particular species settling from pelagic larvae may depend on how these have fared in another part of the species' and individual's range. The smaller pool of species from which to draw on in the Caribbean compared with the Pacific (*ca* 550 species compared with over 1500 species) was thought to result in the more highly structured reef communities in the Caribbean as described by Gladfelter, Ogden & Gladfelter (1980).

Comparison with freshwater communities
Coral fishes of many families differ greatly from cichlids in the African Great Lakes in having a pelagic larval phase away from the reef, eggs and larvae being distributed widely in the open sea. The coral reef community depends very much therefore on which species manage to settle there. Success in one habitat may be due to success in another (see Robertson & Lassig, 1980). Quite different selection pressures (about which we as yet know very little) will be affecting the pelagic phases from those affecting adults on the reef. On the reef the juveniles that manage to settle are unlikely to be closely related to the resident adults.

Coral reef communities are much older than lake communities (*ca* 50 million years compared with but a few million years); in this time specializations have evolved to family level, whole families of reef fish being specialized for herbivory for instance, compared with the adaptive radiations to use all kinds of foods within the Great Lake cichlids.

Barlow (1974*a*), who compared acanthurid and cichlid social behaviour, stressed that in the older family Acanthuridae the species fall into feeding guilds within which there is broad overlap in diets, but the social systems differ radically, both when breeding and not, which can be understood as consequences of their strategies in obtaining food. The Central American cichlids, mostly *Cichlasoma* of *ca* 100 species, with which Barlow compared them, are a much younger group, and they express a spectrum of feeding behaviour ranging from grazing herbivore through omnivore to predator, each with varying degrees of specializa-tion. By contrast, the *Cichlasoma* social behaviour is very conservative (Lake Tanganyika cichlids however show much greater variation). In these *Cichlasoma* species the tendency for division of labour, females doing more direct caring for eggs and fry and males responsible for more defence, leads towards polygyny but is counterbalanced by the need for both parents to defend the fry. Communication in cichlids is based on colour patterns which are based on a common plan. In acanthurids the poster colouration is apparently as important in extraspecific as in intra-specific aggression, but they show pronounced rapid colour changes when fighting intraspecifically.

Populations of small patch reefs seem to be in a state of flux, the species composition of the community changing as different members of a guild of ecologically similar species gain access to the reef. In freshwaters somewhat analogous situations occur on floodplains with, in many cases, the seemingly random retention of ecologically similar but different species in pools which become isolated as the floods recede. The experimental colonization of artificial reefs in Lake Malawi studied over five years (see p. 114) indicated that each habitat developed a different pattern of species accumulation and population density, as in artificial reefs in the sea, although the cichlid colonizers lacked the pelagic phase found in coral reef fishes. The basis of community structure in marine and freshwater communities is compared by Emery (1978).

9

Demersal fish studies

Factors affecting demersal fishes

Information on the ecology of demersal fishes comes mainly from
two sources: (1) line fishing over hard substrates, old dead coral, rock or
other rough ground; and (2) trawl fishing on the continental shelves where
the shelf is wide and there are suitable deposits of sand or firm mud. Some
Neritic pelagic fishes may also be taken in the trawls, particularly fish
which have the habit of forming demersal schools, as do carangid jacks
and scads (*Caranx, Trachurus*), and some clupeids such as *Sardinella*
which congregate near bottom depressions by day moving up to feed by
night. Many of the rough-ground fishes caught by lines may also rest some
way above the bottom but feed on the bottom, often by night.

Shelf faunas are greatly affected by the width of the shelf and nature of
bottom deposits, by oceanographical conditions (salinity, clarity of water,
temperature, water movements and so on), and by the history of the
oceans and possibilities of colonization from neighbouring areas. The
continental shelf may be over 100 km wide, sloping gradually to about
100 m deep before it falls away steeply into deep water (as off Guyana).
The type of bottom is of prime importance in controlling the distribution
of demersal fishes. Soft mud, sand, hard rock and coral, each has a
characteristic community of fishes and of invertebrates so important as
fish food. Open habitats which lack cover, such as bare stretches of sand
or mud, carry very different and much less diverse fish communities than
structured habitats such as coral reefs. Some open habitats may be visited
only under cover of darkness when nocturnal fishes from nearby reefs
move out over the flats to feed on fossorial invertebrates, which also
emerge from the cover of their burrows by night to feed on the sand or
mud surface. Fishes which live in open situations by day are generally
heavily camouflaged if they cannot burrow into the mud or sand (as do
some gobies and *Hemipteronotus* razor wrasses), or else they live in dense

schools, only dispersing under cover of darkness to feed in the water column or over the flats.

The great rivers of the tropics bring down enormous quantities of silts and soft muds which are deposited on the shelf off the estuary mouth and are then drifted along the shelf by currents and tides. Off West Africa thick soft mud lies off the Niger Delta east of Lagos; west of Lagos deposits are harder and sandier. West Africa has only a few pockets of live coral, but thickets of dead coral, relics of reefs which grew here when the sea level was lower during glacial epochs, are found in 60–80 m towards the edge of the shelf, half-smothered in softer deposits. Similar dead coral reefs occur near the edge of the shelf off Guyana in the western Atlantic. The 100 km-wide shelf along the northeast coast of South America has soft mud deposits inshore and by the mouths of the great rivers, Amazon, Orinoco, Essequibo and many others. These deposits are pushed north-westwards along the coast by tides and currents, staining the water a rich opaque brown. Harder sandy mud, sand and shelly sand deposits lie further offshore, generally in clearer greener water. Clear blue ocean water overlies the hard bottom near the edge of the shelf. The sharp interfaces between the different water types can be seen clearly from the air when flying over the Orinoco mouth, the muddy brown water inshore outlined by bright green water where nutrients brought down by the river have stimulated phytoplankton growth, this in turn separated by a sharp interface from the clear blue waters of the open sea. These interfaces are not static but move with the tides and currents, taking much of their associated fauna with them.

The estuarine systems of major continental drainages are rich in organic sediments, often replenished seasonally by the rivers. Broken down by bacteria, fungi and microorganisms, fed on by shrimps and fish, these make a large contribution to productivity. By contrast, the shelf system away from large estuaries depends much more on planktonic primary production, greatest where upwellings occur near the coast.

Off West Africa the tropical surface water is warm and of low salinity (temperature 24°C, salinity 35‰). Tropical conditions are, however, much less extended on the African than on the American side of the Atlantic, both latitudinally and in depth. The cold Benguela Current (Fig. 1.1) flows northwards up the coast of Southwest Africa and its cooling influence is increased by upwelling from the sea floor. Off West Africa the tropical area is also bounded on the northern side by cold currents, a boundary which forms one of the most marked faunal discontinuities in the ocean. Off America, however, the tropical region of the western Atlantic has no such sharp boundaries, as warm currents run northward along the coast of North America and southward along the coast of Brazil. Furthermore, as the thermocline inclines down to the west (p. 8), the warm water forms a considerably thinner layer off the African coast. The average depth of the first thermocline is at 20–35 m over most of the Gulf of Guinea, but as shallow as 12–14 m off Senegal to Liberia, in the Bite of Biafra and south of the equator, and it inclines downwards offshore. The tropical water off West Africa is thus only a thin layer lying over cooler water, whereas in the western Atlantic off America, where the thermocline is much deeper, waters may be warm to the bottom right to the edge of the shelf; the mean annual temperature at 100 m near the edge of the shelf is 26°C off Bahia in Brazil, 24°C off the Antilles in the Caribbean. The presence of a thermocline between the tropical surface water and underlying South Atlantic central water off West Africa for at least part of the year has marked effects on the fish communities. The upper limit of the thermal discontinuity is the sharpest, its average position that of the 25°C isotherm, its lower limit approximately that of the 19°C isotherm.

West Africa has a different fish fauna from that of the Caribbean, which was once in contact with the eastern Pacific. The open waters of the Atlantic have evidently presented a barrier to most shelf-dwelling fish species. Some of those which did reach West Africa, either across the openwater barrier or round the Cape, have in their isolation there undergone adaptive radiations and show convergent or parallel evolution with forms elsewhere; this is clearly shown among the sciaenids, all members of a tribe (Pseudotolithinae) endemic to West Africa.

Shelf communities illustrate very well the rule that faunas at low latitudes have many more species than those at higher latitudes. A 2-yr trawl survey off Guyana by the *R.V. Cape St Mary* took about 200 fish species, representing 72 families (Lowe-McConnell, 1962). Trawl catches off West Africa also took about 200 species (Longhurst, 1963). There were, however, dominants in the Guyana catches, three species of sciaenid making up 62% of the weight of fish landed. Island faunas in the Caribbean, well away from the seasonal inflows of nutrients from large rivers, are more diverse and lack such dominants.

Rough-ground demersal fish studies

Hard rock and rubble and old coral bottoms support important line fisheries in many parts of the tropics: the Campeche Banks in the Gulf of Mexico, Seychelle Banks and rough ground off East Africa, and parts of the Sunda Shelf. As an example, the tropical West Atlantic has much rough ground and here three families, snappers (Lutjanidae), groupers (Serranidae) and grunts (Haemulidae), are the dominant carnivores. Snappers and grunts congregate in large schools, groupers are more solitary and scattered. Species composition changes with depth. In the Caribbean at least 13 species of grouper (*Epinephalus, Mycteroperca*) and 12–14 species of snapper (*Lutjanus*) are commercially important. Red groupers (*Epinephalus morio*) are important members of the benthic sublittoral communities. Growing to over 70 cm long and 90+ kg in weight, they live on rocky bottoms in water from 3 to 100 m deep, lurking below ledges and in caverns. Unspecialized carnivores, they eat small fishes of many kinds – crabs, spiny lobsters, shrimps and other Crustacea, octopuses and squids (an estimated 83% Crustacea, 17% fish) (Randall & Brock, 1960). Immature red grouper (less than 40 cm SL and under 5 yr old) remain associated with particular inshore reefs. At about maturation size they move offshore to join adult populations up to 70 km away. Some groupers make long migrations to spawn in large aggregations at a particular place on the reef; red groupers off Florida have an extended spawning season (March to July). Age and growth rate studies based on otolith rings revealed great variations in growth rate between individuals in each age group. Spawning and associated processes seem to be the primary cause of annulus formation, but age groups 1 to 4 form the annulus in March to May, whereas age groups 5 to 10 form it later, May to July. The red grouper, like many other serranids, is a protogynous hermaphrodite. Sexual maturity as a female occurs when 4–6 yr old and about 45 cm SL; some of the females become functional males after spawning as females. Male red groupers live longer and grow slightly larger than females. Maximum size is generally reached when 13 yr old; fish up to 25 yr old have been recorded (Moe, 1969).

Snappers (Lutjanidae) are very important food fishes throughout the tropics, especially on reefs and rough ground. The family has some 17 genera, one of which, *Lutjanus* (*ca* 70 species), occurs in all tropical oceans. On the basis of their feeding habits the 13(+) western Atlantic snappers fall into several groups (Starck, 1971). *L. synagris* (lane snapper) and *L. analis* (muttonfish) are predominantly invertebrate-feeders in open water, the smaller *L. synagris* (*ca* 30 cm) being common inshore and the larger *L. analis* (60 cm) further offshore where it takes molluscs; both species are active by day and night. Four primarily piscivorous species are all nocturnal: *L. jocu* (dog snapper, *ca* 60 cm) and *L. apodus* (schoolmaster, *ca* 45 cm) which both live in rocky areas; *L. mahogoni* (mahogany snapper, *ca* 25 cm), a large-eyed reef-roving species which lives in open areas between reef corals where it takes crustaceans as well as fish; and *L. cyanopterus* (cubera, the largest species, growing to over 100 cm, 45 kg) which prefers deep channels, ledges and coral patches. *L. griseus* (grey snapper, *ca* 40 cm) is the most generalized in feeding areas and habits and mainly nocturnal; juveniles live inshore in eel-grass (*Zostera*) beds, where they eat mainly Crustacea and may feed by day, while adults live offshore in open areas over various types of bottom, feeding mainly on fishes at night. Another wide-ranging reef and inshore species is the yellow-tail snapper *Ocyurus chrysurus*, which feeds in midwater by day and night.

The fish-eaters have larger mouths than the invertebrate-feeders, and the rock-dwellers have longer pectoral fins than the openwater-dwellers. Fish-eaters have well developed canine jaw teeth, *L. analis* having heavy pharyngeal teeth which can crush molluscs, and the dentition is reduced in *Ocyurus chrysurus* which feeds largely on plankton, small fish and crustaceans. Comparable relations between body form and way of life are shown by Indo-Pacific lutjanids: the secondarily openwater-living fusiliers (*Caesio*) are spindle-shaped compared with the deeper-bodied reef-wandering *Lutjanus* species. The degree of gregariousness is also related to way of life; these openwater species live in schools, whereas the reef snappers form schools when travelling, and nocturnal species form very large aggregations when resting on the reef by day. Colours are also related to way of life: the openwater species are silvery and unpatterned, whereas reef species blend well into the reef environment, yellow colours predominating in *L. apodus* and *L. jocu* which live among rocks and corals encrusted with sponges and algae, and pink or red by day in the deeper-living species (which will look grey-black at depth since red light rays are absorbed by the water). By night the colours change, most snappers developing bars or blotches. Colour patterns also change with age and behaviour, stress and other emotional factors.

Grey snappers (*L. griseus*) change their habitat as they grow, moving out from shallow grass beds into channels when about 7 cm, and out to reefs when they mature at about 19 cm SL. Schooling behaviour is strongest in adults, and greatest in areas of reduced cover. Schools of mixed snapper species are common. Schools break up at dusk, the snappers dispersing 1–2 km along the reef to feed. Foods used change with size of fish; copepods and amphipods used by postlarval fish are generally replaced by shrimps (*Caridina* then penaeid prawns), crabs and fish as adults increase in size. Spawning occurs more than once a year (probably around full moon); a 31 cm female contains half a million eggs.

Scale-ring studies suggested that off Florida *L. griseus* are 3 yr old when they mature, and may live 9 yr, growing to 40 cm. In the Caribbean, lutjanids appeared to spawn throughout the year but the greatest proportion of ripe fish was found in the coolest months (Munro *et al.*, 1973).

Continental shelf communities

The very valuable trawl fisheries in the Gulf of Thailand, off Indonesia and around India have been much studied (see references in Cushing, (1971) and Simpson (1982)), but Atlantic catches are chosen here for consideration; these demonstrate very well how species associations are related to water type.

Western Atlantic

Catches off Guyana during the 1957–59 trawl survey by the *R.V. Cape St Mary* included over 200 species, representing 72 families; these were mainly of sciaenids (21 species), carangids (14 species), haemulid grunts (*ca* 13 species), catfishes (12 species) and lutjanid snappers (*ca* 7 species) (Lowe-McConnell, 1962). In the inshore brownwater zone (Zone I: Fig. 9.1) the 'brown-fish' zone, catches over the very soft mud deposits included numerous small ariid catfishes, stingrays (*Dasyatis* species) and young fishes of many species. The main commercial catches came from Zone II in 20–60 m over firmer mud or sandy mud. Here the invertebrate-feeding croaker *Micropogonias furnieri* predominated (43%

Fig. 9.1. The continental shelf off Guyana, Northeast South America, indicating fish fauna zones, physical characteristics and areas sampled by different types of fishing gear. (After Lowe-McConnell, 1962.)

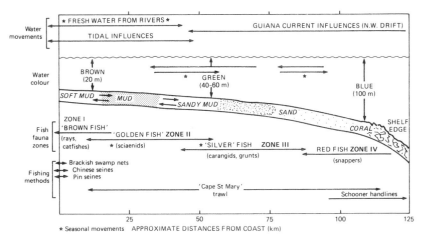

of the catch by weight), accompanied by two predatory sciaenids, the small *Macrodon ancylodon* (18%) and the larger *Cynoscion virescens* (11%). Over hard sand catches were much lower; here the water was clearer and greener and more pelagic fishes were taken, silver-bodied carangids and Spanish mackerel (*Scomberomorus*), pomfrets (stromateid *Peprilus*) and spadefish (*Chaetodipterus*). Towards the edge of the shelf, in the deeper clear oceanic water over rough ground strewn with sponges and dead coral, was a 'red-fish' zone with lutjanid snappers and many spiny fishes characteristic of the clearer more saline water around the Caribbean islands.

Off Guyana there were no regular seasonal fluctuations in fish numbers in trawl catches, but there was a general tendency for fish in sciaenid Zone II to move inshore in June–August, months when Trade Winds ceased to blow and the main rainy season when the rivers discharge most freshwater to the sea, and to move offshore into deeper water in January–March, when northeast Trade Winds blow most strongly, stirring up the bottom mud in inshore waters. There was a great diversity in spawning rhythms. Many species (sciaenids including croaker, carangids, haemulids, gerrids) appeared to be capable of spawning at any time of year, and ripe fish were always present in catches. But some other species (including some sciaenids) had much more restricted spawning seasons. Many species spawned in the main rains and juvenile fishes were abundant in the estuaries from June to September. Many kinds of live-bearing elasmobranchs (sharks and rays) contained young almost ready to be born around March.

Off Guyana, fish food resources, such as shrimps and small fishes, were shared by various sciaenid species. The very common *Macrodon ancylodon* and the larger *Cynoscion virescens* were often caught together, packed with the same kinds of food organisms. Interspecific competition appeared to be lessened by the various species predominating in different places (as at varying distances from the coast), by seasonal movements, which may separate species sharing resources at other times of year, and by changes in feeding habits at different stages of the life history (species overlapping in food habits as adults not necessarily doing so when young). *M. ancylodon* young were common in the estuaries of the large rivers, places where *C. virescens* young were rarely caught.

These Guyana sciaenids fell into three feeding groups. (1) Bottom-feeders, with inferior mouth surrounded by well-developed pores of the acoustico-lateralis (lateral line) system, and generally with barbels on the lower jaw, included *Micropongonias furnieri* which took mainly polychaetes from the bottom mud, and a whole range of species, from

Lonchurus lanceolatus with very small eyes and huge pectoral fins living inshore over very soft mud, to the 'deep-sea' croaker *Menticirrhus* living offshore over hard sand. *Equetus* was the only sciaenid living in the red-fish zone amongst old coral blocks, where it ate small molluscs. Type of bottom rather than distance from shore appeared to be the controlling factor. (2) Pelagic (sergestid) shrimp-feeders, sciaenids with obliquely upturned mouths, large eyes, rather short laterally compressed bodies and silver in colour, included *Larimus breviceps*, which shows many convergent features with the carangid *Vomer setapinnis* which also feeds on these shrimps. (3) Predators feeding on penaeid prawns, mantis shrimps (stomatopods) and fish included all the larger sciaenids: six species of *Cynoscion* which formed a series occupying increasingly offshore zones from inshore brackish swamps to 60 m depth, and *Macrodon ancylodon* common in estuaries and out to 50 m over mud (Lowe-McConnell, 1966). All these species swallow their prey whole, generally headfirst, and include both shrimps and fish in their diet, preying on whatever is most readily available. In one trawl haul individuals of different species were often all packed with mantis shrimps (*Squilla*) or with the same species of prey fish. Fig. 9.2 shows how the predominant sciaenids shared resources with the grunts *Genyatremus* and *Conodon*, the gerrid *Diapterus*, and with more pelagic carangids *Vomer* and *Chloroscombrus* and some herrings, preyed on by *Trichiurus* (cutlassfish), *Scomberomorus* (Spanish mackerel), *Sphyraena* (barracuda) and carcharinid requiem sharks.

Fig. 9.2. Trophic relationships among trawl-caught fishes off Guyana. (After Lowe-McConnell, 1962.)

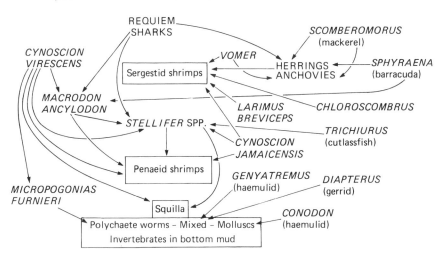

Eastern Atlantic

Many trawl surveys have been made off West Africa, including the International Guinean Trawling Survey in 1963–64 along nearly 2700 miles of coastline from 12.5°N to 6°S. Computer analyses of the catches confirmed earlier subjective groupings of species into six faunal assemblages by Longhurst (1965, 1969). These make up three faunas one above the other in a vertical series: (1) sciaenid-dominated assemblages where the tropical water extends to the bottom, with subcommunities (i) in the estuaries, and (ii) offshore on soft deposits to the base of the thermocline; (2) sparid-dominated assemblages below this where cooler water lies on the bottom, subdivided into (i) a shallow sparid community on sandy corally rocky substrates above the bottom of the thermocline, and (ii) a deep sparid community on both hard and soft bottom deposits below the thermocline to near the edge of the continental shelf; and (3) hakes and related fishes at depth (i) at the edge of the shelf to 200–300 m, and (ii) a deep-slope community below 400 m.

Many families of fishes and numerous species were represented in each of these three main assemblages; the sciaenid community for instance included polynemids (threadfins), haemulids (grunts), carangids (jack and jack-mackerels) and ariid catfishes. The sparid community bears a close relationship to the faunas of the warm temperate region north of Senegal and south of Zaire, while hakes and related fauna, found here at depth, became progressively more important in fisheries at higher latitudes. The outlying banks of old dead coral had a distinct assemblage of lutjanids (snappers) and other fishes of the kind of lutjanid community found on rough ground elsewhere in the tropics, and which appear to be even more restricted to the tropical zone than are the sciaenids.

Studies of the bathymetric distribution of the sciaenid community in relation to the depth of the thermocline between Sierra Leone, Nigeria and Cameroons, showed that where the shallow thermocline is more or less permanent, with little seasonal variation, the lower limit of the sciaenid community corresponded with the lower limit of the thermocline. Below this, well-lit water on the shelf carried sparids. The main occurrence of the deep sparid community was restricted to water cooler than 21°C. The thermocline here thus forms an important biological barrier between fish communities.

The main assemblages of species recognized by computer analyses off West Africa, the sciaenid, sparid and lutjanid communities, are also found in the western Atlantic and appear again in the Indo-Pacific. There are, however, much larger areas of hard bottom in the western than in the eastern Atlantic, and the lutjanid community assumes much greater importance in the Caribbean than off West Africa. Sparids on the other hand were less in evidence in the western Atlantic survey. Sparids are also sparse in Indo-Pacific fish communities. Surveys by the *R.V. Cape St Mary* using the same gear off Guyana and off West Africa allowed direct comparisons to be made of catches in the western and eastern Atlantic (Table 9.1). For comparison, catches from Indian Ocean surveys by other vessels (summarized by Cushing, 1971) are also shown.

Longhurst's (1960) studies of foods used by demersal fish off West Africa showed that most of the species were very unspecialized in their feeding habits, taking food items from many groups: cephalopods were found in 15 species, other molluscs in 22, polychaete worms in 40, small Crustacea in 39, crabs in 40, stomatopods in 14, ophiuroids in 12, other echinoids in 9 and fishes in 50 species. The need to examine samples over a series of years was demonstrated clearly by stomach contents of the flatfish *Cynoglossus senegalensis* from the Sierra Leone estuary where the brachiopod (lampshell) *Lingula* was its main food one year, though this hardly appeared in stomachs in two other years; stomach contents thus reflected the relative abundance of *Lingula* (Fig. 9.3). Comparisons of

Fig. 9.3. Seasonal and year-to-year variations in the diet of the flatfish sole *Cynoglossus senegalensis* from the Sierra Leone estuary (after Longhurst, 1957).

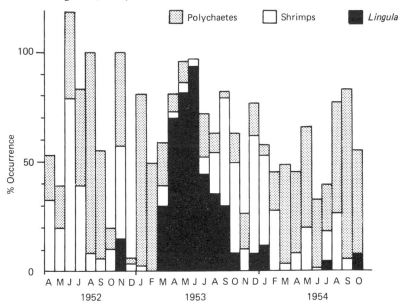

Table 9.1. *Comparative trawl catches of demersal fishes from tropical waters ($kg\,hr^{-1}$). Atlantic data from Lowe & Longhurst (1961); Indian Ocean data from Cushing (1971). (T1–T10 refer to sites on Fig. 1.1).*

	Western Atlantic off Guyana (T1)	Eastern Atlantic off West Africa (T2)
Atlantic Ocean		
Sciaenid fauna	187	178–295
Serranid/sparid fauna	59	21–130
Deep fauna	7.5	14–90

Indian Ocean	**Western**	**Eastern**
	Off Kenya (T3) 10–300	Off Bangladesh (T7) 160–320
	Off Sind (Pakistan) (T4) 40–320	Off Thailand (T8) 100–500
	West coast of India (T5) 100–800	W. Malaysia (T9) 40–320
	Cape Comorin (S. India) (T6) 120	(Sunda Shelf)
		South China Sea (T10) 40

food eaten under estuarine and offshore conditions by seven species showed that empty stomachs were far commoner in offshore than in estuarine fishes. Estuaries seem to have very high densities of suitable benthic foods such as polychaetes, a high percentage of which occurred in fish stomachs, whereas offshore there was increased dominance of Crustacea, fish and molluscs in the stomach contents. The sparid *Pagrus ehrenbergi* showed peak feeding activity at dawn and dusk; cephalopods were present in day-caught fish, whereas relatively sedentary organisms, molluscs, polychaetes, echinoids and ophiuroids, were found in the stomachs of night-caught fish. This West African study emphasized the relatively high level of the fishes in the trophic network. A very high proportion of the diet was of benthic invertebrates, many of which are themselves carnivores. Many other fishes were deposit-feeders. Much of the organic matter in the deposits is probably of terrestrial origin brought down by the many rivers which discharge the runoff from heavy rains onto the shelf. The nutrient cycles of the sea are here intimately connected with the land; few fishes feed directly on primary production by phytoplankton.

Determinations of ages and growth rates of tropical fishes are notoriously difficult to make. The sciaenids off West Africa ceased to spawn when the water temperature fell, and this enabled growth rates to be determined by progression of length frequency modes (Petersen's method) when very large samples of fish were measured (Longhurst, 1963). This showed that fishes here grow very fast and have a very short life span compared with fishes from temperate seas. Catches are therefore composed of very few year classes. For example, 98% of the sciaenids *Pseudotolithus elongatus* and *P. senegalensis* were 2- or 3-yr-old fish (as were 77% of the polynemids, 87% of the spadefish and 81% of the main sparid species); fishes more than 4 yr old were rarely seen. The implications of this for the fishery are discussed below (p. 326).

Estuaries and mangrove swamps

Huge estuaries, with complex delta systems, are a feature of large tropical rivers. Characteristically they are mangrove-lined, the mangroves providing shelter for juvenile stages of many marine species. Estuaries, with their turbulent muddy waters, are, however, hard to sample and few studies have yet been made. Primary production appears to be lacking but estuaries evidently provide rich feeding grounds, food webs being based on detritus from organic matter brought down by rivers from the land.

Estuaries are characterized by varying salinity (from almost freshwater to *ca* 33‰), variable temperature ranges, turbidity, muddy bottoms, strong water movements, freshwater running down over the denser more saline water. Freshwater fishes tend to move down into the estuarine reaches in the flood season and back upriver in the dry season. In some estuaries plankton develops at the end of the rains and marine immigrants are then more abundant (Day, Blaber & Wallace, 1981). In southern African estuaries the largest group of fishes are marine migrants, juveniles of species which breed in the sea, the young entering estuaries to feed and shelter until their gonads start to mature; some adults never return to the estuaries, others move in seasonally to feed. Families represented by such juvenile immigrants include Clupeidae, Engraulidae, Atherinidae, Sciaenidae, Sparidae, Haemulidae. Estuarine residents include grey mullets (Mugilidae) though these spawn in the sea, and many small species: gobies, syngnathids, ambassids, atherinids, stolephorids and some clupeids. Fishes breeding under estuarine conditions often show some reproductive specialization: brood pouches for eggs in male syngnathids, gobies guarding sticky eggs on stones, oral incubation of eggs by male ariid catfishes, live-bearing in the cyprinodontoid 'foureye' fish *Anableps*.

Trophic relationship studies in the Tortuguero estuary on the Caribbean coast of Costa Rica (Nordlie & Kelso, 1975), where the seasonal cycles follow the rainfall pattern with two wet and two dry seasons a year, showed that planktonic primary production was higher during the dry season but input of allochthonous materials by river flow was higher during wet seasons. Numbers of typically freshwater species increased in the wet seasons and typically marine forms increased in number in the dry seasons. The most striking seasonal difference in the estuarine community was the presence during dry seasons of dense aggregations of 'tismiche', composed principally of larval shrimps and larval fishes, food for other species. A preliminary carbon budget indicated that estuarine respiration accounted for more organic material than was produced in the estuary during either wet or dry seasons. The 29 fish species present throughout the year included tarpon, ariids, belonid needlefish, grey mullets, pipefish, centropomids, sciaenids, gobies and cichlids; ten species taken only in the wet season included gobies, a cichlid, a characin and an anchovy; seven species taken only in the dry season included barracuda, *Scomberomorus* and a carangid.

Marine species predominated in the estuaries of Northeast Brazil (over 90% of the 117 species recorded); here species of Engraulidae, Atherinidae, Centropomidae, Gerridae, Mugilidae, Gobiidae, Bothidae and Soleidae were characteristically estuarine forms in their abundance and the presence of both young and mature fishes (Oliveira, 1972; Paiva, 1973). The ecology of fishes in brackish lagoons has also been studied in Mexican coastal lagoon systems (Warburton, 1979). In Lagos lagoon, Nigeria, Fagade & Olaniyan (1973) examined trophic interrelationships of 26 fish species which used a wide variety of food organisms.

Mangrove swamps (mangals), associations of halophytic trees and shrubs growing on tropical coasts, are limited to calm waters and grow best on muddy shores of sheltered bays and estuaries where the rainfall exceeds 1500 mm yr^{-1} and where temperature does not fall below 20°C, but some are associated with coral reefs (Teas, 1983; Por & Dor, 1984). They generally provide very unstable habitats in which temperatures and salinities vary greatly (from 20–39°C, 0–46‰ salinity), with dissolved oxygen also fluctuating greatly. Few fish spawn here, but numerous (reputedly 400) marine species use mangrove swamps as nursery grounds (Gundermann & Popper, 1984). Only very euryhaline fishes can live here. They include freshwater cyprinodonts, poeciliids, *Oryzias* and some tilapia species. The true residents include typically intertidal fishes, basically detritivorous or preying on aerial or subtidal prey (gobies, *Periophthalmus*, eleotrids, blennies), and open lagoon fishes which fulfil their life cycles in or near lagoons; these include ariid catfishes, a centropomid–gerrid association in tropical America (*Centropomus* and *Diapterus* species), sphaeroid pufferfish and syngnathid belonids. Other species include grey mullets (Mugilidae) and sciaenid predators (*Kuhlia* in the Indo-Pacific). Associated fishes include *Chanos chanos*, the fry of which move into lagoons for 10 days, predatory sciaenids in Atlantic mangrove swamps (their niche filled mainly by serranids, lutjanids and some sparids in the Indo-Pacific), with carangids and eleotrids in lagoons, the clupeid *Ethmalosa dorsalis* in West African lagoons; also small pelagic fishes (especially anchovies) and threadfins (polynemids) are found here in the Indo-Pacific (Por & Dor, 1984). Phillips (1981) has described the spatial and seasonal distribution of the dominant ariids, gerrids, sciaenids, several flatfish families and pelagic engraulids and clupeids in a Central American mangrove embayment.

Southeast Asian mangrove swamps appear to have particularly rich estuarine faunas. Amphibious fishes such as the climbing perch *Anabas scandens* occur here, as well as the gobioid mudskipper *Periophthalmus*, widespread along Indo-Pacific coasts, whose territorial habits are described by MacNae (1968). Goby territorial defences are signalled by such means as raising and lowering distinctively marked dorsal fins and by mouth-fighting.

10

Pelagic fishes

Production in the pelagic zone of tropical seas is very patchy, greatest in upwelling areas and convergence zones, often with much variation seasonally and from year to year. The resources are exploited by fishes in two main ways: (1) by small species with short cycles (such as clupeoids) which multiply fast, so producing many mouths to exploit planktonic food when it is abundant; (2) in larger species (such as tuna) by migrating long distances in search of food. Adaptations to life in the pelagic zone include schooling which helps the fishes to keep together when moving in the vastness of the open ocean, as well as providing protection from predation in the small species. Physostome swimbladders enable the fishes to migrate vertically with ease (which physoclistous fishes cannot do). The uniform silvery colouration of the openwater fishes is linked with an environment that is visually uniform except for light intensities varying with depth and time of day. The acoustico-lateralis system of clupeoids, linked with the swimbladder, enhances sensitivity to water vibrations and so aids synchronous swimming (Blaxter & Hunter, 1982).

Clupeoid fishes of upwelling areas

Clupeoids are among the world's most important food fishes; in 1971 they made up 35% of world landings, and the fourfold increase in catches since 1958 was mainly due to increased landings of clupeoids from low-latitude waters, particularly those of the Peruvian anchovy fishery. In recent years this fishery has collapsed, providing an object lesson in what can happen to such an important world food source. Clupeoids are soft-rayed fishes specialized for pelagic life, silvery, 20–39 cm long as adults, mostly with filamentous gill rakers used to strain plankton from the water (Whitehead, 1985). There are three main families: (1)

Clupeidae, herrings; (2) Engraulidae, anchovies; and (3) Chirocentridae, wolf-herrings. Herrings, which include the herrings, sprats and sardines of temperate waters, have many genera important in tropical waters. Anchovies are smaller, characterized by the very underslung lower jaw; they feed mainly on phytoplankton, rather than on the zooplankton used by many herrings. Wolf-herrings (*Chirocentrus*), found mainly in the Indo-Pacific, are larger predatory fishes with canine teeth.

Of nearly 300 known clupeoid species, the majority occur in warm seas, where at least 25 genera are important in the fisheries. They obey the general ecological rule of higher diversity at low latitudes where conditions are relatively stable. This latitude rule is, however, broken where there are seasonal upwellings at low latitudes, as here there are few dominant and extremely abundant species, generally specialized plankton-feeders. The major upswelling areas, where cold currents meet the coasts, each carry a dominant pair of complementary anchovy and sardine species (*viz.* in the Peru Current (Fig. 1.1) sweeping north along the coast of Peru the anchoveta *Engraulis ringens* and sardine *Sardinops sagax*; in the cold Benguela Current sweeping north up southwest Africa, *E. capensis* and *S. s. ocellata* (Longhurst, 1971)). Elsewhere, particular species may support very large fisheries, such as the oil sardine *Sardinella longiceps* which migrates seasonally along the west coast of India, and other *Sardinella* species in Indonesia, off West Africa, and off Venezuela and Brazil. The richest clupeoid faunas are in Indonesian seas where there are many species of *Ilisha* and *Sardinella* and the anchovy genus *Stolephorus*. West Africa has a poor clupeoid fauna compared with other tropical areas (about 7 species). Clupeoids often tolerate low salinities; the *Sardinella* species off West Africa fall into a series, from the very stenohaline *S. aurita* found only in full seawater (35‰ salinity), through *S. maderensis* (syn. *S. commersoni*) which lives in less saline water (20‰), to *S. eba* found in estuaries and brackish lagoons. *Sardinella* species around the coast of India may fall into a comparable series of stenohaline to euryhaline forms. Many euryhaline shad-like forms make anadromous spawning migrations, for example *Hilsa ilisha* which supports important fisheries as it migrates into Indian rivers to spawn.

Clupeoids are preyed on heavily by piscivorous fishes such as scombroid bonitos, and supported huge numbers of guano-producing birds (cormorants, boobies and pelicans) off Peru. Schooling provides some protection to individual fish against predation. In many species the schools make vertical migrations, the fishes moving down and living in dense schools by day, moving up and dispersing to feed at night. They thus follow the vertical migrations of the plankton. At night clupeoids are often drawn to a light source.

Clupeoid growth in tropical waters is often very rapid and life cycles are short. Off Peru, few anchoveta (*Engraulis ringens*) survive beyond 18 months, compared with fish of 5–6 yr which form a significant part of the *E. mordax* populations off California. Peaks of spawning activity follow rapidly on upwellings and their associated plankton outbursts.

The anchoveta *E. ringens*, a diatom-feeder, was the mainstay of the Peruvian anchovy fishery, with lesser numbers of the sardine *Sardinops sagax*. Since *E. ringens* catches have declined, *S. sagax* catches have become increasingly important. These clupeoids were once protected because the birds producing the valuable guano fertilizer depended on them for food. The industrial anchoveta fishery only started in about 1940. In 1957 nylon roundhaul purse seine nets were introduced and the fishery developed until in 1971 these fishes formed 20% of world landings. This very high production was possible as the rate of carbon fixation is extremely high in this upwelling zone, anchoveta are phytoplankton-feeders, and they have a very high growth rate. In 1973 the fishery crashed from 11–12 million tonnes yr^{-1} to zero, following the first major failure of the upwelling since fishing was at the estimated maximum sustained yield of stock. Upwelling failures occur here regularly, though not cyclically, when in certain years warmed water thrusts southwards down the Peruvian coast into the upwelling area (conditions known locally as 'El Niño' since they usually occur around Christmas). The anchoveta disappear at this time, perhaps into deeper water. There was a slight recovery in the late 1970s, when sardines became more important in catches, but there was a devastating El Niño event in 1982–83 (Steele *et al*. 1984). These anchovetas are short-lived (2 yr), growing rapidly to 12 cm long and to 12.5 cm in their second year. They were recruited into the fishery at 5 months. Spawning occurred at low intensity throughout the year with peaks in August to October following upwelling. Year classes (groups of fish born within one year) varied very much in strength, large and small classes often alternating.

In the Indian Ocean the oil sardine *Sardinella longiceps* is the most abundant of the many clupeoid species and supports fisheries from Java to Arabia. Clupeoids can form up to 30% of India's marine fish landings, 80% of them from the west coast; catches show irregular fluctuations of 2–6 yr periodicity. The much larger clupeid *Hilsa ilisha* (growing up to 60 cm) occupies much the same geographical range but is an anadromous species living mainly in coastal and estuarine waters and moving up rivers to spawn, where some remain all year. Eggs are demersal in still water but buoyed up by currents. Scale studies suggest that hilsa live for 5–7 yr.

In the Caribbean off Venezuela the main commercial fishery is for the sardine *Sardinella anchovia* (Simpson, 1971). Scale studies have shown that 50% of the catch is of 2-yr-old fish, 20% are 3 yr old, and there are negligible numbers of 4- and 5-yr-old fish; they grow to 18 cm at the end of the third year, females growing slightly faster than males. Some spawning occurs throughout the year but is most intense in December–April, corresponding with the main upwelling season and peak values for phyto- and zooplankton. Differences in intensity and duration of upwelling are reflected in numbers of sardines which come to an area to spawn. The anchovy *Cetengraulis edentulatus*, also caught here, probably came from a common stock with *C. mysticetus* in the eastern Pacific before the formation of the landbridge divided the stocks.

Epipelagic fishes of the high seas

The epipelagic zone of the high seas has few niches and carries relatively few fish species (*ca* 225, 1.3% of the known total fish species). Most of these are mobile far-ranging fishes, such as scombroids and atherinomorphs (flyingfishes, belonid garfish, saurey skippers). Some species, such as *Scomberomorus* and belonids, also occur in continental waters. Flyingfishes have been studied off Barbados where they support a fishery. Fishes associated with the pelagic *Sargassum* complex in the Sargasso Sea were studied by Dooley (1972).

Scombroid fishes are typically pelagic, schooling, carnivorous, fast-swimming fishes (see Collette & Naven, 1983). They include the large tunas (*Thunnus*), the skipjacks (*Katsuwonus* and *Euthynnus*), dogfish tuna (*Gymnosarda*), wahoo (*Acanthocybium*) and smaller bonito (*Sarda*), frigate mackerels (*Auxis*), Spanish mackerels (*Scomberomorus*), Indian mackerels (*Rastrelliger*) and mackerels (*Scomber*); all these are now included in the family Scombridae. Other scombroid families include the swordfish (*Xiphias gladius*, Xiphiidae), sailfish and marlins (Istiophoridae), and cutlassfish (Trichiuridae). Scombroids fall into two categories: (1) those inhabiting the open oceans far from any land, such as the large tunas; and (2) those frequenting coastal waters or living within a few hundred kilometres of land, generally the smaller species (small tunas, bonitos, various types of mackerel).

The large tunas are migratory fishes ranging over very wide areas of the epipelagic zone, completely pelagic at all stages of life, from pelagic eggs which hatch in about 2 days into 4 mm long pelagic larvae, through rapidly growing young stages, to adults. Circumtropical fishes, their distributions are related to shifting patterns of ocean currents rather than to geographical features. Concentrations of tuna occur where upwelling nutrient-rich water leads to concentrations of food organisms, for instance along the equatorial countercurrent convergences in the eastern Pacific (from where nutrient-rich water moves westwards and northwards, Fig. 1.1). In the Indian Ocean catches vary seasonally with the monsoons.

The very important tuna fisheries have led to a vast amount of research (reviewed by Rosa, 1963; Blackburn, 1965; Nakamura, 1969; Rothschild & Suda, 1977; Sund, Blackburn & Williams, 1981), though research on such far-ranging and fast-moving fishes is costly and difficult. Of the large tunas, the yellowfin (*Thunnus albacares*, growing to 195 cm long, 136 kg in weight) is the most 'tropical' species, occurring mainly in equatorial waters where the surface temperature is high (over 27°C) and where the thermocline is shallow, and in seas adjacent to islands or reefs. The bigeye (*T. obesus*: 236 cm, 198 kg) shares the yellowfin's range but tends to live deeper, rarely being taken in surface waters. The albacore or longfin tunny (*T. alalunga*: 137 cm, 31 kg) chooses homogeneous water of 18°C, so tends to live at higher latitudes, though a few are caught across the tropics on subsurface hooks. The skipjack (*Katsuwonus pelamis*: 100 cm, 18 kg) has its distribution centre in the tropics and does not live as deep as the albacore. The very large bluefin (*T. thunnus*: 260 cm, 454 kg) lives at the highest latitudes (to 70°N and 35°S in the eastern Atlantic); although continuous across the tropics, the few taken there are always in deeper water. Temperature appears to limit tuna distribution, though they are rather eurythermal and range through water with a 10-degree temperature difference (20 degrees for the bluefin). Tuna body temperatures may be as much as 10 degrees above the water temperature, the blood supply to the muscles of these very active fishes making the flesh red in colour.

Tuna are carnivores, feeding partly on the larger zooplankton (such as euphausiid crustaceans) but even more on larger more active animals, the 'micronekton' of crustaceans, molluscs and fishes of 1–10 cm size groups. They appear to locate prey visually at close range (over longer distances by olfactory means) so may not be able to feed efficiently in turbid water, whereas clear water contains little food. Each species eats a great variety of organisms, and different size groups use different proportions of prey species. Euphausiid crustaceans are much used by skipjack, seldom by yellowfin. In two Indo-Pacific studies, medium-sized yellowfin stomachs contained fish (42–46%), cephalopod squids (26–36%) and crustaceans (11–24%); representatives of 30 fish families and of nine invertebrate orders were eaten, but only a few types of forage were important in any one locality. Little information exists on factors affecting the distribution of tuna larvae.

Tuna swim in schools of similar-sized individuals, the number in the school diminishing with the size of fish, from several thousand when small to 6–20 for very large fish. Tuna of different species but comparable sizes may school together (bigeye with yellowfin, skipjack with albacore). Bluefin can travel at 5–14 km h^{-1} in a dense group (pod) just below the sea surface, and may leap clear of the water in a feeding flurry. Sharks, including the huge whale shark *Rhincodon typus*, may accompany them. Size segregation is less distinct in the yellowfin which spends a large part of its life in the tropics, but is very clear in albacore and bluefin which evidently spawn in tropical seas and then spend their immature life in waters at higher latitudes.

Yellowfin and bigeye appear to share habitats in equatorial seas, while albacore and bigeye occur together in North Pacific currents. A particular ocean current may represent a distinct habitat for a tuna population, the boundary between two currents delineating one tuna species from another. There appear to be two types of tuna migration: a rather passive movement within the habitat (current), and an active movement between habitats (currents) following a change in physiological or ecological requirements of the fish (Nakamura, 1969).

Tuna have been tagged for over 30 years off California and skipjack tagged there have been recaptured off Hawaii. Tagging data have revealed that yellowfin grow at a rate of 20–40 cm yr^{-1}, skipjack at 6–17 cm yr^{-1} and albacore at 10 cm yr^{-1}. Such growth data are particularly valuable because determining tuna ages and growth rates has proved difficult as there are no seasonal growth checks and spawning seasons are protracted (see Rothschild & Suda, 1977). Results, though not altogether in agreement, suggest that the large tuna are fast-growing species. In the North Atlantic, bluefin become sexually mature at the end of the third or fourth year (when 100–200 cm long, 16–27 kg). Tuna in tropical waters have prolonged spawning seasons and probably produce more than one batch of eggs per season, but grow so rapidly over their rather short life that it is apparently not too difficult to identify size frequency modes corresponding to age groups.

Some of the other large scombroids, the billfishes (marlins, swordfish and sailfish) feed mainly on smaller fishes including tuna, bonito and mackerels (Shomura & Williams, 1974; Nakamura, 1985). Billfishes probably spawn far offshore and the young develop in the high seas.

Tagging has shown that these fishes undergo extensive migrations. A blue marlin (*Makaira nigricans*) tagged in the Bahamas was recaptured over 1300 km away in the Gulf of Mexico: weighing *ca* 90 kg when released, it had doubled its weight when recaptured 2.5 yr later. In the Pacific, striped marlin (*Tetrapterus audax*) and blue marlin have been recovered over 1800 km from the site of release.

PART IV

Syntheses

11

Responses of fishes to conditions in tropical waters

Ecological attributes of tropical fish communities considered in earlier chapters are summarized here and in Table 11.1. There are of course many exceptions to such broad generalizations, but they are made here as a framework to stimulate further study.

1. **Phyletic patterns.** Phylogenetically different types of fishes dominate the various communities. Riverine communities are dominated by otophysan fishes (primary freshwater fishes), together with other endemic primary freshwater fish families in African rivers; lacustrine communities are dominated by secondary freshwater fishes, cichlids in the littoral zone, clupeids and their centropomid predators in the pelagic zone. In the sea, clupeids and their acanthopterygian predators abound in upwelling areas, while reef communities are dominated by acanthopterygian fishes able to manoeuvre backwards into crevices, with jaws and teeth capable of exploiting the very varied food resources, finspines and poisons protecting the fish from the numerous piscivores. A diverse group, primarily acanthopterygians (sciaenids, grunts, sparids, snappers) live demersally on the continental shelves. There is also an apparent relationship between phylogenetic position and temporal habits (Hobson, 1974): the more generalized teleosts (such as percomorphs) are typically large-mouthed nocturnal or crepuscular carnivores, while many advanced fishes have specialized towards diurnality and feeding on smaller animals and plants. Ontogenetic changes may also occur, for example juveniles feeding diurnally then switching to nocturnal feeding with increased size.

2. **Levels of diversity.** Tropical fish communities show a great range in diversity from simple pelagic communities with few families and species, as in the open ocean, the upwelling zones of the sea

and pelagic zone of Lake Tanganyika, to greatest diversity on coral reefs and the littoral rocky shores of the African Great Lakes.

3. **Levels of environmental complexity.** Community diversity parallels (is causally linked with?) that of habitat structure, which ranges from very little in the pelagic zone (here due to invisible factors such as water temperature and salinity boundaries about which little is known), to the most physically structured habitats on reefs and off rocky shores. A structured habitat provides cover from predators. In pelagic communities 'cover' is provided by schooling, enhanced by vertical migrations into dimly lit zones of sea or lakes.

Table 11.1. *Ecological attributes of tropical fish communities discussed in the text*

Seasonality of environment:	Very seasonal ⟶ Aseasonal	
Examples:	Floodplain Pelagic upwelling zone	Lacustrine littoral Coral reef
Fish population response:	Fluctuates greatly by: (1) migrations (high mobility); (2) rapid multiplication	Remains constant, through year and from year to year
Life cycles:	Short; early maturation; longevity low	Long; delayed maturation (often sex change); longevity higher
Growth rates:	Fast	Generally slower?
Spawning:	Seasonal, rapid response to nutrient supply	Multiple through year
Feeding:	Facultative, or specialized for low trophic levels	Specialized for all trophic levels; adaptive radiations
Production/biomass ratio:	High	Lower
Behaviour:	Simple; uniformity; schooling	Complex, with learning, terri- toriality; symbioses
Predominant selection:	r-type, abiotic & biotic agents	K-type, mainly biotic agents
Diversity:	Less diverse, with dominants	Highly diverse, lack dominants
Community:	Rejuvenated	Very mature
Implications:	Resilient?	Fragile?

4. **Temporal patterns.** The seasonality of the habitat appears to be a key factor affecting many interrelated aspects of community life. This ranges from extremely seasonal in floodplain rivers (amounting to new habitats opening up each year), with biannual floods in some equatorial rivers, to relatively aseasonal conditions in some equatorial lakes and reefs. Life history strategies are geared to these seasonal changes in environment which affect all aspects of the fishes' biology – their foods, movements, growth and breeding seasons. Fish movements to feed or spawn are tied up with seasonal and other changes in the environment. For example, many riverine fishes move upriver (a few down-river) to spawn. In some reef fishes spawning movements fit in with lunar cycles which may be related to optimum conditions for survival of the pelagic larvae. Diurnal feeding movements are known in many species, for example from coral reef hiding places to grassy flats. Diurnal vertical migrations, generally down by day, surfacewards at dusk, follow the vertical migrations of the zooplankton food source but also appear to be antipredation measures.

5. **Behavioural attributes.** Schooling fishes are often migratory, though some may keep within a home range; many make feeding or breeding migrations. For territoriality some kind of cover is generally essential. Species may, however, be territorial for part of the 24 h cycle, moving long distances to feed at other times, for example, many marine reef fish and gymnotoids emerging from holes in river banks to feed nocturnally. In many species the juveniles school and wander to feed, while adults become territorial to breed. Cichlids which nest on open sand gain protection by nesting colonially or in arenas; only very large species (such as *Boulengerochromis microlepis* in Lake Tanganyika) appear to be able to nest in solitary pairs on open sand.

6. **Population dynamics.** Population fluctuations may be caused by movements in and out of the zone considered, or by breeding and mortality cycles affecting the actual numbers of fish. Fish populations fluctuate most under the changing conditions in floodplain rivers, estuaries and pelagic zones, both seasonally and from year to year, and least in the much more stable conditions of littoral regions of the lakes and coral reefs. Not only do the fish 'stay put' more under the latter conditions, but life history strategies

ensure that some young are produced throughout the year, replacing adults that are removed piecemeal, mainly by predation. This is in contrast with, for example, floodplain rivers, where young are produced mainly as the water level rises and mortalities are very high as the environment contracts, affecting the actual numbers of fish present; populations are also augmented by seasonal movements from other areas.

7. **Factors controlling population size.** In rivers, abiotic factors, such as water levels and deoxygenation, may be more important than the biotic factors (with which they interact) in controlling fish populations. In lakes, biotic factors appear to be especially important in controlling populations. In Lake Malawi shortage of suitable food for the smallest larval stages of the pelagic cyprinid *Engraulicypris sardella* may govern sizes of year classes entering the fishery. Competition for food appears to be less intense than competition for living space in many littoral and reef populations. In lakes and reefs predation is a major controlling factor; these very diverse communities characteristically have a very large number and diversity of piscivorous species.

8. **Trophic specializations.** These are less marked in riverine than in lacustrine fishes. In tropical lakes specializations occur in both pelagic and littoral/benthic fishes; these are possible because the food sources are available throughout the year, whereas many riverine fishes have to eat what is available seasonally. The great morphological diversity in feeding structures (jaws, teeth, alimentary canal) within a single family, the Cichlidae, and indeed within the haplochromine group of that family, equals or even exceeds the kind of differences found between families of marine fishes which have had longer evolutionary histories. The pressures leading to these specializations are puzzling, as competition for food does not appear to be severe at the present time.

Some of these attributes are now discussed.

Seasonality in tropical fish communities

Fishes display a whole spectrum of life history responses associated with the varied seasonality encountered in tropical waters, ranging from markedly seasonal conditions on floodplains and in upwelling areas of the sea, to the relatively aseasonal conditions in some equatorial or deep lakes and on coral reefs at low latitudes. These responses have been discussed elsewhere (Lowe-McConnell, 1979).

Although seasonal changes in daylength and temperature are small within the tropics compared with those in temperate regions, seasonal changes in wind and rainfall regimes do cause some seasonality in most tropical ecosystems. Seasonality in rivers is induced mainly by water level changes, in lakes and pelagic zones of the sea by wind-induced upwellings; continental shelves near estuaries have seasonal changes in salinity. Seasonal changes in the habitat affect the fishes mainly through qualitative and quantitative changes in available food, though there may be more direct effects, such as opening up the habitats on the floodplains, or wind turbulence affecting delicate planktonic larvae. Many coral fishes living in relatively aseasonal conditions on the reef have a pelagic phase (eggs and/or larval stages) living in a more seasonal environment in the pelagic zone; does this impose seasonality on reproduction and/or recruitment to the reef communities? Even in some relatively aseasonal habitats, some fishes still have defined breeding seasons (for example, some cichlids in Nicaraguan and African lakes), which may be caused by biotic pressures from associated species.

Case histories in earlier chapters have shown some of the attributes of fish populations in the most and least seasonal communities, as summarized in Table 11.1. When conditions fluctuate seasonally, fluctuations in fish populations are brought about by migrations, seasonal spawning and mortality (as on floodplains and in pelagic zones). In these fishes investment is for rapid multiplication, with short life cycles, early maturity and high fecundity. In fishes living under conditions where the nutrient supplies fluctuate less (both seasonally and from year to year) the investment is for efficiency of resource use; these species generally have longer life cycles, deferred maturity, iteroparous (little and often) spawning, often with parental care of offspring, and more complex behaviour related to homing, territoriality, social behaviour and symbiosis with other species. Fecundity appears to be much reduced in species practising parental care. In marine reef fish, however, it is high, associated with the production of pelagic eggs and larvae and mortality is probably much higher in the juvenile stages than in adults; this is in contrast with other species in which adult survival is very uncertain. Fishes living under seasonally changing conditions have to change their foods to eat what is available at different times of year, so they cannot be great specialists. The predominant type of selection in such communities seems to be for rapid population increase as in pioneer communities. Much of the floodplain is indeed a new habitat opened up for opportunists to exploit every year. The extent and duration of flooding varies greatly from year to year; abiotic factors, such as those associated with the flood regime (stranding, deoxygenation), interact with intense predation, leading to great fluctuations in numbers of particular fish species from year to year. In the pelagic zone with seasonal upwellings, small fishes (clupeoids) are specialized to crop foods low in the trophic web (such as plankton), multiplying fast and thus producing many mouths to do so. These and their predators form the mainstay of the pelagic communities, which tend to be much less diverse than those of floodplains or of bottom-dwelling reef fishes.

Ecological theory is concerned with the predictability of the environment (p. 303), and on the whole the less seasonal environments are more constant and predictable. Where conditions are less seasonal, fish populations remain very stable throughout the year and from year to year, as many tagging experiments have shown, for example on marine reefs. Many of the fish are territorial or have restricted home ranges; repopulation happens in a piecemeal way, vacancies arising as individual fish disappear, removed mainly by predation. In lakes, fecundity of the littoral fishes appears to be low, few young being produced at a time (and given parental care), though this may in some cases be over a relatively long life span. On marine reefs, fecundity also appears to be relatively low among demersal-spawning fishes (small species such as pomacentrids), but is much higher in fishes with pelagic larval stages which have to withstand the exigencies of life in the very different environment of open waters. In these species, numbers on the reef may be limited more by factors affecting recruitment to the reef fauna than by fecundity of the adults. The distribution of fecundity through the year might also be crucial in some cases.

Under aseasonal conditions it is very difficult to assess ages and growth rates, but such information as is available suggests that some of the large reef fishes live many years (large groupers up to 25 yr). The presence of the same food sources throughout the year allows trophic specialization, and spectacular examples of adaptive radiation occur in these relative aseasonal communities, for example among the lake cichlids. What Margalef (1968) referred to as 'stores of information' held by the fish are high: complex intra- and interspecific behaviour, including symbioses with other fishes (such as cleaners) and invertebrates in the sea and social transmission of behavioural traditions. Predators are abundant and anti-predator devices, such as camouflage or possession of poison, are common. These less seasonal communities have many of the attributes of what Margalef (1968) calls 'mature' communities (p. 325), in which K-selection leading to efficiency of resource use appears to be the predominant type of selection. Margalef commented that mature communities often exploit less mature ones, for example when their young stages move into a less mature community in which they develop and grow rapidly, as larval fishes do in estuaries or the pelagic zone.

Thus it seems that in tropical waters, where temperatures do not limit fish spawning (as at altitude), seasonality is imposed mainly by: (1) environmental factors leading to injections of nutrients (affecting the fish through the food webs), though some factors (such as turbulence affecting pelagic fish larvae) may influence species directly; and (2) biotic pressures (such as competition for spawning grounds or living space) which may impose seasonality in the relatively aseasonal environments with their very diverse fish communities.

Life history strategies

Life history strategies (patterns) result from natural selection for a species to produce the maximum number of young surviving to maturity under the conditions imposed by its biotopes (i.e. to maximize its fitness). These strategies have been discussed by Pianka (1970), Stearns (1976) and in Potts & Wootton (1984) among others. Natural selection would be expected to favour non-reproductive activities at the expense of reproduction only when they advance reproduction at a later stage in the life history and thereby maximize overall fitness. Habitat provides a templet in which life histories develop (Southwood, 1977). Changes in allocation of a species' resources from reproductive to competitive activities should only occur in habitats where these will enhance the survival of future offspring. The result of this is that organisms under different selection pressures will have characteristic life history patterns. For example, in ephemeral conditions, such as temporary pools, selection will have been for very rapid growth to maturity and for high fecundity to produce the maximum number of young, a strategy termed 'r-selection' by ecologists. At the other end of the scale, where conditions are very stable, the effective strategy has been to produce relatively few young at intervals throughout the year, in association with delayed maturation, longer life cycles and marked trophic specialization; this strategy is called 'K-selection'. The terms r and K refer to the parameters of the logistic growth curve for populations, where r is the slope representing the growth rate of the population and K the upper asymptote, which represents the carrying capacity of the environment. An *r*-selected species will have life history strategies which tend towards productivity, the *K*-selected species towards efficient exploitation of resources. Fishes in the tropical communities considered here show very clearly the attributes (correlates) of these two types of selection. This has important practical applications, as the two types respond differently to exploitation (p. 326).

Do fish mature at a fixed size or age? Neither, according to Stearns & Crandall (1984), but along a trajectory of age and size that depends on demographic conditions and is determined both by genes and by environment. Ware (1984) held that fish reproductive rate is solely a function of body size in short-lived species, but depends on the supply of energy surplus to maintenance requirements (except when feeding conditions are very poor) in long-lived species. For intermediate life history types the fitness of each strategy depends on values and forms of juvenile and adult mortality rates.

Many fish species are, however, very plastic in their behaviour and can change allocations of resources to growth or reproduction according to environmental and social conditions. Tilapia breed at a smaller size and produce more batches of eggs as waters dry up, growing to a larger size again when better conditions return (see pp. 250, 262). The shift is here to increased reproductive effort with the onset of bad conditions. (This is in contrast to the shift in some small fishes in English streams, which showed a general trend from single spawning, delayed maturity and prolonged generation times in low-productivity streams to multiple spawning and a short period of reproductive life in high-productivity streams, contrary to the predictions of *r* and *K* theory (Mann, Mills & Crisp, 1984).) *Lates calcarifer* also matures at different sizes in a series of rivers draining into the Gulf of Carpentaria, North Australia (Davis, 1984), though the factors responsible are not known. These *Lates*, like many other marine fishes, change sex in response to environmental or social conditions (p. 204). The trade-off between growth and reproduction has been studied in guppies (*Poecilia reticulata*) in Trinidadian streams, where predation appears to affect life history strategies (p. 301); the effects of population density on their fecundity were reported by Dahlgren (1979).

Types of reproduction

A very few tropical fishes spawn once in a lifetime and then die: 'big bang' spawners, such as the catadromous *Anguilla* eels which feed and grow in freshwater then migrate back to the sea to breed. The majority of fish spawn at repeated intervals. Among these the ova may either ripen all at one time of year so that the eggs are produced in one batch, 'total spawners', or eggs may ripen in batches ('multiple spawners' of Bagenal (1978)), the eggs being laid either at intervals throughout a breeding season or aseasonally. Multiple spawners include fishes with a wide range of behaviour, from 'partial spawners' in which perhaps one third of the ova ripen and are released at one time, the other ova remaining in reserve for spawnings later in the season, to 'small-brood' spawners (such as cichlids) which produce batches of eggs at frequent intervals. There is thus a whole range of breeding possibilities. How do these relate to seasonality of spawning and other ecological factors?

Examples of different types of reproduction in freshwater fishes are indicated in Table 11.2, while Table 11.3 lists the fecundity (the number of eggs ripening at one time in the ovaries) of representative species (see also data collated by Albaret, 1979a, b). Graduations between the various categories exist, but on the whole the total spawners have more clearly defined spawning seasons and are more fecund, producing numerous small eggs, many making long migrations to do so, whereas multiple spawners have less clearly defined breeding seasons and make only local movements to spawning places. Small-brood spawners generally establish a territory and often make a nest in which to spawn and guard eggs. Complex ritual breeding behaviour is important for synchronization of spawning in the small-brood fishes whereas total spawners may be brought together by external factors such as floods.

Coral reef fishes with pelagic eggs spawn in pairs or small groups near the edge of the reef, often at a particular time of day characteristic of the species. Reproduction in reef fishes was discussed in Chapter 8; they represent a fascinating array of spawning and social systems, as collated by Thresher (1984). He examined the costs and benefits of their five spawning systems – monogamy/long-term pairing, harems, explosive breeding assemblages, promiscuity and leks – and also of the four major social systems – solitary with individual territories, paired, harems and multi-male heterosexual groups. He also discussed reproductive behaviour in reef fish, looking at spawning height above the reef, diel spawning times (throughout the day in demersal-egg-layers, mainly at dusk in pelagic-egg-producers), lunar spawning cycles (known in 10–15 families, semilunar in some species of seven families, non-lunar in 13 families), and the possible reasons for dimorphism in so many reef fishes.

Table 11.2. *Types of reproduction in representative tropical freshwater fishes*

Type and fecundity	Seasonality of reproduction	Examples	Movements and parental care
Big bang + + + +	Once a lifetime	*Anguilla*	Very long migrations; catadromous; no parental care
Total spawners + + +	Very seasonal with floods: annual or biannual	Many characoids: e.g. *Prochilodus* *Salminus* *Hydrocynus* Many cyprinids Some siluroids	'Piracema' fishes, with very long migrations No parental care
	Extended season	*Lates* (L. Chad)	Local movements: pelagic eggs
Partial spawners + +	Throughout high-water season(s)	Some cyprinids Some characoids: e.g. *Serrasalmus* *Hoplias* Some siluroids: e.g. *Mystus*	Mainly local movements Guard eggs on plants ($\male$; $\male + \female$) Guard eggs on bottom ($\male$) Guard eggs and young ($\male$)

Grades into	*Arapaima*	Guard eggs and young; bottom nest (♂ + ♀)
	Some anabantoids	Guard eggs, surface bubble nest (♂)
Small-brood spawners +	*Hoplosternum*	Guard eggs surface nest (♂)
	Hypostomus	Guard eggs bank holes (sex?)
High-water season; may start at end of dry season or be aseasonal	*Loricaria parva*	Guards eggs under stone (♂)
	[a]*Loricaria* spp.	Carries eggs on lower lip (♂)
	Aspredo sp.	Carries eggs on belly (♀)
	Osteoglossum	Mouth broods (♂)
	Cichlids:	
	[a]Most S. American spp.	Guard eggs and young (♂ + ♀)
	[ab]Most African spp.	Mouth brood eggs and young (♀)
	[b]*Sarotherodon galilaeus*	Mouth brood eggs and young (♂ + ♀)
	S. melanotheron	Mouth brood eggs and young (♂)
	Stingrays	Live-bearing
	[b]Poeciliids	Live-bearing
	Anableps	Live-bearing
End of rains	Cyprinodont annual spp.	Resting eggs in mud through dry season

[a] End of dry season.
[b] Or be aseasonal.

Table 11.3. *The number of ripe ova in the ovaries (fecundity) of representative tropical freshwater fishes (sources given in Lowe-McConnell, 1975)*

Species	Number	Species	Number
Protopterus aethiopicus	1700–2300	Labeo victorianus	40 133
Arapaima gigas	47 000	Catla catla	230 830–4 202 250
Osteoglossum bicirrhosum	180	Lates niloticus	1 104 700–11 790 000
Mormyrus kannume	1393–17 369	[a]Mystus aor	45 410–122 477
Marcusenius victoriae	846–16 748	Hypostomus plecostomus	115–118
Gnathonemus longibarbis	502–14 624	Arius sp.	118
Hippopotamyrus grahami	248–5229	Loricaria sp.	c. 100
Pollimyrus nigricans	206–739	Sarotherodon leucostictus	56–498
Petrocephalus catostoma	116–1015	Sarotherodon esculentus	324–1672
Alestes leuciscus	1000–4000	Pseudotropheus zebra	17–<30
Alestes nurse	17 000	Cichla ocellaris	10 203–12 559
Alestes dentex	24 800–27 800	Astronotus ocellatus	961–3452
Alestes macrophthalmus	10 000	Anableps anableps	6–13 embryos
Hoplias malabaricus	2500–3000		
Salminus maxillosus	1 152 900–2 619 000		
Prochilodus scrofa	1 300 000		
Prochilodus argenteus	657 385		

[a] May spawn 5 times a season; most species above this in table (except *Arapaima* and *Osteoglossum*) are **total spawners**, those below it **multiple spawners**

Spawning seasonality

Spawning seasons may be either (1) restricted, as for many total spawners (such as many riverine otophysans and certain groupers, as well as some small-brood cichlids in Lake Malawi, or (2) extended, sometimes throughout the year (depending on the latitude): (i) in certain local spawners (*Lates niloticus* in Lake Chad (Hopson, 1972), *Clarias gariepinus* in Lake Sibaya (Bruton, 1979), the milkfish *Chanos chanos* in the sea); (ii) in multiple-spawning reef fish with pelagic eggs (many labrids, scarids, acanthurids (Sale, 1978*a*, 1980*b*)), in marine pelagic fish (such as Indian mackerel, *Rastrelliger* species, and the scombroid skipjack *Katsuwonus pelamis*), also in freshwater clupeids (which produce two or three batches of demersal eggs (Reynolds, 1974)); or (iii) in small-brood spawners, with parental care (such as cichlids in freshwater and pomacentrids on marine reefs, broods being produced through all or part of the year – though generally not as often in natural waters as by cichlids in aquaria).

Data on spawning in coral reef species have been collated by Sale (1977, 1978*a*) (see Table 8.2 for examples). They are multiple spawners, many species spawning throughout the year, but how often an individual spawns is less easy to determine. *Labroides dimidiatus* is said to spawn daily over 7 months (Robertson, 1972); some other species spawn every 3 or 4 days, weekly, or once or twice a month, some with a lunar cycle. Pygmy angelfish (pomacanthid) females are said to produce a few eggs every day (Bauer & Bauer, 1981). Patterns of individual spawning within the stock spawning cycle remain obscure for West African sciaenids (Longhurst, 1963) as well as for numerous other types of fish. We do not know the fecundity of a species until we know how many batches of eggs are produced a year; furthermore, the number of batches a year is quite likely to vary from year to year and with size and age of female.

Total spawners are mostly very fecund, fishes producing large numbers of ova at a time; for example, in the sea groupers which collect from a wide area of the Caribbean, and in rivers many species which make long migrations to spawning grounds and spawning tends to be very seasonal. Many otophysan floodplain fishes are total spawners, but a few, such as *Prochilodus scrofa*, should perhaps be regarded as partial spawners, as their numerous eggs (some 1 300 000 in a 4 kg female) are deposited in batches as they move upriver (Godoy, 1959), with many residual ova being resorbed after spawning. This suggests that there is a continuum from total to partial spawners. *Lates niloticus* in Lake Chad is said to be a total spawner (Hopson, 1972) though it has an extended breeding season and spawns in the lake; it is very fecund, producing very small pelagic eggs, each buoyed up by an oil globule; in Lake Victoria this species produces between 1 and 11 million eggs at a time. Many riverine characoids and siluroids produce demersal eggs which sink gradually as they descend with the current until they reach a stable substratum; others lay adhesive eggs amongst water plants.

Tropical freshwaters include numerous multiple spawners. There must be an adaptive advantage in producing several batches of eggs where the first may be endangered by fluctuations in water level. Many of these fishes make only local movements to spawn. The many tropical species which guard their eggs and young tend to have smaller broods than do fishes which leave their eggs unguarded; mouth-brooders have the smallest batches of all. In freshwaters, fishes with parental care often produce a batch in the low-water season before the floods come; this occurs amongst cichlids in both Africa and South America and in some loricariid catfishes in South America. We have seen above (p. 100) that in Lake Victoria the indigenous tilapia produce broods through the rains, the introduced tilapia spawn throughout the year and some haplochromines at the south end of the lake have dry-season spawning peaks (Witte, 1981). In Lake Malawi, although some breeding fish can be found throughout the year in some species, the majority of species display some seasonal breeding activity (Lewis, 1981; p. 103). In equatorial waters in Africa and in ponds it is quite usual to find ovaries starting to ripen again in cichlid females which are mouth-brooding young. These brooded eggs are relatively large. For marine reef fishes, Johannes (1978) suggested that extended spawning seasons may have an adaptive value related to the spread through the year of vacancies in reef communities which are taken up by larval fishes on their return to the reef; vacancies arise as resident fishes disappear from the reef in an aseasonal manner, many picked off by predators. It has been suggested that synchronous spawning and synchronous settling on the reef of some species may help to swamp predators.

The staggering of peak spawning times in sympatric closely related species, or among those sharing the same resources (food or living space), has been noted in many cases: among cichlids in Lake Malawi and Lake Jiloa, Nicaragua (McKaye, 1977), among small characins in Panama streams (Kramer, 1978), *Barbus* in Sri Lankan streams (Schut *et al.*, 1984), *Haemulon* grunts on Jamaican reefs (Munro *et al.*, 1973) and for other reef fish (Smith & Tyler, 1972). These cases all stress the probable importance of biotic pressures in determining seasonality in these tropical communities. They suggest that there must be competition for food or living space. In Lake Victoria, availability of spawning and nursery grounds may limit tilapia numbers.

Selection should determine that the young are produced at the time of year most favourable for their survival, when there is abundant food for rapid growth and cover from predators. Among the total-spawning otophysans in rivers, the start of the highwater season is the main spawning time for fishes in which the young feed on floodplains. Spawning may be stimulated either by local rains or by floods coming downriver from rain higher in the drainage basin (Fig. 1.2). Some species spawn in the riverbed before they move out, as catches between the Kainji coffer dams (p. 55) showed; others wait until they have moved out. Related species may differ in this respect, as for example *Hydrocynus forskahlii* spawning in river channels, *H. brevis* on the floodplain. Should the floods fail, the fish may not spawn that year. Bruton (1979) discussed delays in *Clarias* spawning. Kirschbaum (1984) described aquarium experiments in which, by decreasing the conductivity of the water, increasing the water level and simulating rain, he induced gonad maturation in both gymnotids and mormyrids; changing pH had little effect.

Certain fishes such as *Arapaima* have to await floods to gain access to their spawning places and cannot spawn until the water is deep enough for them to do so. In equatorial areas where there are two rainy seasons and two floods each year (as in rivers at the north end of Lake Victoria and in the lower Zaire) fishes are known to spawn in both floods but it is not known if individual fish spawn twice each year. At higher latitudes in the tropics (as in Lake Nasser/Nubia on the River Nile, Lake Chad and in southern Africa) spawning activity generally ceases when the water temperature is relatively low in winter. The effect of environmental conditions on the number of batches of young produced is well shown in cichlids bred in aquaria and ponds, where they produce many more broods a year than under natural conditions in lakes and rivers. The reasons for this are not clear and may vary from species to species. Townshend & Wootton (1984) found the effects of feeding levels on various aspects of spawning in the Central American cichlid *Cichlasoma nigrofasciatum* to be complex. A social factor, the removal of broods earlier in aquaria than in the field, may affect spawning frequency in *Etroplus maculatus* (Ward & Wyman, 1977; Ward & Samarakoon, 1981). In tilapia (see p. 262) the ability to breed at a small size with increased frequency according to the circumstances has an adaptive value in enabling them to survive in small pools and then to repopulate lakes and rivers when conditions permit. This adaptive switch has been observed many times in a number of tilapia species, for instance in Lake Rukwa in Tanzania and Lake Chilwa in Malawi when these lakes almost dried up, the fish reverting to normal size when the lakes refilled. The mechanism for this switch is not yet known, though understanding it might be very useful for fish culture.

Where breeding habits are known, characoids and cyprinids leave adhesive demersal eggs unguarded among plants or on the bottom. The eggs of many stream fishes need to be kept free of silt and need water with a fairly high oxygen content. The spatial distribution of the spawning grounds determines both the distance travelled and the timing of fish movements. Fishes living in short streams have short compact migrations; for those in large rivers the ascent is often long and entry into the river from a lake may continue over many weeks. The migratory categories of fishes ascending rivers from Lakes Victoria and Malawi were discussed earlier (p. 97). Species ascending the Mwenda River from Lake Kariba varied very much in their behaviour, as described by Bowmaker (1973). In South America some fishes which make long migrations upriver, in the Pilcomayo for example, are trapped in feeding lagoons until the river starts to rise and only reach their spawning grounds far upstream when the floods are no longer at their height (Bayley, 1973). Many lake fishes retain the habit of moving into rivers to spawn, which they do seasonally. But amongst those which spawn in lakes breeding may continue more or less throughout the year, or be affected by hydrological and biotic events in the lake. In the sea some groupers congregate in special places in the Caribbean, their numbers suggesting that they must come from a wide area.

Fecundity

Fecundity, defined by Bagenal (1978) as the number of eggs in the female prior to the next spawning period, varies very much in individuals of one species of the same weight, length and age, but in general it increases in proportion to the weight of the fish, hence with the cube of the length. Where fecundity is at a premium, as in most total spawners such as characoids and cyprinids, the females are generally larger than the males, the egg number increasing with the size of the female. Where eggs are guarded by one sex, the guarding sex (often the male) is the larger fish. Males are also larger than females in many cichlids, as in tilapias (p. 261); where breeding is very frequent this size discrepancy increases, probably for bioenergetic reasons, more weight being lost with egg than with sperm production, retarding the growth of the female compared with that of the male. Also the female feeds little, if at all, while mouth-brooding. Egg number increases with weight of female in the substratum-spawning *Tilapia zillii*, but Welcomme (1967*b*) found that in the mouth-brooding *Oreochromis leucostictus* the 'fertility' (number of young reared) increased with the square of the total length, the numbers of young being limited by mouth size which increases in linear relation to body length, so brooding efficiency decreases as the parent increases in length. In *Alestes macrophthalmus* in Lake Bangweulu, Bowmaker (1969) noted that lake fish appeared to be more fecund than swamp fish; he thought that individual females bred twice a year.

Individual tilapia in equatorial lakes, dams and ponds may spawn three or more times in one season. The number of eggs generally diminishes at each spawning, but in some Central American cichlids Townshend (personal communication) found that the number of eggs may increase in successive spawnings. In aquaria, individual *Oreochromis esculentus* spawned seven times in 24 months, with broods as close as 39 days and at average intervals of 2 months. These small fishes were as fecund as lake fish, producing as many but smaller eggs (Cridland, 1961). Peters (1963) found egg number to be related reciprocally to egg weight, tilapia producing many small or few large eggs. The maximum size to which tilapia young are brooded varies with the species and its normal habitat: in inshore-dwelling species it is smaller (to *ca* 12 mm long in *O. leucostictus*) than in openwater-dwelling species, such as *O. lidole* in Lake Malawi which may continue to brood young that are 50 mm TL (though rather few of them (Lowe, 1952)). The haplochromines of the African Great Lakes, many of which produce young throughout the year, produce large and few eggs (less than 50 per brood in *Pseudotropheus zebra* (Balon, 1977)) (see Table 11.3).

The South American substratum-spawning *Cichla ocellaris* in ponds has a minimum of 22 days between spawnings; Fontenele (1950) recorded 20 spawnings in 4.5 yr. Eggs hatch in 78 h and the young are guarded by both parents until they are *ca* 35 mm long. In Gatun Lake, Panama Canal, where this species is now established, it grows to *ca* 20 cm in 8 months; both sexes first mature here at 32–33 cm SL, but after they mature the male increases in size relative to the female (Zaret, 1980); in Guyana all individuals over 38 mm were males (Lowe-McConnell, 1969*b*).

For reef fish, Thresher (1984) maintains that fecundity per size of fish is much the same in demersal and pelagic spawners, but as most of the pelagic spawners are larger they produce more eggs.

Skewed sex ratios

Attempts to produce sterile hybrid tilapia for fish culture led to the unexpected discovery at Malacca (Malaya) that crossing certain species led to skewed sex ratios (Hickling, 1960). When female *Oreochromis mossambicus*, a species from southern Africa acclimatized in Malaya, were crossed with male *O. hornorum* imported from Zanzibar, the offspring were all males. The reciprocal cross produced about three males to one female. Since then, skewed sex ratios have been found in a number of other tilapia crosses (see discussion of possible mechanisms in Pullin & Lowe-McConnell, 1982). These skewed sex ratios in tilapias are not due to sex changes.

Parental care

Some form of parental care is now known in 89 fish families, 50 of them freshwater families (see Blumer (1982) who gives full references). Blumer defined forms of care ranging from prefertilization activities such as substrate cleaning and nest building, to guarding eggs or fry attached to the body or in the mouth (oral-brooding), and internal gestation (livebearing). Grouping these families into categories based on the sex of the carer revealed male parental care to be as common or commoner than female parental care. Male-alone caring is known in 36 families, female-alone in 16 families, with biparental care in 10 families, and various combinations within the other families.

Of the families in tropical waters considered here, male-alone care is found in Lepidosirenidae, Notopteridae, Erythrinidae, Lebiasinidae, Clariidae, Plotosidae, Callichthyidae, Loricariidae, Batrachoididae, Antennariidae, Syngnathidae, Synbranchidae, Sparidae, Nandidae, Labridae, Blenniidae and Tetraodontidae, with male-alone and biparental care in Pomacentridae, and male-alone or female-alone in Bagridae, Ariidae and Apogonidae. Families in which the female alone gives care include Aspredinidae and those with internal gestation, such as various cyprinodontoid groups. Families with biparental care include the Heteropneustidae, channidae, Osphronemidae and Balistidae. Families with male-alone, female-alone or biparental care include the Osteoglossidae, Cichlidae and Gobiidae.

The rapid development of parental investment (PI) theory based on

Trivers' (1972) ideas has led to much interest in parental care patterns in fish (Keenleyside 1979, 1981; Perrone, 1978; Perrone & Zaret, 1979; Baylis, 1981; Gittleman, 1981; Blumer, 1982). Trivers analysed the evolution of sexual strategies in terms of PI, defining PI as any investment by the parent in an individual offspring that increases the offspring's chance of surviving (and hence of reproductive success) at the cost of the parent's ability to invest in other offspring. He suggested that selection will tend to favour desertion by whichever parent has invested least in the offspring up to that moment, and to favour continued investment by the other parent. Great care is, however, needed in extending this new behavioural ecology to the entire class of fishes, as reproductive patterns (including mating systems, courtship and breeding behaviour and parental strategies) are more diverse among fishes than in any other vertebrate group (see Keenleyside, 1981).

Balon (1984) classified the reproductive guilds of fishes, differentiating between: (1) 'open substratum-spawners', producing floating eggs (e.g. *Ctenopoma murieri, Lates niloticus*), laying eggs on the bottom which later become buoyant (*Prochilodus*), or scattering eggs on many types of bottom or among plants; (2) 'brood hiders' putting eggs in gravel, caves, mussels or with dormant eggs (annual fishes); and (3) those with parental care. These latter include: (i) guarders, substratum-choosers attaching eggs to rocks (some loricariids), to plants (*Polypterus*) or leaves overhanging the water (*Copeina*), or guarding floating eggs (some *Channa* and *Anabas* species); and (ii) nest-spawners, preparing an area of rock or gravel (many Neotropical cichlids, some characins such as *Leporinus*), constructing a nest of vegetation (Asian *Clarias bratrachus*), making shallow nests in sand (*Tilapia zillii*), with floating bubble or froth nests (as in *Hepsetus odoe, Hoplosternum*, some anabantids), depositing eggs in cleaned holes or crevices (several cichlids), building nests with any materials (*Notopterus chitila, Hoplias malabaricus*), or even nesting in sea anemones (*Amphiprion*). Amongst 'bearers' he differentiated between external bearers, notably mouth-brooding cichlids, *Osteoglossum*, ariids, and those carrying eggs attached to the parent's skin (as in some loricariids, *Bunocephalus* and other South American catfish); and internal bearers, (a) those with internal fertilization but eggs expelled immediately afterwards (*Pantodon*) or almost ready to hatch (*Tomeurus*), (b) with ovoviviparous eggs incubated in the body cavity and fed by yolk (as in most elasmobranchs), or (c) viviparous, with eggs in the body cavity fed by absorptive organs (as in poeciliids).

Oral incubation has evolved independently in many fish groups (listed by Oppenheimer, 1970): in the sea in apogonids; in freshwater in ariid catfishes, various anabantoids, osteoglossids, and in cichlids, in which the mouth-brooding habit has evidently arisen several times independently from substratum-spawning species, as it has in the tilapiine fishes (Trewavas, 1983). Bearing live young has also evolved many times independently, even amongst the cyprinodontoid fishes (Parenti, 1984b).

There are variations within families as to whether eggs are guarded by one or both parents, whether parents continue guarding young after they are hatched and for how long they do so (see data collated by Breder & Rosen, 1966). Such differences are well illustrated in the anabantoids. In Asia many of these species which live in deoxygenated water make floating bubble nests, generally guarded by the male (as in *Osphronemus goramy*) or by both sexes (as in *Macropodus* species), though very small pelagic eggs are left unguarded in *Anabas testudineus* and *Helostoma temmincki*. In the African *Ctenopoma* there are two ethological groups – *C. muriei* which does not care for its floating eggs and *C. damesi* in which the male guards a surface foam nest; while *Sandelia capenis* produces adhesive eggs guarded by either sex.

In the Channidae eggs and young are guarded by both parents in *Channa striatus* in the Philippines and by *C. punctatus* in the Punjab, but not in *C. asiatica*. *C. striatus* produces 100–1000 eggs at a time which take 3 days to hatch in a clearing in the vegetation; breeding occurs in every month and individuals may breed twice a year. *C. punctatus* in the Punjab spawns between April and July; the 500-strong brood is guarded for over a month, until the fry are 10 cm long. *Notopterus chitila* attaches up to 20 000 small eggs onto posts, which are then guarded by the male. The nandids all show some form of parental care. The Indian *Mastacembelus panacalus* leaves 10–20 demersal eggs in demersal masses which are not guarded.

Among the characoids which guard nests (an unusual feature in this group of fishes), *Serrasalmus* species (*S. nattereri* and *S. rhombeus*) spawn in the main rains in Guyana, where they are reputed to guard eggs laid on tree roots trailing in the water; guarding behaviour by the parents has been observed in aquaria. In South American streams *Hoplias malabaricus* produces batches of 2500–3000 eggs at 15-day intervals; these are laid in a small depression in bottom shallows where they are guarded by the male; they hatch in 4 days and take about 10 days to absorb the yolksac (von Ihering, 1928; de Azevedo & Gomes, 1943). Nestmakers in South American waters include the callichthyid armoured catfish *Hoplosternum littorale*, in which the male makes a nest in flooded trenches during the rains, buoying up grass and other vegetation with bubbles passed into it from his intestine; the male guards the nest and several females may lay in it. In *H. thoracatum* eggs are laid under a large leaf buoyed up by bubbles. The female *Aspredo* catfish carries eggs adhering to her belly. The male guards the eggs in seven of the eight species of *Loricaria* studied (the female in *L. macrops*). Some male *Loricarichthys* species carry the eggs adhering to the specially elongated lower lip. Male *Loricaria uracantha* guard batches of eggs from several females in hollow pieces of wood in Panamanian streams. Moodie & Power (1982) speculate that the egg-carrying habit in some loricariids has evolved as a means of circumventing intraspecific competition for nesting cavities.

The guarding fish eat little, and not at all in carnivorous cichlids, a behaviour which may prevent the young being eaten by the parents. In some cases (as in some anabantoids) the habit of the male being the guarder is thought to be correlated with the greater need for the female to recover after spawning, as more body weight is lost in egg than in sperm production. Some male cichlids develop nuchal humps before the breeding season, possibly used in display or as a fat store over the guarding period. In some cases the male appears more aggressive and thus a more effective defender of the young. In species in which the male alone guards the nest, he may induce several females to lay in his nest in succession; he then rears a mixed batch of young, of different genetic constitution and of slightly different ages. This occurs in such diverse groups as *Protopterus* (Greenwood, 1958), *Hoplosternum* and *Loricaria uracantha*, as well as in some cichlids. Communal broods guarded by two sets of parents are known in *Etroplus* and *Tilapia rendalli* (Ward & Wyman, 1977; Ribbink *et al.*, 1981), and brooding by several parents in some pomacentrids on coral heads (Fischelson *et al.*, 1974). McKaye (1984) thought about 50% of the cichlid broods he observed in Lake Malawi were multispecific and many were observed by Ribbink *et al.* (1980). Cichlid young were also found in *Bagrus* nests in Lake Malawi (McKaye & Oliver, 1980), and with *Synodontis* in Lake Tanganyika (Brichard, 1979; Sato, 1986).

The mouth-brooders include most of the African cichlids but only a few Neotropical ones (species of *Geophagus* and *Aequidens*). Among tilapias the male broods in the *Sarotherodon* species, found mostly in West Africa (see Trewavas, 1982, 1983); *S. galilaeus* is unusual in that either sex may brood eggs and young. In *Oreochromis* species, as in most other mouth-brooding cichlids in the African Great Lakes for which breeding habits are known, it is the female which mouth-broods. Males remain near their nests on the spawning arenas (leks); females pick up the eggs immediately (they are in many cases fertilized in her mouth) and move to special brooding grounds, eventually leaving the young in nursery areas near the shore. Nest forms often have specific patterns (for example domed, or with accessory pits around a central plaque) resulting from the movements used by the male to make the nest. Courtship displays are species-specific, using the colours of the breeding fish to full advantage, though how they manage to display in waters with very low visibility is still a mystery. Some male *Oreochromis* have long genital tassels with egg-simulating blobs of tissue on them, which may enhance the chances of the female picking up sperm in her mouth together with the eggs (Trewavas, 1983). Many haplochromines have 'egg-dummy' spots on the anal fin (Wickler, 1962). Sperm and eggs carry adhesive filaments in some cases and their structure varies from species to species (Peters, 1963). There is a considerable literature of experimental work on the breeding behaviour of cichlids; it seems that both visual and chemical (pheromones) means may be involved in keeping the young near to the parent (references in Barnett, 1982). In *Pseudocrenilabris multicolor* the young learn the mother's territory and stay within it (Albrecht, 1963).

Predator pressure would appear to be an important factor in the evolution of some form of parental care for eggs and young. It is interesting that so few South American cichlids have evolved oral incubation, especially as predator pressure appears to be particularly intense in South American waters. Perhaps the presence of both parents to guard the young fishes when they are first free-swimming is of greater value; also more eggs can be laid at a time by substratum-spawners. It is interesting too that it is in Lake Tanganyika with its many piscivores that substratum-spawning cichlids are found in Africa. McKaye (1984) has compared reproductive behaviour in substratum-spawning Nicaraguan cichlids and mouth-brooding Malawian cichlids. Piscivores appeared to influence the location of spawning territories in both cases; in Lake Malawi the multispecific spawning arenas were considered to be an antipredator response to the six species of specialized egg-eaters in this lake. The monogamous pairs of substratum-spawners in Nicaraguan lakes stayed on their territory for up to several months, whereas the polyandrous and polygynous Malawian males were generally on territory for less than 2 weeks. This suggests that availability of food in the potential territories must also be an important consideration. Roberts (1972) considered that physicochemical factors may also have been important in the evolution of parental care, for such care occurs in many fish in which the adults spend at least part of the time in swamps or other oxygen-deficient habitats. *Protopterus, Lepidosiren, Notopterus, Arapaima, Heterotis, Gymnarchus, Hoplosternum* and anabantoids all live in rather deoxygenated waters and care for their young, helping to aerate the water for them. There are, however, other air-breathing fishes which lack parental care (such as *Polypterus*), while all the cichlids provide examples of non-air-breathing fishes with parental care. So it would seem that physicochemical factors may have been important in the evolution of parental care in ancient groups of fishes, and those that had it survived to tell the tale, while in cichlids (probably a young group) predator pressure may have been more important.

A number of fish groups produce live young. Viviparity has evolved repeatedly and independently from oviparity in many fish groups. Wourms (1981) cites 54 families of extant fishes bearing live young. These include 40 chondrichthian families (sharks and rays), *Latimeria*, and 13 teleost families with over 500 species, including some Neotropical cyprinodontiform groups and some species of about eight marine families (also listed by Blumer, 1982). The many types of live-bearing are discussed by Wourms. In live-bearers such as *Poecilia* and *Anableps* the male is generally smaller than the female, unlike the males of egg-laying cyprinodonts, which are generally larger than the females. Cases of internal fertilization leading to the production of fertile eggs are also known, for example in glandulocaudine characids (Nelson, 1964).

Annual fishes

Certain cyprinodontoids – *Cynolebias, Astrofundulus* and *Rivulus* in South America, *Nothobranchus* and *Aphyosemion* in Africa – are 'annual fishes' which spend the dry season as resting eggs. These annual species are often readily recognizable by the large dorsal fin of the male with which he holds the female when spawning. The anterior parts of the anal fin of both sexes are folded into a tube, used to ensure proper fertilization and to bury eggs in the bottom mud. Air is brought down to the eggs by expansion and contraction of the dry mud caused by diurnal temperature fluctuations. Annual rivulins are generally very fertile and once mature spawn ever day until they die. Some eggs remain transparent while others develop and aestivate as 'resting embryos' (diapause II); still others develop ripe embryos and aestivate as 'resting fry'. There are generally three phases of diapause in the African rivulins. The adaptive advantages of this are obvious, for when rain falls and the oxygen supply is cut off from the eggs, those with ripe embryos have to hatch or die, but earlier diapause eggs remain if the pool should dry up again after the first showers. The eggs of the South American *Cynolebias nigripinnis* can remain viable even after 3 yr under anaerobic conditions in soil below water (Scheel, 1968). Simpson (1979) gives a good general account of these fishes, with references.

Fish growth under tropical conditions

As already mentioned, it is more difficult to determine fish ages and growth rates under tropical conditions where seasonal checks in growth are not as regular and apparent as are the winter checks in growth which mark skeletal structures (scales, bones, otoliths) in temperate zone fishes, and extended spawning seasons complicate tracing length frequency modes (Petersen's method). Growth studies and age determinations are, however, essential for estimates of production, stock sizes, recruitment and mortalities, so much effort has been put into trying to determine them (for methods see Bagenal, 1978; Pauly & Murphy, 1982; Pauly, 1983). Growth data have been collated by Pauly (1981) in relation to environmental temperature in 125 fish stocks (mainly marine), and by Merona (1983) for African freshwater fishes. General points which arise are as follows.

1. At the high temperatures prevailing in tropical waters growth rates are faster, the fish mature at a younger age and the life span is shorter than in temperate waters. For example, Chinese grass carp (*Ctenopharyngodon idella*) grew at up to 10 g day^{-1} in ponds at Malacca in Malaysia compared with 3.5 g day^{-1} in South China, and at Malacca they matured in 1 yr compared with 3–5 yr in South China. Similarly, the North American bluegill (*Leopomis macrochirus*) and largemouth bass (*Micropterus salmoides*) acclimatized in Cuban lakes matured earlier (1 yr or less compared with 2 or 3 yr), grew faster, and had a shorter life span in Cuba than in the USA (Holčík, 1970). Conversely, tilapia grow more slowly in temperate zone ponds and at high altitudes where temperatures are lower than in their tropical home waters.

2. The species considered in earlier chapters show a wide range of longevity, from annual fishes completing their life cycle in less than 1 yr to 25 yr in groupers. Small pelagic clupeoid fishes, both clupeids in Lake Tanganyika and anchoveta off Peru, have very short life cycles, rarely longer than 1.5–2 yr. In the Niger and Lake Chad the small characids *Alestes leuciscus* and *Micralestes acutidens* were never more than 2 yr old (Daget, 1952; Lek &

Lek, 1977), and in Lake Kariba five small species (including *Alestes lateralis*) did not exceed 3 yr. In trawl catches West African sciaenids and polynemids have very short life spans (Longhurst, 1963), rarely more than 4 yr. In Lake Kariba most species evidently reached maturity in their second growth season (Balon & Coche, 1974); the majority of the 22 species studied here had a life span of 7–9 yr (including *Tilapia rendalli, Mormyrops deliciosus, Mormyrus longirostris, Hydrocynus vittatus, Labeo altivelis, Synodontis zambesensis*) or 4–6 yr (as in *Oreochromis mortimeri, Hippopotamyrus discorhynchus, Marcusenius macrolepidotus, Eutropius depressirostris, Schilbe mystus, Syndontis nebulosus*), though four species lived longer than 10 yr (*Heterobranchus longifilis, Clarias gariepinus, Malapterurus electricus, Anguilla nebulosa*). In aquaria, individual *Serrasalmus* have been kept up to 25 yr and *Bagrus bayad, Clarias lazera, Barbus bynni* and *Loricaria parva* have all survived for over 16 yr.

3. The size to which a species grows is partly genetically determined and partly modified by prevailing conditions. The North Atlantic tuna *Thunnus thynnus* becomes sexually mature when 100–130 cm long (16–27 kg) at the end of the third or fourth year, the time taken for cichlids to mature at a very much smaller size in Lake Malawi. Though many large fish are piscivores, other foods can support large fish: the giant catfish *Pangasianodon gigas* of the Mekong, for example, is a vegetarian, and the microphagous osteoglossid *Heterotis niloticus* in West Africa grows fast to over 1 m long.

4. In many fish species growth rates vary greatly with environmental conditions, the available food, and factors which affect its utilization such as temperature and crowding. This is very clearly shown by variations in growth rates with food provided in fish ponds; in natural waters, growth rates must also be influenced by many conditions interacting in their effects. In Lake George (Uganda) the maturation size of the tilapia *Oreochromis niloticus* was found to be lower following years of intensive fishing (Gwahaba, 1973), and lowering of maturation size in response to intensive fishing is known in other commercial species (Garrod & Knights, 1979; Garrod & Horwood, 1984). Edwards (1985) considered the growth rates of three lutjanid snapper species living for 10–14 yr in tropical Australian waters to be lower than that of cod in the North Sea, but thought this might reflect the

lower rate of exploitation in the tropical waters; data collated by him suggested they grow faster in reef areas than on open bottoms, which he considered might be related to the availability of food.

5. Growth rates change through the life history. Though they often slow down with increased age and after maturation, they may also change with change of diet, as Hopson (1982) found for *Lates* in Lake Chad (p. 60). Changes in growth rates with available prey have been explored in temperate fishes (see for example Kipling, 1984).

6. Growth rates also vary seasonally. In the Middle Niger, Daget (1952) found that *Alestes* of various species which fed on the floodplains at high water then retired to the river in the low-water season, grew in length and laid down fat stores only in the highwater season, subsisting on their fat reserves and losing weight and fat content in the dry season. In *Alestes baremose* and *A. dentex* the females grew larger than the males. Such sex differences in growth rates are often found when fecundity is at a premium, the egg number increasing with the weight of the female.

7. Growth rates also vary with habitat. Comparable data for *A. baremose* growth in the Bandama River, Ivory Coast (Paugy, 1978), in Lake Chad (Durand, 1978) and in Lake Turkana (Hopson, 1982) show that growth is faster and to a larger size in the lakes than in the rivers, supporting the view that lakes can provide better feeding grounds than rivers.

8. Sexual dimorphism in size and/or colour, is widespread in reef fishes (see tables in Thresher, 1984), e.g. in most scarids, labrids, epinephaline groupers, some acanthurids and pomacanthids, while most chaetodontids, lutjanids, mugilids and many pomacanthids are permanently monomorphic. Some (e.g. labrids and scarids) have small and large forms of breeding males, often with distinctive colours. Hypotheses to account for dimorphism are discussed by Thresher (1984).

The most reliable growth data for tropical freshwater fishes in natural waters come from waters where seasonal differences are marked, such as on floodplains. In the Zambezi system on the Kafue flats it proved possible to study variations in tilapia growth rates from year to year in association with the prevailing water levels and other conditions (Dudley, 1974, 1979; Kapetsky, 1974). In the Rupununi, Guyana, in floodplain pools, length frequencies and scale rings evidently caused by dry season growth checks suggested that most of the cichlids matured in less than one year, ready to spawn in the next annual flood, though the larger fishes such as *Osteoglossum* probably took 2 yr to reach maturation size. On floodplains there are probably strong selection pressures for fishes to grow fast to avoid being eaten by the many piscivores which feast on the fishes as the waters contract. Many cichlids in the African Great Lakes on the other hand may take several years to attain maturity (Fryer & Iles, 1972).

Growth rates of tilapia have been much studied as these are important food fishes throughout the tropics (see Pullin & Lowe-McConnell, 1982). Summarizing the available data brings out the great variations in growth rates, maturation sizes, and relative maturation and final sizes of males and females. Some species, such as *Tilapia sparrmani*, never grow large; others like *Oreochromis niloticus* and *O. mossambicus* vary their growth rates very much according to the conditions (Fig. 11.1). *O. niloticus* grow

Fig. 11.1. Variations in length distributions and maturation sizes (m.s.) (arrowed) of male (×· · ·×) and female (●——●) tilapia *Oreochromis niloticus* from various East African waters (after Lowe-McConnell, 1958). (A) Lake Turkana, seined fish, m.s. 39 cm TL both sexes; (B) Lake George, gillnetted fish, m.s. 28 cm both sexes; (C) Lake Kijanebalola, gillnetted fish, m.s. 17 cm both sexes; (D) Buhuku Lagoon, Lake Albert, seined fish, m.s. 14 cm male, 12 cm female; (E) Kijansi Pond, Uganda, males of 18 cm in breeding colours.

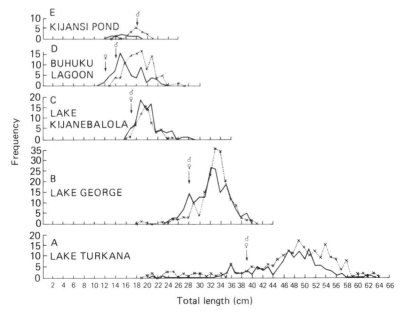

large when in good condition (high weight for length), and males and females then grow to, and mature at, comparable sizes. In 'dwarfed' populations the males are generally larger than the females (Lowe-McConnell, 1959, 1982).

In most populations of tilapia maturing at a small size (as on Fig. 11.1) there has been a switch in life history strategy from growth to reproduction, a switch which Iles (1973) and Fryer & Iles (1972) considered to be unique, but which is now known to occur in some other fishes (e.g. *Lates calcarifer*). Several, perhaps most, tilapia, including *O. mossambicus* which now has a circumtropical distribution in fishponds and the widely distributed *O. niloticus*, are able to breed when a few months old, instead of 2 yr or more, and when about 8 cm long compared with over 20 cm in the lakes. They also produce small eggs, relatively more in relation to body weight than are produced by large individuals. Iles (1973) stressed that this was a form of miniaturization and not 'stunting'. In some tilapia populations growth of adult tilapia may, however, be stunted, as Bowen (1979) found for *O. mossambicus* in Lake Sibaya (southern Africa), where food available to the adults was less nutritious than that used by juveniles living in shallower water.

Communication between fishes

In these complex communities, with so many species crowded together, biotic interference is continual. For each fish there is a conflict between the need to make itself known to its conspecifics, for schooling or territory maintenance and for successful breeding, while at the same time avoiding the attention of the numerous predators.

The senses used depend very much on the usual habitat of the group of fish involved. Fishes living in very clear waters, such as diurnal fishes on coral reefs and cichlids in lakes, are mainly sight-orientated, though other senses (hearing, olfactory senses) may be used too. In open waters, where the light changes are mainly of intensity with depth and diurnal, colours tend to be uniformly silvery in both prey and predatory fish, counter-shaded for camouflage. In these open waters the acoustico-lateralis system, which detects water pressure changes, plays a vital role in schooling behaviour. In nocturnal fish and those living in turbid water, sound production is often most important. Chemical communication is particularly significant for fishes living in confined spaces (moray eels in reef crevices, some catfishes in freshwaters). Electrical signals are used by some nocturnal freshwater fishes. Each of these methods of communication has a large and growing literature. Useful review papers include

Woodhead (1966) on behaviour in relation to light in the sea, Levine & MacNichol (1982) on colour vision in fishes, Partridge (1982) on the structure and function of fish schools, Schwarz (1985) on sounds, and Bardach & Todd (1970) on chemical signals.

Significance of markings and colours

Colours in fishes include fixed colours present throughout a phase of life, and chromatophore-influenced colours which change with the emotional state of the fish and the light intensity. The former include the 'poster' colours of many reef fishes; these may change as the fish grows (as in the angelfish *Pomacanthus paru*). The vivid colours and stripes of some of the Great Lake rockfishes are probably of significance in the same way for species which are territorial. Permanent colours also include the shining neon colours of small forestwater species (such as neon tetras prized by aquarists); these may have a role in enabling fishes to build up viable populations when they have been scattered by the oscillating water conditions. In darkwater streams, black patterns surrounded or supported by white, yellow, orange or red show up most clearly. These are generally found as vertical, longitudinal or diagonal stripes on the body, in caudal and humeral spots or blotches, or on the caudal or adipose fin, in contrast to the brilliant spots on the head or opercular region of greater significance for close-up head-to-head encounters (as during sexual displays).

Colours which change with the emotional state include those depicted for *Hemichromis fasciatus* in Fryer & Iles (1972) to illustrate the language of the cichlids: a neutral livery means 'I am quietly going about my business'; vertical stripes and darkening colours (with reversed lateral spots in very dark fish) show increasing readiness for combat to defend territory; the fright pattern is mottled with the development of a longitudinal stripe if there is nowhere to hide; brooding fish carry a series of lateral spots; spawning fish lose their spots.

In many cichlids, including mouth-brooding tilapia, male breeding colours advertise spawning grounds. Such colours may be kept throughout the spawning season. In tilapias and other cichlids in which both parents guard young, there is less difference in colour between the sexes, and colours change with fish mood, playing an important part in the prespawning displays which ensure the synchronous ejaculation of eggs and milt.

Colour patterns and markings come to have behavioural uses, as shown by aquarium experiments (collated by Keenleyside, 1979). For example, in the small characid *Pristella riddlei* the conspicuous black patch on the dorsal fin (Fig. 11.2), jerked rapidly when the fish is alarmed, has been shown to be a 'social releaser', leading other *Pristella* to follow the school (Keenleyside, 1955). In many haplochromines the male has 'egg-dummy' orange spots on his anal fin, to which the female reacts by trying to take them into her mouth, thereby sucking up sperm with her eggs (Wickler, 1962). In many cichlids (e.g. *Pterophyllum, Mesonauta, Etroplus*) the

Fig. 11.2. Communication systems in freshwater fishes: (A) *Pristella riddlei* in which the black spot on the dorsal fin acts as a social releaser for schooling; (B) *Astyanax bimaculatus* with dark humeral and caudal dark markings found in many midwater schooling fishes; (C) *Corynopoma riisei* a glandulocaudine characid in which the male wafts a pheromone from his caudal gland to the following female (note opercular extension paddles of the male); (D, E) fishes utilizing electric discharges (note convergent evolution in body form in the African mormyroid *Gymnarchus niloticus* (D) and the South American gymnotoid *Gymnotus carapo* (E)).

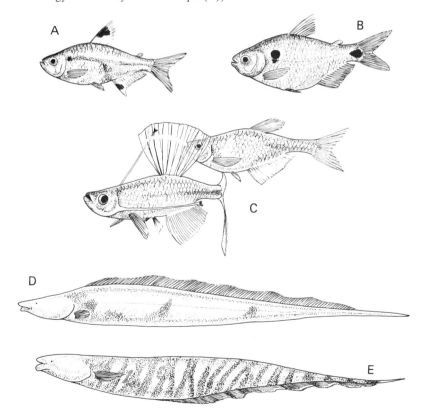

pelvic fins become conspicuous in colour and are jerked in characteristic fashion when 'calling' young to follow the parent. The caudal ocellus developed in many cichlids which lead their young (e.g. *Cichla ocellaris, Mesonauta festivum*) may help the orientation of the young to the parent. Changes in eye colour often have social significance (as in gulls); their importance for dominance and territory establishment has been demonstrated in *Pomacentrus* (Rasa, 1969). As we learn to decipher fish language we can guess a great deal about how a fish uses its markings, guesses which call for field observations and experiment.

In Mato Grosso streams small characoids schooling in midwater, in a band within 2 m of the bank, mostly had a dark spot at the base of the caudal fin and often a second dark humeral spot. Mixed species (a dozen or so belonging to several genera) were found in these bands, all heading into the current. Selection here, as for schooling species in lakes, is likely to be for uniformity in appearance, any individual differences making it easier for predators (of which there are many) to pick out an individual. This must help to explain the many cases of mimicry, fishes in different genera being so similar in appearance. The small bottom-pecking fishes here lacked the caudal blotches and were mainly uniformly sandy-grey making them very inconspicuous (compare fishes on sandy bottoms in lakes and reefs). Some species living solitary lives here carried very distinctive specific recognition marks; these included the small pufferfish *Colomesus ascellus* belonging to the family Tetraodontidae which con- tains many poisonous species.

The roles of sound production

The behaviour of fishes in their acoustic environment has been discussed by Schwarz (1985). Bioacoustic studies of reef organisms have related daily cycles of sound production to the activities of particular fish species; holocentrid squirrelfish, for example, have an evening chorus (Winn, Marshall & Hazlett, 1964). Sounds may also be produced during aggressive behaviour, or escape, or be noises made while feeding, for example when crunching coral. Playing back sounds to reef fish has demonstrated their significance for courtship in the male damselfish *Eupomacentrus partitus* (Myrberg, 1972).

Sound travels much faster in water than in air (see Schwarz 1985). The acoustico-lateralis system of pelagic clupeids is sensitive to sounds as well as other changes in water pressure (Blaxter & Hunter, 1982). Many freshwater fishes transmit and receive sound. Otophysan fishes have a Weberian apparatus connecting the swimbladder with the inner ear, which aids the detection of sound and possibly helps to locate the direction of the sound source. Many fishes produce specific frequencies which may be used for communication. This is still a little-explored field of research, though sounds of catfishes such as *Aspredo*, nicknamed the 'banjoman', and of brackish and marine fishes such as sciaenid croakers, pomadaysid grunts and toadfish have long been known. In South American freshwaters *Prochilodus* sound like an outboard engine when on their upriver spawning runs at the start of the rains, a time when underwater visibility is much reduced by the rapidly rising water swirling debris and mud into the river.

Serrasalmus of all sizes produce specific grunts when caught, perhaps with a warning function. The cichlid *Geophagus surinamensis* grates its pharyngeal teeth audibly when caught. Spawning *Curimata elegans* are said to sound like treefrogs and glandulocaudine males also croak. In aquaria clicking sounds made by *Pterophyllum* are connected with aggression in the male. The African *Synodontis* are known as 'squeakers' from the sounds they produce. There is here a huge field for research.

Electric signals

Many fishes are highly sensitive to electric currents (a ray can detect the minute electrical discharges in the resting plaice on which it feeds, thus helping it find its prey), but the production and detection of special electrical signals are confined to a few groups. The gymnotoids of South America and mormyroids of Africa provide striking examples of convergence in body shape, habits, ecology and types of electric impulse produced (see Fig. 11.2). Mostly insectivorous, they all appear to be nocturnal. They all emit and receive electric impulses (Lissmann, 1958, 1963; Kirschbaum, 1984). These electric fishes fall into two groups: (1) those emitting Type I 'tone' discharges of very regular sequences of continuously emitted monophasic impulses, which vary in frequency from species to species, and within narrower limits from individual to individual; and (2) those emitting Type II 'pulse' discharges which are less regular in frequency and with a much smaller pulse duration (0.2 s in *Petrocephalus*). Type I tone discharges occur in *Gymnarchus niloticus* in Africa (a large species which lives solitarily along the fringes of swamps and near river mouths), and in *Hypopomus* and *Eigenmannia* in South America; frequencies range from 50 to 1600 pulses s^{-1} according to the species and do not alter with the state of excitement of the fish, and the duration of the individual impulse is relatively long. Field observations in South America showed that Type I tone discharge fishes live in fast-flowing streams and open water, those with the highest discharges (up to 1600 s^{-1} in *Porotegus*) living in the fastest-flowing water (Lissmann, 1961). Type II pulse discharges are found in mormyrids in Africa and *Gymnotus carapo* and *Steatogenys elegans* in South America; basic discharges are lower, 1–6 pulses s^{-1} in resting mormyrids but increasing up to 130 s^{-1} in excited fish, and they can be inhibited for long periods. Type II discharges are common among sluggish and bottom-living forms, with basic discharges of 5–190 pulses s^{-1} in undisturbed fish but increasing when fishes are excited.

Some of these gymnotoids (*Sternopyus macrurus, Gymnotus carapo, Apteronotus albifrons, Hypopomus* species) live solitary lives. Others, such as *Eigenmannia* species, live in small schools. Both solitary and gregarious species show homing tendencies. *Eigenmannia* hide away by day, under a bridge for example, then fan out to feed at dusk with clockwork precision (Lissmann, 1961). Individuals of the solitary *A. albifrons* are belligerent in their defence of daytime refuges, crevices in rocks. Young *Gymnorhamphthichthys hypostomus*, transparent elongated fishes *ca* 12 cm long, live buried in the bottom sand of shallow streams by day; their seasonal biology has been studied by Schwassmann (1976).

Conditioned reflex experiments in tanks showed that *Gymnarchus niloticus* and *Gymnotus carapo* use their weak electric fields to sense their environment (Lissmann, 1958). Their prey is not affected by their discharges, unlike the prey of the 'electric eel' *Electrophorus electricus* which in addition to an involuntary slow regular discharge has a voluntary one of up to 550 V used to stun prey and deter enemies. These weak discharges provide a locating mechanism, an adaptation to nocturnal habits and life in turbid water, and they also have a social significance. In Rupununi pools gymnotoids could be seen backing into crevices at great speed; in tanks it has been shown that they also use the electric field to detect food, and could use it to detect predators. Gymnotoids are palatable to other fishes (they make good bait for hooks); they have good powers of regeneration, and specimens with regenerating tails are often caught.

More recent work has stressed the social significance of these discharges. In tanks *Gymnotus carapo* use weak electric displays agonistically to determine dominance hierarchies (Black-Cleworth, 1970). Sexual differences in signalling, analogous to bird song, have been demonstrated for *Sternopygus macrurus* (a Type I fish) in Rupununi streams (Hopkins, 1974*a*,*b*). The Moco Moco creek here had 11 species of gymnotoids all producing electric discharges (four of Type I tone and seven of Type II pulse discharge). Playback experiments in the field indicated that *Sternopygus* males are able to distinguish between their own species and sympatric species that have tone discharges on the basis of frequency alone. Males are also able to distinguish the sex of a conspecific fish, as they send out courtship signals to passing females but not to males. The species specificity of electric discharges in the many sympatric gymnotoids in the Rio Negro has been studied by Heilgenberg & Bastian (1980). Among the mormyrid species living in large schools over mud banks in the Zaire River, the electric signals may have social use in keeping the schools together and may help to explain how so many sympatric species manage to coexist (Poll, 1959*b*).

Chemical communication

The role of fish pheromones (chemical substances produced by the fish which have a signal function to other members of the species) has been reviewed by Bardach & Todd (1970). Though as yet little work has been done on pheromones in tropical fishes, it seems likely that these will prove to be of vital importance in the social behaviour of many species, especially nocturnal ones such as the numerous catfishes, many of which produce mucus with a characteristic strong smell. An immense and exciting field is here being opened up to investigation.

Cyprinids have long been known to produce 'alarm substances' when damaged which warn other members of the school of danger; fishes of the African families Kneriidae and Phractolaemidae also produce them. In glandulocaudine characids (such as *Corynopoma*) the caudal glands of the males appear to direct a pheromone towards the female during courtship (Nelson, 1964) (Fig. 11.2). In *Arapaima gigas* a pheromone exuded from the parent's head attracts the young to keep near the parent. The role of chemoreception in the care of young by cichlid parents has been investigated experimentally. Some cichlid parents respond to water in which their own young have been living and can distinguish this from water inhabited by conspecific broods of the same age or by other species. When reviewing the roles of vision and chemoreception in parental recognition of cichlid young, Myrberg (1966) concluded that chemoreception is important when the young are at the wriggling stage but that vision is more important when they become free-swimming. As species vary very much in their behaviour, generalizations are dangerous at the present state of knowledge.

In North American bullhead catfishes (*Ictalurus* species), in which the entire body is covered with taste buds, the sense of taste leads them to their food and the very keen sense of smell, through olfactory receptors in the nose, is used for social communication (Todd, 1971). The clustering of bullheads appears to produce a concentration of pheromones which inhibit aggression. When deprived of nose tissue they fight viciously, and the loser's growth is stunted. Through pheromones these catfish can tell not only the species and sex of another bullhead, but also whether it is a dominant or subordinate individual. In the Rupununi remarkable concentrations of catfishes (15 species with auchenipterid *Tatia* and *Trachycorystes* predominating) were found packed into crevices in submerged logs (Lowe-McConnell, 1964); such behaviour calls for studies on pheromones in these catfishes.

12

Trophic interrelationships

Trophic interrelationships and community structure

Although food webs in tropical waters are often very complex, they may be based on relatively few sources, for example the aufwuchs which supports numerous species on the rocky shores of Lake Malawi (Fig. 4.2), or the benthic algal flora that supports herbivorous fishes on marine reefs. In any food chain there are rarely more than four or five links; long chains are energetically expensive as a large proportion of the potential energy is lost at each successive stage. In freshwaters alternate chains run: (1) from bottom detritus, through microorganisms, to detritus-feeding invertebrates or fish, to several levels of piscivore; or (2) in the pelagic zone from phytoplankton to zooplankton, to zooplankton-feeders, then to one or more levels of piscivore. In river systems the detrital chain is more important, based largely on allochthonous materials in the headwater streams while in lower reaches detritus comes mainly from decomposition of aquatic macrophytes (see the 'river continuum concept' of Vannote *et al.*, 1980; but see also Rzoska, 1978). As a lake forms, the pelagic plankton chain becomes increasingly important, as seen after the filling of man-made lakes, and also in lakes such as Chad, populated with riverine fishes such as *Alestes baremose* which changed its riverine diet of insects and seeds to exploit zooplankton in the lake.

In these tropical communities many kinds of fishes have specialized on foods at the lowest trophic levels: organic mud with its microorganisms, aufwuchs, phytoplankton, and even forest debris, are all utilized by fishes able or specialized to do so, instead of these sources being converted into fish foods by invertebrates, as is more often the case in temperate waters. Tropical seas, too, have numerous fishes specializing in algae-feeding: acanthurids, siganids and pomacentrids. This has important fishery implications as food fishes can be cropped low in the trophic chain, avoiding losses in production inherent in moving from one trophic level to the next. There are also specializations at all trophic levels, even such extremes as feeding on the fins or scales of other fishes. Despite these specializations, however, most fishes show a considerable amount of plasticity in the foods used.

Limnological conditions in tropical waters differ from those in temperate waters in some fundamental ways, as yet little explored, as evidenced by the importance of detritivorous fishes, which can dominate ecosystem biomass in tropical waters, and the relatively large biomass (size and numbers) of predatory species. Comparing diets of fishes in the freshwaters of the three main land masses shows: (1) the importance of allochthonous vegetable material as direct food for many fish species; (2) the important role that insects have as fish food, both aquatic stages of Odonata, Ephemeroptera and Diptera and terrestrial insects stranded on the water surface; (3) the importance of mud and detritus as food for fishes which have become specialized to strain large quantities of these for their contained microorganisms; and (4) the large numbers of individuals and kinds of piscivorous fishes present in the communities.

The partitioning of foods in the African Great Lakes and on coral reefs has been discussed in earlier chapters. For riverine fishes, the case histories considered here have shown clearly that there is a linear succession of dominant food sources in streams and rivers (Fig. 12.1).

Fig. 12.1. The linear succession of dominant food sources for fishes in a tropical river system.

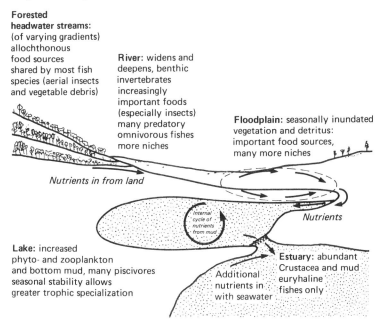

Forested headwater streams: (of varying gradients) allochthonous food sources shared by most fish species (aerial insects and vegetable debris)

River: widens and deepens, benthic invertebrates increasingly important foods (especially insects) many predatory omnivorous fishes more niches

Floodplain: seasonally inundated vegetation and detritus: important food sources, many more niches

Nutrients in from land

Internal cycle of nutrients from mud

Nutrients

Lake: increased phyto- and zooplankton and bottom mud, many piscivores seasonal stability allows greater trophic specialization

Additional nutrients in with seawater

Estuary: abundant Crustacea and mud euryhaline fishes only

Fishes in shaded headwater streams depend primarily on allochthonous foods (terrestrial insects and vegetable debris); fishes here are mostly euryphagic, taking whatever food drops into the water and apparently rarely using the feeding specializations that they may possess. Where the streams are open to the sun, grazing fishes such as loricariid catfishes specialize in grazing algae. As the stream enlarges and deepens, predatory omnivores using benthic invertebrates become more important in the fauna, and laterally flooded areas with their adjacent pools provide more niches. Feeding specializations appear to become increasingly important, though their importance may fluctuate seasonally. In the lower reaches, detritus and soft mud accumulates and supports species specialized to use these foods (*Prochilodus* and curimatids in South America, *Labeo*, *Citharinus* and some tilapia in Africa and certain cyprinids in Asia (see Bowen, 1983)). The extensive growth of macrophytes in the large swampy areas of tropical rivers leads to large accumulations of detrital mud and very large populations of mud-feeding fishes can be supported (e.g. *Prochilodus* in the La Plata, p. 158), which may, however, have to move long distances to find suitable places in which to spawn.

The headwater streams carry fewer species than the lower stretches. Are these mainly endemic species, or ubiquitous riverine species, or both? The mere dozen species in forested headwater streams in the Mato Grosso (p. 145) were nearly all ubiquitous species or their young stages, hardy fishes which could withstand poorly oxygenated acid water and relatively low temperatures and which were able to utilize many kinds of allochthonous and autochthonous foods, such as aerial and aquatic insects, vegetable debris and small fishes. Knöppel (1970) recorded a similar situation in other Amazonian streams. In Zaire, Gosse (1963) found many small species and the young of widely distributed large species in headwater streams. In these cases and in the Gombak River in Malaysia (Bishop, 1973), the numbers of species increased very much in the lower stretches of the river where there were many more biotopes. Some species may have evolved in the lower stretches, others probably evolved in discrete tributaries, then dropped downstream and became more widely distributed. Species would thus accumulate in the lower stretches.

Resource (food and habitat) partitioning among tropical stream fishes has recently received considerable attention (Gorman & Karr, 1978; Angermeier & Karr, 1983; Power, 1983, 1984*a*,*b*; Ward & Stanford, 1983; Moyle & Senanayake, 1984; Schut, de Silva & Kortmulder, 1984; Watson & Balon, 1984). In their studies of fish community structure in relation to food availability in nine (1–6 m wide) Panamanian streams, Angermeier & Karr (1983) assigned the fishes to seven feeding guilds (algivores, aquatic insectivores, general insectivores, piscivores, scale-eaters, terrestrial herbivores, omnivores). Species richness of feeding guilds increased with stream size and canopy openness. They concluded that trophic diversity may be related to the reliability of available food resources, but terrestrial predators appeared to be more important than food availability in determining distributions of fishes among habitats. Power (1983) found that the loricariids in Panamanian streams tracked variations in algal productivity very closely, moving from pool to pool, which led to higher densities of them in sunny pools; this affected fish growth rates, which were then similar in sunny and shaded pools. But within the pools avoidance of avian and terrestrial predators outweighed foraging considerations, the larger fishes avoiding water less than 20 cm deep. Overlaps in foods used by loricariid species were greatest in the dry season when food was most limiting; seasonal variations in growth rates reflected these constraints. These Panamanian streams provided a dynamic environment, as habitats could be created, destroyed or drastically altered in hours.

In Sri Lankan streams Schut, de Silva & Kortmulder (1984) described the typical habitats of eight *Barbus* species and concluded that due to differential food choice and/or spatial segregation direct competition for food appeared negligible, though it was impossible to say whether these differences were themselves the results of interspecific competition (either through present-day character displacement, or through divergent evolution resulting from character displacement in the past). Moyle & Senanayake (1984) made a thorough study of a fish assemblage typical of small rainforest streams in Sri Lanka to see if the fishes were characterized by a high degree of specialization (expected of 'equilibrium' communities), or if they were relatively unspecialized (expected of more stochastic, non-equilibrium, communities). The 20 most abundant fish species examined included 11 cyprinids, two gobies, two channids and one each of five other families. Morphologically the fishes showed a high degree of specialization, especially in structures related to feeding. Although several species were habitat generalists, most species occurred in distinct habitats. Within habitats, microhabitat overlap among co-occurring species was low, particularly in relation to position in the water column (depth distribution). Fishes not segregated by habitat or microhabitat tended to show low dietary overlaps. Overall, the fish assemblage had the characteristics expected of an equilibrium (deterministic) assemblage, but long-term studies would be needed to test its persistence and resilience to perturbation. Specialization in feeding habits was more pronounced here than reported for fishes in other small rainforest streams (e.g. in Amazonia), partly because of the greater reliance of the Sri Lankan fishes on autochthonous foods. These streams were in secondary forest, not as heavily shaded as the Amazon streams, and algae were important foods. Furthermore, this study was made at the low-water time of year (January to March) and climatic conditions in the streams were relatively constant (contrast the rapid changes in conditions in the Panamanian streams noted by Power). In the Borneo stream communities studied by Watson & Balon (1984), community structure appeared to be determined by the range of available resources and associated specializations of the existing species, rather than by the number of species present. But in forest creeks and pools in the Ivindo Basin (Gabon, Africa), where environmental conditions were very uniform, the spatial distribution of eight cyprinodont species was not dependent on preferences by participating species but on characteristics of the habitats, the fishes living in multispecies size groups. Here sudden changes in habitat, caused by random factors such as local storms or disturbance by elephants, completely changed the species composition of the community at a given locality within a very short period (Brosset, 1982). These 'random factors' caused an interspecific lottery for space, as reported for some coral reef communities. This case thus appears to fit in with the 'intermediate disturbance' hypothesis (Connell (1978), discussed for lotic systems by Ward & Stanford (1983)). The species composition of some floodplain pool fish assemblages in South America (Lowe-McConnell, 1964; Bonetto et al., 1969) also appears to be largely stochastic. Some Zaire River communities may also be stochastically determined (see p. 48).

Plasticity in fish diets

Most fishes show considerable plasticity in their diets. Predators change their preferred prey as they grow and change biotopes, or with whatever foods are available seasonally, or with lunar cycles, or from year to year, or by active selection of preferred foods according to individual choice. Variations in diet from year to year were very clearly shown by the flatfish *Cynoglossus senegalensis* in the Sierra Leone estuary (Fig. 9.3) where the main food one year was the brachiopod (lampshell) *Lingula*, an organism that hardly appeared in the diet among the polychaetes and shrimps the preceding and subsequent years. In freshwaters plasticity is particularly marked in riverine fishes, especially in the more seasonal rivers where many of them feed on benthic invertebrates and fish. In lakes it is more marked in the fishes of riverine origin (as in the non-cichlids in Lake Victoria discussed on p. 80). The cichlids of the African Great Lakes appear to be more specialized in their feeding habits though, as recent work shows, even fishes with extreme morphological specializations in their feeding apparatus will use a variety of foods as occasion offers. Specialization could otherwise lead to stenophagy and the inability to change feeding habits should environmental conditions change; for this reason the African Great Lakes had been regarded as 'evolutionary traps' (Briggs, 1966).

Widely distributed riverine species, such as *Schilbe mystus*, *Aucheno-glanis occidentalis* and *Heterobranchus longifilis*, tend to be bottom insectivores that can change easily to other diets. Euryphagy, the ability to use many different foods effectively, is an important characteristic of ubiquitous species, and omnivores have a better chance than specialists of becoming widely distributed. The food varies with what is available in each habitat and also depends on interactions with other species present. De Kimpe (1964), who examined data for widely distributed species that occurred in Mweru-Luapula and elsewhere in Africa, concluded that (1) a large number of these freshwater fish do not have a strict food regime; (2) a species may use different foods in different places (*Alestes macro-phthalmus*, for example, eats plants in the Malagarasi swamps but is mainly carnivorous in Lake Mweru); and (3) there are individual differ-ences within a species, an individual selecting a particular food at one time. The food regime can vary with season, abundance of food organisms, activity of fish, change of biotope and other fish species present.

Ivlev (1961) stressed the choice of food by a fish, comparing what is eaten by the fish with what is actually present in the water column. This relation called by him the electivity (E), ranges from $+1$ to -1, an electivity of zero indicating complete lack of selection by the fish. (Selection refers to the fish's choice given many objects equally available.) The electivity of planktonic organisms and other items by '*Haplochromis*' *squamipinnis* in Lake George was studied by Moriarty *et al.* (1973). Factors influencing the size of food niches in *Clarias senegalensis* in Ghana were discussed by Thomas (1966) who stressed that intraspecific compe-tition will tend to enlarge the food niche, while interspecific competition will restrict it.

Numerous studies have shown seasonal differences in diet (for example, Daget's studies of *Alestes*, p. 260). In Lake Victoria (p. 80) diets of several species changed with the phase of the moon, which affected the emergence of their insect prey. Numerous fishes take zooplankton when young, whether they turn to a herbivorous diet in adult life (as in tilapias and some haplochromine cichlids), or to a piscivorous one (as in *Cichla ocellaris* (Zaret, 1980), *Lates* and many others). In Lake Bangweulu in Zambia, Bowmaker (1969) found that *Alestes macrophthalmus* fell into four trophic groups entering into different trophic relationships with other species in the lake, juveniles taking zooplankton, adults up to 14 cm (70% of them males) living inshore eating insects, and larger adults (mainly females) preying on '*Engraulicypris*' and other small cyprinids offshore. These and numerous other cases all stress that it is impossible to draw conclusions about overlaps in foods used from small samples. Many marine fishes also show the same kinds of variations and individual differences in their diets.

The community and evolutionary effects of 'prey switching' are considered later (p. 289). Piscivorous fishes living in the pelagic zone may specialize (as van Oijen found among Lake Victoria's cichlids, p. 83) but most predatory fishes take a varied selection of prey; the size of prey and the way it moves may be more important than the species (tuna, for example, take neuston including cephalopods and other invertebrates as well as fish). Data on foods used by *Hydrocynus vittatus* from various waters, by predatory fishes on the Kafue flats and by *Lates* in different lakes, all show this clearly. In Lake Chad, Hopson (1972) concluded that it was impossible to assign *Lates* to a particular niche because of the marked seasonal changes in size and types of prey used which result in fluctuations of the trophic level of food intake. During summer months when prawns were the dominant food of the offshore population the food chain was very short (bottom algae/detritus → Macrobrachium → *Lates*) compared with occasions when there were three or even four links between *Lates* and primary producers (phytoplankton- → *Entomostraca* → small characins → *Hydrocynus* → adult *Lates*). Thus, the relationships between *Lates* and smaller predators such as *Hydrocynus* and *Eutropius* alternated between those of competition and of predator–prey. This change complicated attempts to quantify transfers of energy from one trophic level to the next in this lake.

Transfer efficiencies

Production estimates are considered in Chapter 14, but what proportion of the solar energy can be cropped as fish from tropical waters? In the sea, Bardach (1959) estimated the energy production of coral fishes on a Bermuda reef as 9.24×10^5 kJ ha^{-1} yr^{-1}, considered by him to represent 0.0014% of the annual solar energy there available (estimated as 6.72×10^{10} kJ ha^{-1} yr^{-1}). In freshwaters, various attempts have been made to relate fish production or yield to primary production (see papers in Kajak & Hillbricht-Ilkowska, 1972). Gross primary production is fairly easy to measure, and if this could be used as a reliable indicator of likely fish production or yield this could be a useful tool in fisheries management. A major problem is that the net primary production, the amount of carbon available to be passed through the food webs to the fish, is not a constant proportion of the gross primary production (since it depends on respiration needs of both autotrophs and bacteria and other small heterotrophs in the plankton or other samples). Discrepancies in estimated transfer values stem from such problems.

In five Malaysian experimental ponds, conversion of incident solar energy to primary production ranged from 3.98% to 7.04% (compared with a terrestrial agricultural average of 7%), and conversion efficiencies from net primary production to fish production (of tilapia and some vegetarian carps) ranged from 1.02% to 1.79% (Prowze, 1972). High-protein pelleted fish foods gave a conversion efficiency of 6%. Data for Indian waters (some six impoundments, seven ponds and lakes) collated by Ganapati & Sreenivasan (1972) and Sreenivasan (1972) indicated that the photosynthetic efficiency depended on the biota and was higher in confined waters than in reservoirs with water renewal (e.g. 4% in a pond compared with 1.4% in Amarvarthi Reservoir). The highest conversion value of photosynthetic efficiency *versus* fish yield was 0.774% in Chetpat Swamp (which had a high fish yield of 2200 kg ha^{-1} yr^{-1}), compared with 0.136% in Amaravathi Reservoir (where fish yield was 160 kg ha^{-1} yr^{-1}).

In Lake George (Uganda) the ratio of production of algae-feeding tilapia to net primary production was estimated to be less than 1% (Burgis & Dunn, 1978). But in this lake the tilapia production represented only part of the total production of herbivorous fish, so fish production (*ca* 1.1%) here was considered to be comparable to the 1.02–1.79% obtained by Prowze (1972) from the Malaysian experimental fish ponds.

In the equatorial Lake George, soupy with the blue-green algae on which the tilapia feed, probably well over 90% of the energy fixed by the planktonic algae is dissipated in respiration by the plankton community. Thus, though the gross primary production is extremely high, the net primary production is less than 10% of this (probably nearer 8%). In this lake the system as a whole was shown to be inefficient in the transfer of solar energy from primary to other trophic levels. Nevertheless, enough is passed on to support a fishery of some 3500 tonnes yr^{-1} (Burgis & Dunn, 1978), equivalent to *ca* 138 kg ha^{-1} yr^{-1}, much the same level as in unfertilized fish ponds.

Burgis & Dunn (1978) compared production at different trophic levels in the equatorial Lake George and the temperate Loch Leven in Scotland (Table 12.1). In Loch Leven secondary production was about 8% of net primary production, but in Lake George total secondary production (herbivores fish and invertebrates) was estimated to be only 3% of primary production, and it would appear that much of this net primary production is not being converted into secondary production. Possible reasons for this were discussed by Burgis & Dunn who suggested that the size of the tilapia population may be limited by lack of suitable nursery or breeding grounds, or that juvenile tilapia may suffer heavy losses from predation or be limited by the supply of the zooplankton food used by the young stages. Although the yield of fish, as a proportion of net primary production, was about ten times higher in Lake George than in Loch Leven, Burgis & Dunn questioned why it was not even higher in this equatorial lake with the added advantage of harvesting a herbivore. In both lakes the yield of the exploited species was about 25% of the total fish biomass. Assuming that 46% of the solar radiation is available for photosynthesis the fish yield was equivalent to 0.0014% of solar radiation from Lake George and 0.00028% from Loch Leven. (Agreeing well with Henderson & Welcomme's (1974) conclusion that fish yield from temperate lakes tends to be about one tenth that from comparable waters in tropical Africa.) Bardach (1959) had concluded that Bermuda coral fish production (equivalent to 92.4 kJ m^{-2} yr^{-1}) represented 0.0014% of solar energy there.

When collating data from International Biological Programme studies at different latitudes, Morgan *et al.* (in Le Cren & Lowe-McConnell,

Table 12.1. *Comparative estimates of production in Lake George (Uganda) and Loch Leven (Scotland)*

	Lake George		Loch Leven	
	kJ m^{-2} yr^{-1}	%	kJ m^{-2} yr^{-1}	%
Net primary production	23 200	100	20 000	100
Secondary production	650	3	1 614	8
Tertiary production	150	0.65	150	0.75
Yield	50	0.21	5	0.025

1980) reached a 'tentative conclusion' that fish production varies between 0.1% and 1.6% of the gross primary production in lakes and reservoirs and that it can reach 4% in fish ponds. The fish yield is likely to be not more than half of this (even an order of magnitude lower in some cases).

The use of primary production estimates to predict fish yields in tropical lakes has been explored by Melak (1976) who developed equations describing the relation between commercial fish yields and gross photosynthesis from data for eight African and 15 Indian lakes. The percentage conversions, based on carbon of net photosynthesis to fish yield, in the African lakes ranged from 0.1% in Lake Victoria offshore (0.35% in gulfs) to 0.72% in Lake Baringo (other values were: Chad 0.18%; Albert 0.21%; Tanganyika, north end, 0.41%; George 0.48%; and for man-made Volta Lake 0.21% and Kainji 0.48%). Comparable rates came from data for the Indian lakes. Estimation of net (PN) and gross primary production (PG) was, however, a major problem. Melak assumed PN = 0.25 PG in most cases but in Lake George, where net production was only *ca* 8% of gross photosynthesis, taking this figure raised the estimated percentage carbon transfer from 0.48% to 1.5%. The data suggested that percentage conversion increased as primary production increased, but this increase was probably partly due to increased fishing efficiency and a greater proportion of herbivores in the catch.

For pelagic fishes in Lake Malawi and Tanganyika, transfer efficiencies from gross primary production to pelagic fish yield have been estimated as 0.15% for Lake Malawi, compared with estimates of 0.25% and 0.12% for Tanganyika (FAO, 1982). Lake Tanganyika is unusual in having a high transfer efficiency (0.45% averaged over the lake) with very low phytoplankton stocks; rates of growth and recycling are extremely high, giving the pelagic waters a deceptive appearance of oligotrophy (Coulter, 1981). The Tanganyika fishery is based mainly on pelagic species but Hecky *et al.* (1981) thought Coulter's (1981) estimates of potential clupeid yields of 350 kg ha^{-1} yr^{-1} from this lake to be too high, even though production of heterotrophic bacteria equals or exceeds primary production by algae in this lake and they had suggested that the immense volume of anoxic deep water might provide the source of energy allowing such high rates of bacterial production. Clearly there is a great deal still to be found out about the limnological cycles in these tropical waters.

For Amazon fishes, Bayley (1982) estimated production of fish in the smaller size groups (15–60 mm) to be *ca* 1.6% of previously determined estimates of primary production in the area. A large quantity of piscivores is a feature of Amazonian waters (as for many tropical waters); piscivores made up 35% of the total biomass and 25% of the production in Bayley's study area. Bayley looked at transfer efficiencies between piscivores and their prey for two size groups of fish (predator sizes 60–239 mm and 240–960 mm, prey sizes 15–59 mm and 60–240 mm). These gave production ratios of 37% and 39%, respectively, suggesting an ecological efficiency of 30–40% for piscivore–prey interactions, i.e. that the system is very efficient for energy transfers at these levels. Amazonian fishes take relatively large prey; high efficiencies have been noted elsewhere when fishes are used as prey. Large differences did however occur from year to year (from 27% to 60% production ratios from 1977 to 1978 for the larger size groups), fluctuations which were probably related to hydrological variations.

These values for tropical waters can be compared with carbon transfer efficiencies from primary production to fish of 0.002–0.04% for the Laurentian Great Lakes in North America and 0.02% for Lake Baikal (USSR) (Oglesby, 1982). For marine fish, values for the North Sea are quoted as 1.2% by Melak (1976), as 0.6% for pelagic fish and 0.8% for the total fishery by Hecky *et al.* (1981, based on Gulland, (1970)), and as about 0.6% for the pelagic anchoveta in the Peruvian upwelling area.

The ratios of fish yields to primary production are much lower. Data collated by Marten & Polovina (1982; based on carbon yield of fish, assumed one tenth of wet weight, divided by 'primary productivity' in carbon units), gave calculated means ranging from 0.000045 for open oceans, 0.0004 for coral reefs, 0.00095 for continental shelves, 0.0028 for lagoons and estuaries, to 0.0081 for coastal upwellings, and in freshwaters from 0.00063–0.0008 for reservoirs and lakes, to 0.0032 in ponds and 0.0071 for rivers.

Predation and antipredation devices

The suggested roles of predation in the maintenance of community diversity and on speciation are discussed in later chapters. Piscivorous fishes such as *Hydrocynus*, *Lates*, *Cichla* and *Boulengerochromis* generally take prey of less than about one third of their own length. Prey species are relatively safe from predation once they are above the critical size, so predation probably provides a powerful selection pressure for rapid growth in prey species.

Antipredation defences include structures such as armour and stout spines, poisons and electric shocks, combined with behavioural devices such as hiding in crevices and mimicry. Many reef fish have venomous spines or poisonous flesh (see Halstead, 1953; Melzak, 1981). In freshwater many catfishes have locking mechanisms to keep their fin spines erect, and crevice-dwellers use their spines to jam themselves into crevices. The triangle of erect spines also increases the effective size of the fish, making it more difficult to swallow and only available to the larger piscivores (Fig. 12.2). Hopson (1972) showed that *Synodontis* found in *Lates* stomachs in Lake Chad never exceeded 18% of the predator's length, compared with 30–40% of its length in the relatively unarmed schilbeids also taken as prey. A crocodile (*Crocodylus niloticus*) has been found dead with a *Synodontis* jammed across its throat. Nevertheless, the cormorant *Phalacrocorax africanus* does manage to swallow some *Synodontis* species (*S. polystigma* from lakes Bangweulu and Mweru and *S. macrostigma* from the Kafue River); *S. macrostigma* was also eaten by *Hepsetus* and *Schilbe* on the Kafue Flats. In some South American catfish the spines appear to be venomous; freshwater stingrays (*Potamotrygon*) have long venomous spines. Torpedo rays (*Narcine*) can give a powerful electric shock. In the gymnotoid electric eel *Electrophorus electricus* in South America and the electric catfish *Malapterurus electricus* in Africa the electric organs have developed a voluntary discharge which is a formidable weapon, as well as being used to stun prey. Freshwater pufferfishes (*Tetraodon*) found in rivers of all three tropical areas inflate themselves when attacked: these boldly striped fish may also be unpalatable, for many marine tetraodontoids are poisonous. We know very little about palatability of prey fish to predators. The dermal armour carried by certain catfishes in all three main tropical areas (p. 22) would appear to deter predators in addition to any other uses; the flesh of doradids and *Sisor* is palatable enough to be eaten by man. In the sea, tetraodontoid pufferfish inflate themselves to increase their size and to wedge themselves into crevices. Concealment in crevices or by diving into the bottom sand (e.g. *Hemipteronotus* razor wrasses) or protective resemblance to some natural object all appear important for survival. Only fishes which are protected in some way or which live in schools (as even the piscivorous barracudas *Sphyraena* do when young) are able to survive in open water.

Tropical fishes living in clear water present many fine examples of protective resemblances to natural objects and of mimicry of one fish by another, both probably antipredator devices (see Randall & Randall, 1960). In New World streams the catfishes *Farlowella* and *Agmus* resemble dead twigs, and the nandids *Monocirrhus* and *Polycentropsis* resemble dead leaves which may help to protect them, though this disguise is used by them to snap up unsuspecting prey fish. The nocturnal behaviour of cichlids which lie completely motionless amongst tree litter or against the stream bank must help them to avoid revealing their presence to nocturnal predators such as large catfish, which probably detect prey chemically or by water vibrations. At least one cichlid has been observed to produce a mucous sleeping cocoon (Mary Bailey, personal communication) as do some marine scarids (p. 186). In brackish lagoons a juvenile tripletail, *Lobotes surinamensis*, resembles a dead mangrove leaf, a resemblance enhanced by its behaviour as it drifts on its side to the bottom.

Among South American characoids associations of species and mimetic resemblances often occur between members of closely related genera. The significance of this type of mimicry is not clear but it has characteris-

Fig. 12.2. Some antipredation devices in freshwater fishes. (A) The African catfish *Synodontis* in which the triangle of locked erect spines increases effective size of the fish. (B) Protective resemblances to gnarled wood in the South American catfish *Agmus lyriformis* and (C) the twig-like *Farlowella*. (D) South American *Centromochlus* catfishes packed into crevices in submerged logs by day.

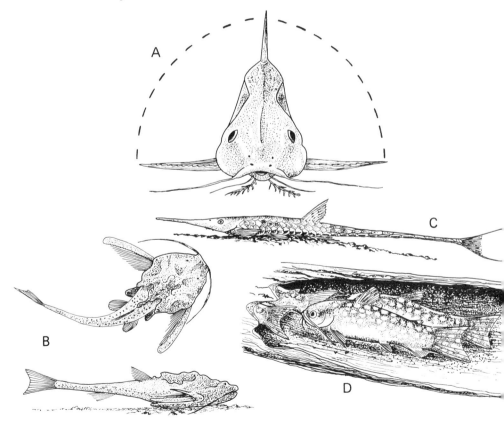

tics of Müllerian rather than Batesian mimicry (Gery, 1969). Possibly the way of life has led to convergence in appearance in these midwater-dwellers which would gain protection from piscivores by all looking alike (see also Dafni & Diamant (1984) on 'school mimicry' in marine fishes). In Lake Malawi, mimicry is used to obtain food by *Corematodus* species which scrape scales from the caudal fins of other cichlids; only when *C. shiranus* dons its breeding colours does it cease to resemble its tilapia 'host'. In this lake, *Haplochromis livingstoni* plays dead to ambush its prey (McKaye, 1981).

Lakes Tanganyika and Victoria also have cichlids specializing on a diet of fish scales; convergence in these cichlids was discussed by Fryer & Iles (1972) and scale-eating by Liem & Stewart (1976). Data on scale-eating in South American fishes, generally characoids, have been collated by Sazima (1983); about four genera and 10 species of scale-eaters are known from the Amazon (Roberts, 1972). Teeth are specialized for removing scales and these fish attack fishes larger than themselves. In Guyana many fish had caudal fin damage attributed to bites from *Serrasalmus*. In Africa the Zaire River has a group of fin-biting characids (*Ichthyborus*) (Matthes, 1964). In Lake Malawi the cichlid *Docimodus* is a fin-biter, one species taking mainly catfish fins, another species mainly cyprinid scales and catfish skin (Eccles & Lewis, 1976; Ribbink, 1984*a*); *Docimodus evelynae* also acts as a cleaner. The victims of these scale- and fin-eaters do not appear to be killed by the attacks. In the Indo-Pacific many teraponids are also scale-eaters (Vari, 1978).

The various forms of parental care (guarding, oral incubation, live-bearing) appear to act as antipredation devices. These appear commoner in freshwater than in the sea and have evidently arisen independently in many fish groups, as mentioned above (p. 252). Paedophagy, feeding on eggs and young extracted from the mouths of brooded females (see p. 82), is a specialization found only, as far as we know at present, in African lake cichlids. Predators have, however, been observed to affect reproductive behaviour and community structure in both New World substratum-spawning and Old World mouth-brooding cichlids (McKaye, 1984); in Lake Malawi multispecific spawning arenas were considered to be helpful in combating predation by six species of specialized egg-predators.

Escaping predation: schooling and the role of cover

Underwater observations in both freshwaters and the sea have shown the very delicate balance between predator and prey, both represented by many fish species. In general, prey fish cannot exist in open water unless they live in schools (shoals). Their protection then appears to derive from the difficulty that the piscivore has in focusing on any one target as it dashes at the school (see Partridge, 1982). There is thus strong selection pressure for these prey fish to look alike, for uniformity. In open waters the surest way to get eaten is to look slightly different; this is in contrast with the situation in a featured environment, where predation may lead to aspect diversity in cryptic species. In open water this leads to schools of similar-sized fishes, and of mixed species resembling one another (as among *utaka* and small tilapias in Lake Malawi) and also to look-alike characoid species of different genera swimming together in South American rivers. Such fishes may only develop specific colours in preparation for spawning, when species recognition is important to achieve success.

The openwater fishes feed visually. Prey fish may gain some protection by descending to more dimly lit zones during the day. This may have been one of the main selection pressures leading to the evolution of diurnal migrations in these pelagic fishes. Marine fish studies have shown how piscivores have special feeding times, often at dawn and dusk, times when the prey is in greatest danger (Hobson, 1968, 1972). The foods available to the openwater prey fishes – phytoplankton or zooplankton – present rather uniform types of food: pelagic fishes have become specialized to utilize one or other of these (for example, by developing long sieve-like gill rakers) or to feed on the small fishes or other planktivorous creatures.

Quite different types of communities live where there is cover. Substrate complexity permits diversity and appears to be a key factor in allowing it. This is true both in freshwater, for example among rock-dwelling fishes in the African Great Lakes, and in the sea, where the innumerable crevices in coral reefs have allowed this most diverse of aquatic communities to evolve. Full advantage of cover can only be taken by fishes that have reached a certain stage of morphological evolution; in many acanthopterygians the pelvic fins have moved forward, making the fish manoeuvrable, and their dorsal fin spines, important fin strengtheners contributing to manoeuvrability, also deter predators. Protection is important for the prey fish when they emerge from their crevices to feed, for speed has generally been lost at the expense of heavy spines. Their protrusible jaws enable these fish to exploit the very diverse foods offered by structured biotopes, for the law that structurally complex cover permits diversity holds for the invertebrate prey too.

Substrate association often leads to territoriality in these fishes. Cover may be provided in many ways: by crevices in rocks or tree litter, in beds of aquatic plants, or by gasteropod shells – used by gobies in brackish water and small cichlids in the African Great Lakes (*Lamprologus* in Lake Tanganyika, *Pseudotropheus* in Lake Malawi). Where plant growth is seasonal, the great losses of young fish to predators when the plant cover dies down are very striking, for example to *Hydrocynus* in African rivers and to *Hoplias* in South American savanna pools. In Lake Turkana, when unusually high rainfall in 1974 increased shoreline vegetation, providing cover for young fish, this was followed by a dramatic increase in tilapia (*Oreochromis niloticus*) production (Bayley, 1977). In the sea the construction of artificial reefs has been found to increase the local standing stock of fish (Randall, 1963).

Tree debris acts as an important substitute in providing hiding places for light-shy nocturnal fishes in streams without rocks, and where bottom water is deoxygenated. Marine fishes hiding in coral or rock reefs by day move out to feed on invertebrates over sandy bottoms under cover of darkness (Hobson, 1965, 1972); these same invertebrates are hidden by day through burrowing in the bottom sand. The hypostomine catfishes in South American streams appear unusual in grazing algae by night; on marine reefs most of the algae-grazers appear to be diurnal feeders. This difference might be related to the different selection pressures: Power (1984*b*) found that aerial predators (mainly birds) active by day influenced the distribution and behaviour of Panamanian stream loricariids; on marine reefs predation from fishes, many nocturnal, is probably greater than from diurnal aerial predators.

Evolution in pelagic/openwater and in littoral/benthic communities thus appears to lead to different results. Pelagic fishes become uniform in appearance, specialized to use a particular food, and their numbers may be limited by that food source as well as by predation pressure. Fishes living with cover become more diverse in appearance, having undergone adaptive radiations to use various kinds of food and fill different niches in a diverse biotope. Many of these fishes appear to be predator-limited rather than food-limited. Predation pressures appear to have contrary effects in these two types of community. In pelagic and openwater communities they appear to lead to uniformity, whereas predators patrolling areas with cover, off rocky shores and along reefs, may help to split up and isolate the populations of fishes, so promoting speciation and contributing to diversity in littoral/benthic communities. The same would appear to hold in rivers and streams, though the distinctions between pelagic fishes and those using cover are not as marked as in lakes. The evolutionary effects of predation pressures are discussed further in Chapter 13.

13

Diversity: its maintenance and evolution

High diversity at low latitudes is a characteristic of both plant and animal communities, and fishes are no exception to this generalization, for among both freshwater and marine tropical fishes greater diversity is shown both within taxa and within communities than at high latitudes. This chapter examines first how species are packed into communities and coexist, and then how these very diverse communities have evolved.

Diversity maintenance

Previous chapters have indicated that diversity is greatest in relatively aseasonal communities, both in freshwaters and in the sea (Table 11.1). Coral reefs carry the most diverse of any fish communities; upwelling areas in the sea have far less diverse faunas than those on reefs. Diversity in rivers becomes less: (1) as the latitude increases and seasonal fluctuations in water level become more marked; (2) with increases in altitude, for example in Andean and Kenyan streams, where this is probably a temperature effect; and (3) towards the headwaters of streams where physicochemical factors, obstructions causing water falls, high speeds of flow, and the size and conditions of dry-season refuges, may be more limiting than food resources.

Fauna lists are very large for many tropical water bodies, but how many species actually share biotopes? The long species lists for a river system could result from each section and the innumerable tributary streams each having its own fauna. What is the distribution of species within the river or reef system? As touched on for Amazon fishes (p. 148) we are here concerned with 'alpha diversity', the number of species within a community (its richness) resulting from diversification in a variety of ecological niches, in contrast with 'beta diversity' of habitat-orientated species added along environmental gradients. For coral reef communities Sale (1980b) concluded that data were still insufficient to answer conclusively whether alpha, beta or both components of diversity were greater in regions rich in reef fishes.

In freshwater biotopes, the use of chemofishing, or some such method which takes all the fishes present, has shown that high numbers of species occur together in many freshwater biotopes. But in some cases these are merely aggregations of fishes brought together by waters contracting in the dry season, when fishes feed little, rather than true communities of feeding fishes. In Guyana, rotenone samples produced 69 fish species from a small brook on an island in the Essequibo River, 54 species from a 30 m diameter pool on the same river and 29 species from under rocks at Warraputa Cataract on the Lower Potaro River (Eigenmann, 1912). A Venezuelan lagoon only 1225 m² in area contained 25 fish species (Mago, 1970). In Argentina the Don Felipe Lagoon (p. 153) had 52 species (Ringuelet et al., 1967). In Amazonas (p. 131) Lac Redondo had 47 species, only six in common with the Preto da Eva which had 49 species, even though these two waters were part of the same river system and less than 100 km apart (Marlier, 1968). In Lake Tanganyika, a 20 m × 20 m quadrat contained over 7000 fishes of 38 species (Hori, Yamaoka & Takamura, 1983). A 300 m stretch of rocky shore on Lake Malawi (p. 86) included over 30 species (25 of them cichlids) and an adjacent sandy shore 23 species (15 of them cichlids, mostly different from the rocky-shore species) (Fryer, 1959). Chemofishing in Lake Chad produced 40–50 species in 0.5 ha (Loubens, 1969). On Lake Kariba chemofishing in a cove produced 28 species (Balon, 1974) even though this was a new lake. The crater lake Barombi Mbo (Cameroons), only 22.5 km² in diameter, had 17 species, though these did form more or less distinct inshore and offshore communities (Trewavas, Green & Corbet, 1972). In Nigeria dry season pools in the Sokoto River fished by Holden (1963) had up to 27 species in a pool. A headwater stream of the Zaire carried 16 species (Gosse, 1963). Over 80 species were taken from a stretch of river 1.7 km long between coffer dams at Kainji (Table 2.4); these fishes were, however, probably transients, unlike the lacustrine littoral fishes which probably spend most of their lives in their restricted habitat, forming true and very stable communities. The same story holds true in Asia: Vaas (1952) for example listed over 17 species from a deep pit in West Borneo.

Thus there is no doubt that alpha diversity is very high in many tropical freshwater fish communities. How do they share resources, and what controls the numbers of species and individuals? Predators? Food supplies? Parasites, disease?

Role of predators

High numbers of predatory species and individuals appear to be a feature of tropical communities, and are certainly found in both freshwater and marine fish communities. By taking as prey the more abundant species, and switching to other prey species as the numbers of prey are reduced, predators may have an important role in permitting the coexistence of prey species by keeping their numbers below the level at which they would compete with one another for food or space (Paine, 1966, 1969; Murdoch, 1969; Glasser, 1979) and as suggested for haplochromines in Lake Victoria by Greenwood. It is obviously not in the interests of the predator to eliminate or endanger its prey species. Most of the piscivorous fishes, particularly those feeding on littoral and benthic prey, include fishes of many species in their diet; those feeding on pelagic fishes often have fewer species on which to prey. The role of predators on speciation is considered later (p. 299).

Packing and resource-sharing

Observations on fish behaviour where the water is clear, as in streams in the Rupununi area, Panama, Borneo and elsewhere, in lakes Tanganyika and Malawi, and in the sea on coral reefs, show that the three-dimensional world is utilized by certain species living always just below the surface, others in midwater and yet others on the bottom; body shape and colouring fit in with way of life and position in the water column. Whether the fish live solitary lives or in pairs or small schools is also quite characteristic. Many species live in schools when small and vulnerable to predation, taking to solitary or paired existence later as they mature. Time-sharing of resources, diurnally for feeding, seasonally by migrations, is also marked. In South American streams the cichlids and most characoids are active by day, the gymnotoids and siluroids by night. The presence of a nocturnal fauna appears to be dependent on cover where these fishes can hide by day; many of the nocturnal species are larger than the diurnal ones. In the forest waters in Zaire the mormyrids and siluroids are nocturnal, using tree litter as a substitute for rocks in which to hide themselves by day. Of fishes migrating downstream at the end of the rains certain species move down by day, others by night. Diurnal changes in activity are also important for resource-sharing in the sea (see Hobson, 1965; Collette & Talbot, 1972); on reefs different faunas are active by day and by night.

The richness of a community in any one place is also affected by ecological conditions in adjacent areas, since fishes are so mobile. Reef faunas are, for example, much richer where there are adjacent grassy flats where some species can move to feed by night. In rivers, flooding drastically alters the faunal composition. Fish movements up- and down-river and out onto the floodplain mean that riverine communities are always changing their species composition with time and water level. In any one area the resident community may be augmented by immigrant fishes, which may join them for some time to feed or spawn, or may just pass through. In Japan the arrival of the fry of the catadromous osmeroid *Plecoglossus altivelis* each spring has been shown to displace the resident cyprinids, leading to changes in the cyprinids' diets (Miyadi, 1960). Comparable interspecific social interference may be important in tropical communities, but this needs investigation. The mobility of riverine fishes makes these riverine communities very unlike the stable ones of littoral fishes in the African Great Lakes, where both species composition and numbers of individuals stay constant throughout the year and from year to year.

The young stages of many fishes live in different biotopes from the parent fish. Most reef fishes have pelagic larval stages. Many riverine fishes move upriver to spawn and the young often remain upstream, where they may have a better chance of escaping predation, though many of them perish as pools dry up or fall prey to birds and piscivorous fishes which frequent such pools. Many of the larger river-dwelling fishes move long distances (see p. 156). The smaller, less strongly swimming mid-water-dwelling tetragonopterids would not be expected to move such long distances, but many of these species are very widely distributed in South America, with many closely related species living together in one biotope or stretch of river. Shorter movements over long periods of time and river captures could account for such distributions. Once such species are widely distributed, oscillations in conditions would appear to help the continued coexistence of the species, rather than leading to its total extinction (see below p. 296), and an area can become repopulated after any local extinctions.

Recent studies in Panamanian streams have shown how algae-grazing loricariids redistribute themselves in the streams according to available food (Power, 1983). Some of these armoured catfishes appear to be territorial. Types of fish which have localized distributions tend to have numerous species. There are, for example, over 600 loricariid species, but among the very large pimelodid catfishes (such as *Pseudoplatystoma* and *Brachyplatystoma* species) which make long migrations relatively few species are known, and these few species are widely distributed through-

out much of South America. Small trichomyterid catfishes are represented by large numbers of species, but some of these (such as the
parasitic candiru *Vandellia*) are widely distributed; the 'hitch-hiking'
habit of living in the gill cavities of larger catfishes has probably extended
their distribution.

Overlaps in foods used by sympatric species

Studies in Amazonian streams (p. 143) showed that the majority
of the species present shared whatever food sources were available, much
of it allochthonous vegetable matter and insects dropped in from trees.
This apparent lack of specialization in foods eaten by many of the fishes
was surprising in view of the adaptive radiations of the characoids. Why
this apparent contradiction? Part of the answer would seem to be that
only eurytrophic fishes can penetrate into shaded headwater streams. The
species which do so are mostly widely distributed ones and their feeding
specializations are probably of greater use elsewhere in their range,
perhaps seasonally. Or they may have been of greater use in the past, as
the fish communities were evolving, in the early days of community
development or when conditions were different. Today such specializations appear to be of greater use in the middle and lower reaches of rivers
where so many species coexist. Whether resource-sharing is greater at
high water, when foods are abundant, or at low water, when they are
scarce, appears to depend on the particular conditions.

The sharing of food by several sympatric species appears to contradict
the competitive exclusion principle – Gause's hypothesis – the tendency
for competition to bring about an ecological separation of closely related
or otherwise similar species. Birds obey this principle well, but among
insects many species may share a food source such as a host plant;
conditions under which competitive coexistence can occur were reviewed
by Ayala (1970). Among Malawi cichlids, Fryer & Iles (1972) suggested
that the many species among the *mbuna* rockfish and zooplanktivorous
utaka sharing the same food are behaving as a 'condominium', members
of an association not in competition with one another as a species but only
as individuals belonging to a mixed indivisible society (see also Wynne-
Edwards, 1962; Margalef, 1968).

In previous chapters we have seen many examples of sympatric fishes
sharing food sources. But when do these overlaps occur? Throughout the
life cycles? Seasonally? If seasonally, are overlaps greater when food is
abundant, specializations then coming into play when foods are scarce, or
are overlaps greatest when foods are scarce and the fishes all have to share

the same few sources? Various studies have led to different conclusions. Overlaps were greatest when food was abundant: (1) in Lake Malawi, where several tilapia species shared the phytoplankton when this was abundant, but diverged in their feeding places when it was scarce; (2) in Lake Victoria, where the food overlap in non-cichlids was greatest when they were exploiting a particularly abundant food source (as during an insect emergence); and (3) in a Central American stream in Costa Rica, where Zaret & Rand (1971) found that among nine sympatric fish species food overlaps were at a minimum in the dry season, the time of least abundant food when competition might be expected to have increased – this was interpreted by them as being in accordance with Gause's competitive exclusion principle. However, overlaps were greatest when food supplies were scarce: (1) in the Rupununi, where fishes were driven to sharing the only few food sources available in the dry-season pools, and it was suggested that trophic specializations would here come into play in the highwater season, the main feeding time, although feeding levels were reduced; (2) in Zaire, where Matthes (1964) also found fishes reduced to sharing the same few foods in the dry season, mainly bottom debris as in the Rupununi; and (3) among loricariids in Panama streams, where food overlaps again appeared greatest in the dry season (Power, 1983). In the Amazon floodforest Goulding (1980) concluded that high water is the period when fishes are most ecologically separated in terms of feeding behaviour and it is at this time that their adaptations are put to their fullest use. During low water these fishes shared the little food that was available, but the main adaptations for survival were the fishes' stores of fat built up by heavy feeding during the floods. Goulding pointed out that the size of the fish is an important factor influencing feeding behaviour, as size determines the amount of fat that can be stored. The fishes studied in Costa Rican streams by Zaret & Rand (1971) were small, so might well have been forced to specialize at low water to ensure their survival in the habitat. The radical differences in conclusions from these different studies indicate that community structure varies greatly with the prevailing conditions.

    Studies on feeding behaviour are much needed, and on how social interference affects this in these polyspecific communities. For example, teeth and mouth shape not only affect the type of food used but also how this is obtained and the attitude of the fish in the water while feeding. Whether a fish 'head stands' or not while grazing algae off rocks will not only affect the food-gathering process but may also affect the fish's vulnerability to a passing predator while feeding (see also costs of obtaining 'safe' and 'unsafe' food (Jones, 1982) and enemy-free space (Jeffries & Lawton, 1984)).

Not all the foods eaten are digested, and one fish species may help provide food for others. For example, in the Malagarasi swamps the characoid *Distichodus* feeding on water lily leaves ejected a flocculent material which became rich in microorganisms and algae utilized as food by tilapias and other fishes living in these clearwater swamps, thus increasing the carrying capacity of the swamps for fish. This is the principle on which polyspecific fish culture is based: for example, the macrophyte-feeding grass carp *Ctenopharyngodon idella* fertilizes the pondwater for plankton-feeding species such as the phytoplanktonivorous silver carp *Hypophthalmichthys molitrix* and the zooplanktonivorous bighead *H. nobilis*.

Competition for living space

Competition is not only between predator and prey but also between species needing the same resources. Examples in previous chapters have shown that very complex food webs may be based on relatively few food sources, such as aufwuchs on the rocky shores of the African Great Lakes or exogenous foods in forest streams. In these circumstances many species share the same foods, not only omnivores and facultative feeders but also some specialists. Such cases indicate that factors other than food are likely to control the numbers of individuals in these species.

Many authors have suggested that competition for food is not the controlling factor governing numbers and distribution of fishes: e.g. Malawi cichlids (Fryer & Iles, 1972), the non-cichlids of Lake Victoria (Corbet, 1961), Lake Chad (Carmouze *et al.*, 1983), and Amazon, Malayan and Sri Lankan stream fishes (Knöppel, 1970; Bishop, 1973; Schut *et al.*, 1984). These examples are all from places where conditions are fairly uniform throughout the year. This raises two questions. First, if foods do not limit the numbers of the various species in these communities, then what does limit them? The total food supply will, of course, put a ceiling on the fish biomass, but the allocation of food amongst diverse fish species is what concerns us here. Secondly, what trophic specializations have evolved and how?

In both freshwater and the sea, living space may be more limiting than food supply in littoral rock and marine reef communities, but food supply may be more limiting in the pelagic system, both in prey species, such as *Engraulicypris sardella* in Lake Malawi, and possibly in their predators too. Food may be limiting at one particular stage of the life cycle (as for the youngest larval stages of *E. sardella* (FAO, 1982)). On marine reefs all stages are shown from fine resource-partitioning, indicating competition has occurred in the past, to overlaps with several species sharing foods, indicating little or no competition – though the same foods may be utilized in different ways – to the presence of alternative species utilizing the resources, resulting from stochastic recruitment to the living space on a patch reef or in a floodplain pool.

Living space concerns protection from enemies and facilities for spawning as well as for food collection. For demersal-spawning freshwater and smaller marine fishes (pomacentrids), space is crucial for spawning. But for larger reef fishes with pelagic eggs, space is needed mainly in the form of shelter sites, as cover from predation. It may be occupied only at the times of day when the fish is relatively inactive; another species may use the same space at other times of day. Space is limited for littoral and reef fishes, and obtaining a suitable territory may in many cases keep the numbers below the levels at which individuals would compete for food (see for example Shulman, 1985). Tagging experiments have shown that rock and reef fish can maintain territories for long periods; some home to their reef when displaced (Bardach, 1958). Fryer and Iles suggested that certain cichlid displays may enable the fishes to assess their own numbers and regulate their breeding accordingly (what Wynne-Edwards (1962) terms epideictic displays). This is still an open question, but the availability of suitable spawning grounds, and in some cases of nursery grounds, does appear to limit populations of certain cichlids.

Territorial behaviour is undoubtedly important in limiting fish populations, especially of crevice-dwelling fishes such as gymnotoids and catfishes in South American streams, among littoral fishes in African lakes and in marine reef fish. In Lake Malawi this appears to keep numbers below the level of competition for food, as has also been found for algae-grazing fishes on the Great Barrier Reef.

Fishes demonstrate many kinds of territorial behaviour (see Keenleyside, 1979). Some species defend feeding territories, and this restricts the number of fishes that an area can support (as for small trout in streams or damselfishes on coral reefs). Territories may be established only for spawning, as in many Malawi cichlids, or for the protection of the family group, as in substratum-spawning cichlids. In the female cichlid *Pseudocrenilabris multicolor* the young learn the mother's territory and this keeps them near to her (Albrecht, 1963). Yet another pattern of territory-holding is shown by the South American dwarf cichlid *Apistogramma trifasciatum* in which the male has a large territory wherein several smaller females each have their own nest (harem system (Burchard, 1965)). In *Gymnotus carapo* electric displays are used to establish dominance hierarchies, territories or intermediate types of organization, according to the population density (p. 267). Mormyrids also show agonistic behaviour for food and territory. Rank order may also be established in some schooling fishes, for instance in some chaetodontids (Zumpe, 1965), a situation in which members of the group might be helped to find food by the more experienced fish. Chaetodontid and pomacanthid reef fish are territorial when young.

Even for pelagic fishes the environment may be more structured than we at present comprehend; fishes may be kept within certain water strata by physical and chemical preferences (as studies in the Gulf of California seem to suggest (Robison, 1972)), and by social interaction. These are exciting fields for observation and experiment in which marine studies are leading the way. The space-sharing mechanisms of coral reef fishes were examined during 60 days of underwater living in the Virgin Islands (Collette & Earle, 1972). These mechanisms included diurnal/nocturnal activity cycles, separate feeding and hunting areas, shelter sites, territoriality, seasonal cycles and symbiotic relationships, leading to the conclusion that a combination of interspecific and intraspecific competition for space may play a major role, if not a decisive role, in maintaining numerical stability in coral reef communities (Smith & Tyler, 1972).

Evolutionary aspects of community development
Evolutionary aspects call for investigations on two levels: (1) on factors affecting the evolution of new species (speciation), the building blocks of new communities (most studied in the African Great Lakes as described in Chapter 5); and (2) on the evolution of the communities themselves (see Odum, 1969; Whittaker & Woodwell, 1972) as conditions change and species come and go (for example, the change from riverine to lacustrine communities in the new man-made lakes).

The many reasons put forward to explain the very diverse plant and animal communities at low latitudes include extrinsic factors such as: time (are tropical communities older?); vicariance (have they been more influenced by geomorphological changes, continental drifting, river captures?); climatic stability (are conditions in tropical regions more stable, or predictable, than those in temperate regions?); spatial heterogeneity (do they contain more niches?); and productivity (are communities living at high temperatures more productive and if so does this help them to support more diverse faunas?). The many intrinsic factors include the roles of predation and competition, whether these are more intense in tropical than temperate communities and, if so, their possible effects on speciation and community development.

Many of these aspects have been discussed elsewhere for tropical freshwater fish communities (Lowe-McConnell, 1969a, 1975) and for coral fishes (Sale, 1980b). Case histories considered in earlier chapters of this book lead to the following comments. Regarding **time**, some coral reef communities may be very long established, but diversity is also high in the African Great Lakes which are geologically very much younger. **Climatic stability** appears to allow specialization, enabling numerous species to coexist, but patch reef studies suggest that local perturbations may also have a role in maintaining diversity (see also the 'intermediate disturbance' hypothesis discussed for riverine fishes by Ward & Stanford (1983)). Reef studies have shown that **spatial heterogeneity** permits high diversity, as it does also for rock faunas in the African Great Lakes. Regarding **productivity**, the most diverse communities may be more efficient but they are not necessarily the most productive; productivity is governed by input of nutrients and these are generally greater in the more seasonally affected biotopes, some of which (such as pelagic systems) do not have very diverse faunas.

A community has been called a record of accumulations of species through geological time, edited by extinctions. Many of the tropical ecosystems appear to be very old. Conditions also appear to be favourable for speciation, both in the African Great Lakes (as discussed in Chapter 5) and in the vast river systems whose numerous tributaries offer plenty of geographical isolation thought to be conducive to species formation. The wide dispersal of riverine fishes, and of larval stages of reef fishes, mitigates against total extinction of a species. Local climatic variations, for example of heavy rainfall, are a feature of tropical regions, but in equatorial rivers most species are widely distributed through a vast system of anastomosing channels and local variations in conditions are unlikely to affect a whole species. The short life cycles, with maturity at 1–2 yr old, means that fish populations can build up rapidly when conditions are good. Furthermore, fishes are able to survive adverse times by reducing their growth rates. The predominant type of selection in the tropics (see below), in which the agents of selection are principally biotic (predators, competing species and food organisms) and density-dependent, means that selection is less likely to lead to extinctions of prey species than where abiotic density-independent selection (such as in Ice

Ages) is dominant. Local extinctions do occur in the tropics (as witnessed by the loss of *Lates* from Lake Edward and proto-Victoria); Roberts (1972) reported for Amazonian species collected by rotenone that only one third or less were common, and a dozen or so were very rare and in danger of local extinction. It has been suggested that adaptation to life at a very low population density might assist invertebrates in tropical rainforest to escape the attentions of the numerous predators, that low density may be the end product of a long evolutionary interaction between predators and prey.

In the most tropical regions it seems that a long period of relative stability prior to the pluvial and interpluvial periods and sea level changes associated with the Quaternary glacials, would have allowed wide dispersal of many species, aided by vicariance events, including continental drift, tectonic changes and river captures. The pluvials and interpluvials leading to expansions and contractions of freshwater environments, far from obliterating the fish populations (as the Ice Ages did for northern temperate regions) could have aided freshwater fish dispersals. These climatic fluctuations could have contributed to speciation by dividing and recombining fish populations, rather than leading to their wholesale destruction. In South America expansions and contractions of the forest environment in wet and dry periods have been invoked to explain bird distributions, forest species being confined to forest refuges in periods of low rainfall; comparable changes are likely to have affected fish populations but evidence for changing fish distributions is much less complete (Weitzman & Weitzman, 1982). The effects of dry epochs may have been more drastic in Africa where the general level of the land is higher than in South America.

Climatic stability is not always associated with high diversity. In the climatically very stable equatorial Lake George in Uganda the relatively low diversity of the flora and fauna for a tropical lake was attributed partly to lack of seasonal succession, and partly to the few types of food available and the uniformity of habitat conditions over most of the lake (Burgis *et al.*, 1973). It appears likely that long-term stability in conditions may have permitted diversity to develop on coral reefs but, as discussed above (p. 207), there is some evidence that perturbations can enhance local diversity of the fish fauna.

Selection pressures

Any organism has, basically, to obtain resources for maintenance at all stages of its life history, for growth, reproduction and avoiding destruction by predators, parasites, disease or other causes. Selection pressures may operate at any point in the life history of an organism and on one, or very few, gene(s) at a time; different types of pressure are likely to operate at different stages of the life history. Dobzhansky (1950) pointed out that biotic selection appears to be of special importance in tropical communities and, being density-dependent, can be more mould-ing in its effects than the abiotic (density-independent) selection encoun-tered in temperate (seasonal) environments; such biotic pressures could be responsible for the subtleties of protective resemblances and mimicry displayed in tropical faunas. The continual interplay between predator and prey, with the development of devices to avoid being preyed upon countered by adaptations in the predator, and these in turn countered by further antipredation devices and behaviour, is a feature of South Ameri-can insect communities, and the same situation appears to hold for tropical fishes.

Among tropical freshwater fish communities the main selection agents appear to be biotic effects of competition for resources (food and living space) and predation, interacting with abiotic (density-independent) effects of drought, desiccation and deoxygenation. The very high (35% per annum) mortalities on the Zambezi system Kafue Flats were thought to result from the combined effects of drought and predation. Interacting factors were also well shown in coastal pools in the Netherlands Antilles where the cyprinodontoid *Poecilia, Rivulus* and *Cyprinodon* fill almost the same niche; in landlocked pools their relative abundances were controlled by interspecific competition, influenced by environmental factors such as plant or algal growth, salinity, oxygen and size of food available, rather than by predation (Kristensen, 1970). Amongst these fishes the maintenance of very small populations was helped by ovoviviparity (in *Poecilia*) and hermaphroditism (in *Rivulus*), and repopulation could also occur from the sea.

The role of competition, itself the product of diversity, on both speciation and community development needs further analysis. Competi-tion for living space (cover from predation and for spawning places) seems more important than competition for food in reef and rock-dwell-ing fishes, but competition for food is more important than for space in pelagic fishes. There are many cases of sympatric, and often closely related species, utilizing apparently similar resources. Predation appears very important in controlling numbers in almost all the tropical fish communities examined. Studies in Lake Tanganyika (p. 116) suggested, however, that predation has opposite effects on speciation in open waters (selecting for uniformity) and in the rocky littoral where prey has cover (where it probably selects for diversity). Again, the nature of selection appears to be different in the seasonal and relatively aseasonal communities (r-selection predominating in the former, K-selection in the latter).

The kinds of changes in selection pressures that occur with changing environmental conditions and with time as communities evolve are indicated in Fig. 13.1, summarizing events that occur after a new lake is colonized by riverine fishes. In the floodplain rivers abiotic selection, interacting with biotic selection, is of great importance seasonally, pressures being greatest as the water level falls. Of the riverine fishes which colonize lakes some will be lost to the fauna (as mormyrids were from Volta Lake) and others will gain access and become established when conditions suit them (as observed in Lake Kariba). Fishes which thrive under the new lacustrine conditions will multiply fast (as pelagic fishes did in Volta, cichlids in many other new lakes, *Alestes lateralis* in Kariba). Intraspecific competition will tend to push them to widen their food niches, but the presence of numerous species (interspecific competition) will tend to lead to niche differentiation and specialization (character displacements and adaptive radiations). With time, aided by environmental changes such as changing lake levels and biotic pressures affecting fish behaviour (processes still under investigation), speciation may occur. Fryer & Iles (1972) maintained that predators facilitate speciation, rather than controlling it (as Worthington p. 116 had suggested and see p. 301); as the numbers and variety of predators increase these may help to preserve the diversity of the community (see p. 289).

Studies of Neotropical birds have suggested that the most potent force affecting the direction in which a particular group will evolve is the associated fauna in the habitat. This may also apply to many tropical fish groups. Riverine fishes are, however, continually changing their habitat, with its associated species, as they move with fluctuating water levels. Foods are considered to be very important in regulating bird populations, operating by bird parents failing to rear the usual number of young (as evolved by natural selection) in adverse times (Lack, 1954). Many young fishes appear to be predator-limited rather than resource-limited, but hard data are required. Exceptions to this occur in the pelagic communities (African Great Lakes and sea) where reproduction and survival of the young is intensified following plankton blooms. This has adaptive value in enabling these fish to exploit such blooms whenever and wherever they occur. In Lake Tanganyika, however, the numbers of clupeids rose

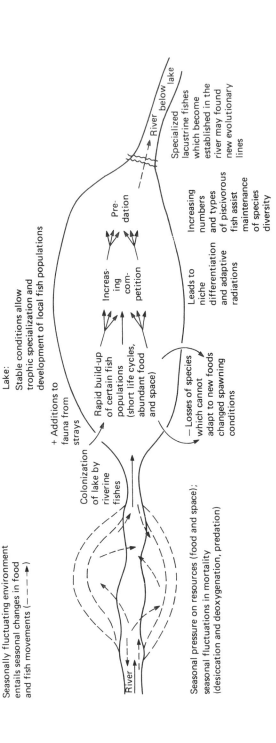

Floodplain:

Seasonally fluctuating environment
entails seasonal changes in food
and fish movements (– – →)

Colonization
of lake by
riverine
fishes

Seasonal pressure on resources (food and space);
seasonal fluctuations in mortality
(desiccation and deoxygenation, predation)

Lake:
Stable conditions allow
trophic specialization and
development of local fish populations

+ Additions to
fauna from
strays

Rapid build-up
of certain fish
populations
(short life cycles,
abundant food
and space)

Increas-
ing com-
petition

Pre-
dation

– Losses of species
which cannot
adapt to new foods
changed spawning
conditions

Leads to
niche
differentiation
and adaptive
radiations

Increasing
numbers
and types
of piscivorous
fish assist
maintenance
of species
diversity

River
below lake

Specialized
lacustrine fishes
which become
established in the
river may found
new evolutionary
lines

River

Fig. 13.1. Summary of some pressures and events which affect the development of tropical fish communities as discussed in the text.

significantly after the fishery for their predators was developed, indicating that predation as well as food supplies is important in controlling clupeid numbers. Zooplankters of a suitable size as food for young fish may limit fish populations in Amazonian lakes (Zaret, 1984*b*).

Riverine studies have shown that there are great seasonal variations in pressure on resources; throughout the tropics these appear to be greatest at the end of the rains and in the dry season. High water is the main feeding and growing time for most floodplain species, except for some planktivores and for piscivores which may be able to catch prey fish more easily when these are less dispersed. As the water level falls, food supplies for most non-piscivorous fishes become restricted in a rapidly contracting environment. Most of these riverine prey species spawn early in the rains and the young grow fast; the ichthyomass appears to be at its greatest at the end of the rains before the enormous losses resulting from predation and stranding as the water level falls. Hence pressures on food supplies are likely to be at a maximum at the time when resources are shrinking. Is this a 'crunch' time when specializations are of most benefit to their owners? As the dry season advances, fish numbers are reduced and many species cease to feed; in some South American fishes parts of the alimentary canal are used as accessory respiratory organs at this time (p. 134). Where fishes do not cease to feed in the dry season, this may be another time of intense competition for food in confined pools. Fluctuations in pressure on food resources are likely to have affected fish evolution greatly. In equatorial waters where there are two floods a year, increased pressures on resources may occur twice a year, instead of annually, thereby perhaps speeding the course of evolution.

The role of predators in fish speciation has been hotly disputed, as reviewed at length by Fryer & Iles (1972). The significance of schooling and cover, discussed above, leads to the conclusion put forward here that predation has contrasting effects on speciation in pelagic and in littoral/benthic communities (see p. 285). A strong case for the direct role of predators in moulding life history parameters of a prey species comes from Reznick & Endler's (1982) observations and experimental manipulations in Trinidadian streams, where they showed that *Poecilia reticulata* makes a higher investment in reproduction (as opposed to growth) in streams subjected to predation by *Crenicichla*.

Zaret (1979) stressed the value of experimental manipulations, by predator removal or following introductions, in evaluating the extent and influence of predation pressure. The study of the introduced *Cichla ocellaris* in Panama showed that this predator had different effects on the indigenous fish communities living in Gatun Lake and in the inflowing Chagres River. The 12 common native species were reduced by 99%, with many local exterminations, in the lake, but a stable predator–prey community developed in the river. Why? One suggestion was that *Cichla*, which hunts prey visually, was able to feed throughout the year in the clear water of the lake, whereas seasonally turbid flow in the river allowed the prey fish a respite (Zaret, 1979). The river also appeared to have more cover, in the form of littoral vegetation, than in the lake. All trophic levels were affected by the decline of the native fish species in the lake. In this case the lake community appeared to be less stable, its fauna less resilient than that of the river; Gatun Lake is, however, a young man-made lake, formed as part of the Panama Canal. The effects of the introduced *Lates* in Lake Victoria have unfortunately not been monitored in detail.

Competition, for living space as well as food, by non-predators can also affect community development. For example, in Lake Victoria where introduced tilapias displaced endemic species (p. 100), *T. zillii* by competing with *O. variabilis* for nursery areas (Fryer, 1961), *O. leucostictus* thriving inshore where dissolved oxygen is lowest, *O. niloticus* replacing *O. esculentus* offshore (Welcomme, 1964, 1966, 1967a, 1970; p. 315).

There has been much discussion on diversity and stability in ecosystems (symposia edited by Woodwell & Smith, 1969; Van Dobben & Lowe-McConnell, 1975; Regier & Cowell, 1972; Zaret, 1984c). Types of stability were discussed by Orians (1975). The former view that diverse systems are more 'stable' than simpler ones is no longer held by most ecologists; it is now considered that a predictable (stable) environment may permit a complex and delicately balanced ecosystem to evolve, but that such complex ecosystems are generally dynamically fragile, that their stability may be easily upset and local extinctions may then occur. Zaret proposed tests of hypotheses on the stability/diversity controversy which led him to conclude that there was support for the hypothesis that stable environments, associated with a high species diversity, will have lower faunal stability.

The removal of biomass from tropical systems, as well as reducing their diversity, is likely to affect their productivity. For instance, the effects of removing littoral plant matter (a density- and diversity-dependent process) from Amazon waters, discussed by Marlier (1968), and Fittkau's (1973) observations on the effects of caiman removal on the fishes, suggest that these impoverished systems cannot give up biomass or diversity without severe disturbance. For African lakes it has been suggested that the removal of large numbers of herbivores may also slow down production, as plant material is made available for renewed plankton growth much more rapidly through the digestive systems of herbivores than when locked away in the bottom mud (Beauchamp, 1964).

Many of the same factors will operate in the sea as in lakes. However, the presence of planktonic stages in the life histories of most marine fishes, and their absence in freshwaters, must have affected evolution in these two types of community. For example, among the lacustrine cichlids juveniles are likely to be related to the sympatric adults, thus forming discrete populations (as are known to occur around rock outcrops in Lake Malawi), whereas most marine fish populations probably come from a much wider gene pool. Such effects have not yet been analysed.

Deterministic or stochastic assemblages?

To sum up, what do the tropical fish communities considered in this book contribute to recent ideas on deterministic or stochastic processes in the evolution and maintenance of communities? Ecologists are concerned about whether communities are primarily 'deterministic' communities, in which partitioning of resources through competition (aided by predation, which alleviates competition) is the driving force affecting coexistence within the community, or whether they are 'stochastic' (non-equilibrium) assemblages, the abundance of species determined largely through unpredictable (random) environmental changes rather than through biological interactions. These 'random' changes either reduce populations to levels at which competitive exclusion do not occur, or cause a limiting resource (such as space) to become available in an unpredictable manner. The classification of systems as either deterministic or stochastic is obviously an oversimplification, as both types of properties may influence the abundance of species in either type of assemblage, but the majority of the inhabitants may appear to be regulated by one or other mechanism. (Whether the deterministic communities are in 'equilibrium', as maintained by Grossman, Moyle & Whittaker (1982), is disputed by Yant, Karr & Angermeier (1984).) A third, intermediate, view – the 'competitive crunch' hypothesis of Wiens (1977) – suggests that interludes of intense competition may be sporadic, say once in many generations, due to exceptional climatic events (such as floods, droughts, lake level changes (p. 88)), when only the individuals that partition resources efficiently survive, and at other times partitioning may be relaxed. Ideas on 'punctuated equilibrium', thought by Fryer, Greenwood & Peake (1983; 1985) to apply to the African Great Lake fish communities, accord with this view.

Key factors appear to be predictability of the environment (conditions in aseasonal environments being on the whole more predictable than in seasonal ones) and the mobility of the fish populations. For reef fishes, we have seen (p. 209) that there is evidence both for resource-partitioning on large reefs, and for stochastic recruitment on patch reefs, indicating that the difference here may be mainly one of scale (as suggested by Smith, 1978; Rahel, Lyons & Cochran, 1984). Some stream studies (in Costa Rica by Zaret & Rand (1971) (see p. 292) and in Sri Lanka by Schut *et al.* (1984) (see p. 273)) indicated that these stream communities were deterministically structured, but other stream studies (in Gabon by Brosset (1982) and on the Rupununi and Paraná floodplains) suggested that stochastic processes were involved in the composition of populations in floodplain pools (see Fig. 6.5). Data from Zaire River studies (p. 48) also suggest that stochastic processes come into play in river communities where the faunal composition is continually changing as the fishes are so mobile. In the littoral communities of the African Great Lakes, where adaptive radiations have led to fine resource-partitioning, many fish are territorial and stay in one area, making for very stable communities; here, deterministic mechanisms may operate throughout the whole life histories of the fishes. In most coral reef fishes, on the other hand, pelagic larvae provide mobile stages which are subjected to stochastic processes during larval life, as well as when settling on the reef.

14

The exploitation and conservation of tropical fish stocks

The aim of fisheries management, to obtain the maximum (or optimum) sustained yield of fish from a water body, involves removing fishes equivalent to the amount of fish flesh produced each year (the production) without making inroads into the capital (standing stock or biomass). When comparing data from different sources, a clear distinction has to be made between biomass (stock of fish present at any one time), biological production (a rate) and the catch (yield). The yield represents only a proportion of the production and can range from almost the entire production in fishponds, to a very small proportion where fishing conditions are difficult, or where there are few fishermen, or where natural mortalities are great. As very few estimates of biological production have yet been made in natural waters (freshwater or sea), the yield (expressed as kg ha^{-1}), is often used as an index of production, and in fisheries literature this yield or catch, i.e. the proportion of the production cropped by man, is often loosely called the 'production'.

Comparative yields

Tropical fishes from freshwaters and the sea within easy reach of land have been fished from time immemorial by subsistence fishermen, and such fishing is still very important in many parts of the tropics. It is only in the last half century that fishing has been heavily mechanized. In the 1950s and 1960s the world catch rose very fast, more than trebling from around 20 million tonnes in 1948 to a record 71.3 million tonnes in 1979. The tropical contribution rose from *ca* 20% in 1948 to 36% in 1972; since then it seems to have fluctuated and was only *ca* 20% again in 1979. The Food and Agriculture Organization of the United Nations, which coordinates world fisheries statistics, estimated a potential yield of about 100 million tonnes (FAO, 1973), predicting that the major part of this increase was likely to be of small species living at low latitudes, as the high-quality food fishes, such as large tuna and the fishes on established fishing grounds, were already heavily exploited (Gulland, 1971; Suda, 1973). It was thought that the best hope for future increase lay in farming small areas of inland or brackish waters where factors governing fish production could be controlled. Since these predictions were made, several extraneous factors have affected catches. The freshwater proportion in fact dropped from 16% in 1972 to *ca* 10% in 1979, partly because China's formerly large production of freshwater fish dropped to about half as their wetlands were 'reclaimed' for agriculture. Furthermore, the acceptance of Exclusive Economic Zones (EEZ) out to a 200 mile (320 km) limit, under the Law of the Sea Convention, means that countries bordering tropical seas are now developing their own fisheries, whereas non-local vessels formerly landed 30% of the catches. Many of these tropical countries have not yet developed their full fishing potential (FAO, 1979, 1981).

By 1979, when world landings had risen to 71.3 million tonnes, about 10% of the catch came from inland waters. Additionally, unrecorded catches of fish for domestic consumption by subsistence fishermen are considerable, particularly from freshwaters. Of the recorded catch, *ca* 15 million tonnes came from the tropical waters considered here (including the Peruvian upwelling of cooler water). Over half the total catch weight was of small clupeoids, another quarter was of demersal fishes from the continental shelves, about 10% was of tuna from the open oceans, and of the remainder a significant amount (*ca* 6%) was of small pelagic fishes such as mackerels and carangids (FAO, 1981). Catches from the west coast of South America still dominated world clupeoid catches; anchoveta catches were no longer of prime importance but catches of sardine (*Sardinops sagax*) and mackerels had increased. Egg and larval surveys and acoustic surveys confirmed that these changes reflected real changes in the sea (probably related to the decline in the huge flocks of piscivorous birds resulting from the El Niño). The influence of El Niño events (p. 228) on the fisheries here stresses the difficulties of trying to predict clupeoid cycles. In the western Indian Ocean, the small pelagic fish catch included a large component of oil sardines (*Sardinops longiceps*); of the mackerel catch, over half was of Indian mackerel (*Rastrelliger* species); of the demersal catch here, a high proportion was of 'Bombay duck' (*Harpadon nehereus*). In the western Atlantic the demersal catch was mainly of sciaenids, mullets, groupers, snappers, grunts, sharks and rays. All tuna stocks except skipjack were considered to be heavily exploited.

When assessing a fishery, it is necessary to know whether the fish population is underexploited, in which case the fishery can be developed, or overexploited, in which case measures must be sought to protect the fish. Decades of experience of controllable fishing are needed to establish sustainable fisheries. As we have seen, fisheries research in tropical waters is complicated by the large numbers of species present and the difficulties in determining ages and growth rates necessary for calculating the biological production. Pauly (1983) gives simple methods for the assessment of fish stocks under tropical conditions (see also Bagenal, 1978; Gulland, 1978, 1983).

Reliable statistics of catches are vital for fisheries research: those from remote tropical areas are of very varying reliability. To compare catches from different places and times, catch per unit effort has to be used. Trawl catches are generally given as catch per fishing hour, but factors such as the size of the trawl and speed of vessel affect fishing efficiency. Hookline catches are quoted as catch per set number (usually per 100) hooks, but these catches vary very much with conditions, the bait used and how long the hooks have been down. Gillnet catches per standard length of net of a particular size of mesh are more comparable with one another, but mesh size is highly selective of fish size and shape.

As a fishery develops, the catch per unit effort inevitably declines as more fishermen share the catch. But at what point does 'overfishing' occur? To determine this it is necessary to be able to assess stock size and recruitment rates. Overfishing may be one of three kinds (Pauly, 1983): (1) 'growth overfishing', when the young fish, recruits to the fishery, are caught before they reach a reasonable size (age); (2) 'recruitment overfishing', when the parent stock is so reduced that not enough young are produced for the fishery to maintain itself; and (3) 'ecosystem overfishing' (not yet clearly defined), in a mixed fishery when the catch decline (through fishing) of the originally abundant stock is not fully compensated for by the contemporary or subsequent increase of other exploited animals, i.e. the transformation of a relatively 'mature' efficient system into an 'immature' (stressed) less-efficient system. This seems to have occurred in the very heavily fished Gulf of Thailand, where a high-biomass system dominated by fish has been turned into an inefficient low-biomass one in which the role of invertebrates has markedly increased.

Changes in trawl catches with time have been well illustrated during intensive surveys in the Gulf of Thailand. Here the catches of preferred species have suffered a decline from ca 100 kg h^{-1} in 1963 to less than 20 kg h^{-1} in 1978, though catches of 'trash fish' have remained more constant. The catches of red mullets (Mullidae), ponyfishes (Leiognathidae), monocle breams (Scolopsidae) and sharks and rays, have fallen much more markedly than those of bigeyes (Priacanthidae) and lizardfishes (Sauridae). Not only has the species composition changed, more smaller species now being caught, but the average size of fish of a particular species has also declined. Leiognathids were a major component of the preferred fish catch in early years but the small ones now landed are considered trash fish (Beddington & May, 1982). It appears that predation by man has reduced these fish populations to below the levels at which they were regulated by competition for resources, and this predation is selecting for species that can reproduce quickly and abundantly ('r-strategists').

Comparative trawl catches in the western and eastern Atlantic and from Indian Ocean surveys were given in Table 8.1. The Cape Comorin catch off Sri Lanka (120 kg ha^{-1} h^{-1}) was equivalent to about half a good haddock catch from north temperate seas off the Shetland Islands (Cushing, 1971). The sciaenids and other fishes trawled off West Africa were of very few year classes compared with catches from temperate waters. Longhurst (1963) pointed out that this would lead to rather large-scale fluctuations according to brood strength, and that the failure of a single year class would greatly affect the adult populations only a year or so later; the rapid growth implies high productivity in relation to standing crop (high productivity/biomass ratio).

Fish yields from various tropical and temperate waters are shown in Fig. 14.1 and Table 14.1. Yields from natural waters range from an estimated 0.02–0.5 kg ha^{-1} yr^{-1} mean for the oceans of the world to yields of ca 1000 kg ha^{-1} yr^{-1} for some estuarine lagoons; but data collated by Marten & Polovina (1982) indicates that yields from natural waters generally fall between 0.1 and 30 tonnes km^{-2} (i.e. 1–300 kg ha^{-1} yr^{-1}, and most of them between 10 and 100 kg ha^{-1} yr^{-1}), though very high yields can be obtained under management in freshwaters. Marine values cited by Pauly (1983) ranged from 36 kg ha^{-1} (demersal fishes) from the heavily fished Gulf of Thailand, through 40–50 kg ha^{-1} (all fishes) from coral reefs, to 120–150 kg ha^{-1} from shallow lagoons and places where estuarine conditions boost productivity.

In African lakes yields range from ca 1.0 to ca 500 kg ha^{-1} yr^{-1}, being on the whole inversely proportional to the lake depth and higher in the more eutrophic lakes, as described below. Floodplains can be expected to yield 40–60 kg ha^{-1} yr^{-1} on a sustained basis (Welcomme, 1979), though yields may be much higher where lagoons retain water through the year.

Floodplain figures cannot be exact as it is very difficult to estimate the number of hectares contributing to fish production in a seasonally expanding and contracting environment; also there is much variation from year to year. The heaviest yields without artificial feeding so far reported appear to be those for the *acadja* fish parks in West African coastal lagoons (Dahomey), planted with brushwood stakes which increase the surface area for growth of periphyton, the food of tilapias and other fishes. Fished twice a year, *acadjas* are reputed to yield the equivalent of 8000 kg ha^{-1} yr^{-1} of mixed species (Welcomme, 1972). But here again the area producing these fish is uncertain, as these *acadjas* also attract fishes from the larger area of the lagoon.

Fishpond studies have shown how much fish production and yield depend on prevailing conditions. Zaire ponds without management produced between 100 and 1000 kg ha^{-1} yr^{-1} of tilapia, but with management (fertilizing the ponds and feeding the fish) produced 1000–4000 kg ha^{-1} yr^{-1} of these same fishes. In Israel, experimental ponds which produced less than 100 kg ha^{-1} yr^{-1} of carp (*Cyprinus carpio*) produced 300–400 kg ha^{-1} yr^{-1} when fertilized; additional feeding with cereals pushed up fish production to 900–1500 kg ha^{-1} yr^{-1}, and the addition of another species, tilapia, augmented it even further to 2500 kg ha^{-1} yr^{-1} (Hepher, 1967). More recently, yields of up to 10 660 kg ha^{-1} have been

Fig. 14.1. Comparative fish yields from tropical ecosystems.

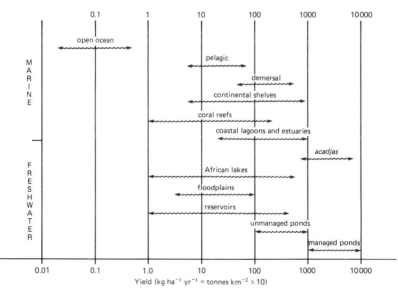

Table 14.1. *Representative yields from tropical waters*

Ecosystem	Location	Yield (kg ha^{-1})	Reference
Marine			
Tropical pelagic	13 sites	5.5–60.2	see Marten & Polovina (1982)
Tropical demersal	16 sites	4.0–67.0	
Coral reef	15 sites	0.9–180	
Estuaries and coastal lagoons	27 sites	17.0–940+	
Coral reef	Jamaica (all fish)	40	see Pauly (1983)
	West Indian Ocean (all fish)	50	
Shelf	Gulf of Thailand (demersal)	36	
Estuaries and lagoons	Philippine, S. Miguel bay	150	
Shallow lagoons	Gulf of Mexico	120	
	Ghana	150	

Freshwaters			
Africa			
Lakes (40+)		~1.0–500	Henderson & Welcomme (1974)
Reservoirs (7)		7.6–127	
Floodplains	Barotse	6.8	see Welcomme (1979)
	Kafue	15–19	
	Shire	118–143	
South America			
Man-made lakes	NE Brazil (30 lakes)	7–882	Paiva (1982)
Amazon floodplains	Itacoatiara	9	Smith (1981)
	Madeira tributaries	52	Goulding (1981)
	Rio Negro	0.5	Petrere (1982)
	(Colombia)		
Magdalena floodplain		32.5	in Welcomme (1979)
Asia			
India	13 lakes and reservoirs	5.3–2200	Sreenivasan (1972)
Rivers	Lower Mekong	40.7	
	Ganges–Brahmaputra	78.2	
Reservoirs	India (8)	2.1–187.7	see Welcomme (1979)
	Thailand (8)	16.8–135.6	
	Indonesia (8)	21.7–356.6	
Floodplains	Worldwide (maximum flood)	40–60	

produced from ponds in Israel using a polyculture of all-male hybrid tilapia (Hepher & Pruginin, 1982), and yields of up to 6183 kg ha^{-1} of herbivorous fishes have been reported from ponds in India (Ganapati & Sreenivasan, 1972). When fish are fed, however, yields are not strictly comparable, since the areas used in the production of the food subsidies are not taken into account. The extremely high yields of fish from running water fish culture and aquarium-type circulating tanks, used mainly in Japan, depend on very heavy feeding with exogenous foods (grains or offal), as well as on running water to remove waste products. Yields of up to 1.5 kg m^{-2} (15 000 kg ha^{-1}) are reported from these systems.

Catches from African freshwaters

In Africa, where both lakes and rivers have important fisheries, an estimated 30% (440 000 tonnes) comes from rivers (FAO, 1981), particularly rivers with extensive floodplains, the total riverine catch reflecting the size of the drainage basin. In addition, a lot of fishes are cropped from lower-order streams and most of these are probably not included in the catch statistics. The seasonal rivers of Africa carry a very varied fauna of food fishes, particularly across the Sudanian region; rivers flowing into Lake Chad produced about 36 kinds of fish. Statistics of catches are more readily available for Africa than for other tropical regions. Catch statistics from rivers are often sparse, but for moderately to heavily fished African rivers Welcomme (1976, 1979) related catch to area of drainage, treating separately rivers with or without extensive floodplains. Rivers with a greater area of floodplain not unexpectedly produce more fish. African river catches ranged from 6.8 kg ha^{-1} from the Zambezi's Barotse floodplain to 118 kg ha^{-1} on Malawi's Shire floodplain. Welcomme's analyses suggested that floodplain catches should increase until there are about 10 fishermen per km^2, after which the total declines as fishermen then interfere with one another. Catches from reservoirs are generally comparable to, or slightly higher than, catches from floodplains.

On the whole, catches from the numerous lakes and reservoirs of Africa correlate well with a morphoedaphic index (MEI) of total dissolved salts (expressed as conductivities, μ mhos cm^{-2}) divided by mean depth of lake (in metres), bearing out the well-known impression that shallow, nutrient-rich lakes are more productive of fish than are deeper lakes low in nutrients (Fig. 14.2). This index, originally developed for homogeneous sets of North American lakes by Ryder, has been applied to African data, taking into account the number of fishermen, by Henderson & Welcomme (1974). The limitations of the index were discussed by Oglesby (1982) and Ryder (1982).

Lake Victoria had important gillnet fisheries based mainly on the two endemic tilapias *Oreochromis esculentus* and *O. variabilis*. This fishery provided a classic example of a decline in catches following excessive fishing in inshore waters (Fig. 14.3). Mesh size was originally limited to a 5-inch (stretched diagonal) mesh, but when fish became scarcer the fishery biologists' advice was ignored and a reduction in mesh size was permitted which led to a further decline in catches. Non-cichlids taken in the gillnets included *Bagrus, Clarias, Protopterus, Labeo, Schilbe* and, later, the introduced tilapias – *O. niloticus* offshore in gulfs and bays, *Tilapia zillii* and *O. leucostictus* in inshore waters – and *Lates niloticus*. There were also fisheries for potamodromous species, especially *Labeo*, large *Barbus* and mormyrids, as they migrated up affluent rivers on spawning runs. The *Labeo* fishery in the Kagera River mouth declined

Fig. 14.2. The relationships between morphoedaphic index (conductivity/mean depth) and recorded catch from African lakes (after data from Henderson & Welcomme (1974), adapted by Toews & Griffith (1979). ● = 17 lakes with more than one fisherman km^{-2} (solid line), × = lakes with less than one fisherman km^{-2} (dashed line).

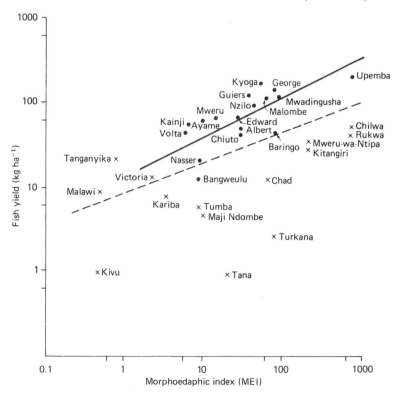

dramatically after the introduction of small-meshed nylon nets which were floated downriver (Cadwalladr, 1965). Marten (1979) suggested that overfishing in Lake Victoria could be remedied by increased fishing effort directed towards the large predatory fishes, particularly *Bagrus*, using hooks.

By 1982 Lake Victoria's fisheries had changed dramatically as the tilapia-dominated fisheries had given way to a number of different ones based on other species – using larger-meshed gillnets for piscivorous *Bagrus, Clarias, Protopterus* and introduced *Lates*, bottom trawls for haplochromine cichlids, and small-meshed ringnets with lights at night to attract the small pelagic cyprinid *Rastrineobola argenteus* (Okemwa, 1984; FAO, 1985). These changes followed (1) the effects of the intro-duced tilapias on the indigenous ones (largely through competition for nursery grounds, see p. 302), and (2) the spread of the introduced piscivorous *Lates niloticus* (see Barel *et al.*, 1985; FAO, 1985). First established in northern waters in 1960, *Lates* spread clockwise round the lake, reaching Tanzanian waters last of all. By 1981 *Lates* made up 50–60% of the total catch (compared with 5% pre-1976). The *Lates*

Fig. 14.3. The decline of the tilapia fishery in Kenyan waters of Lake Victoria (after Fryer, 1972).

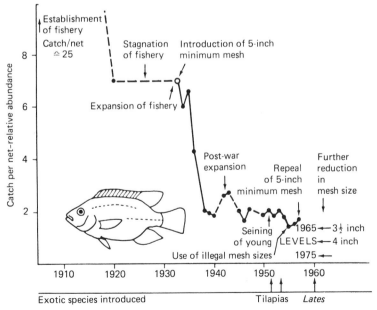

appear to have more or less cleaned out haplochromine stocks from Kenyan waters; by 1984 they were feeding mainly on *Caridina* shrimps and cannibalistically on smaller *Lates*. *Lates* up to 2 m long and 200 kg were recorded; they are not popular locally, and most are exported to the Kenyan coast. Meanwhile catches of other piscivores, *Bagrus*, *Clarias* and *Protopterus*, in Kenyan waters declined (Fig. 14.4). Since larger-mesh nets (153–305 mm) are now used in Uganda waters for *Lates*, tilapia stocks have increased here, but these tilapia are now mainly *Oreochromis niloticus* and *Tilapia zillii*, species which co-evolved with *Lates* in Lake Albert from where they were all originally stocked. The endemic *O. esculentus* has vanished from catches. *Lates* populations have since exploded in Tanzanian waters of the lake.

Following exploratory bottom trawl surveys from the East African Freshwater Fisheries Research Organization's laboratory in 1968–71 (see Bergstrand & Cordone, 1971), commercial trawl fishing developed in Kenyan and Tanzanian waters. A rapid decline in total catches and changes in species composition have occurred in Kenyan waters. In the initial surveys over 80% of the weight was of haplochromine cichlids,

Fig. 14.4. Changes in species composition of annual catches from Kenyan waters of Lake Victoria: note rise in *Lates* catches (data from FAO, 1985).

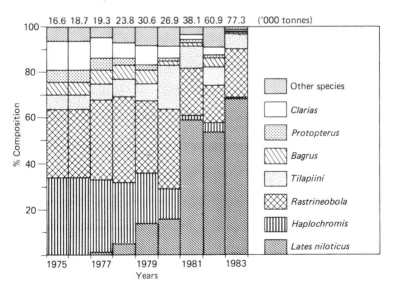

catches falling off with depth from over 900 kg h^{-1} in 10–20 m to 196 kg h^{-1} in 60–69 m (only 29 kg h^{-1} in water deeper than this). In 1976, 68% of the weight landed from Kenyan waters was of haplochromines (from all depths, mostly 40–50 m), included in a mean weight per haul of 137 kg h^{-1} of all species (Benda, 1981). By 1981, haplochromines were no longer taken in the trawl here; mean catch rates included 169 kg h^{-1} *Lates niloticus* and 15.6 kg h^{-1} *Oreochromis niloticus*, with very few *Bagrus*, *Clarias* and other species. As stressed by the HEST team (p. 101), numerous female haplochromines mouth-brooding young are destroyed in trawl catches. Furthermore, the effects of dragging the trawl across the bottom nesting sites is likely to be severely detrimental. In Lake Malawi, where trawl fisheries for small cichlids have also been developed, the species composition in catches has also changed, reducing the numbers of the larger species (Turner, 1977*a,b*).

When reviewing the status of the multispecies fisheries in the African Great Lakes, Turner (1978) considered a fishery to be moderately fished if the 1977 catch rate was within 15% of half the initial catch rate. On this basis, all the gillnet and longline fisheries were classed as heavily fished, and the purse seine, liftnet and bottom trawl fisheries as lightly or moderately fished. The species composition of catches had shifted drastically towards smaller species in all fisheries except the ringnet fishery for tilapia in Lake Malawi. In Lake Tanganyika, however, the catch per unit effort of the pelagic clupeids in purse seine catches increased following the intensive fishing for their centropomid predators (Coulter, 1970).

Catches from South American freshwaters

In South America the food fishes come mainly from rivers and their floodplains. In the forest zone the human population is so low that the rivers and their adjoining lakes are not yet heavily fished, except locally near centres of human population. The human population is, however, now rising very rapidly and some of these centres are growing explosively. Ecological studies (e.g. by Goulding, 1980, 1981) have shown how dependent are these fishes on forest products for their food, and clearing the forest is bound to lead to diminishing fish stocks, even though many fishes migrate long distances up- and downriver. As nutrients are low in these forest rivers it would be very easy to overfish them once the human population of an area had increased. The long estuarine reaches of rivers such as the Amazon and Orinoco (Novoa & Ramos, 1978) should be among the most productive waters, as the river water is fertilized by salts from the sea, but published data from these estuarine reaches are scanty.

A recent survey of the very scattered literature on South American food fishes (Lowe-McConnell, 1984) indicated that over much of the continent the main food fishes appearing in markets are: (1) deep-bodied serrasalmine characoids (such as *Colossoma, Mylossoma*); (2) migratory prochilodontid characoids; (3) other large characoids such as *Brycon, Salminus* (where this occurs), ctenolucid and erythrinid species (the latter on floodplains), anostomids and hemiodids; (4) the smaller but very numerous characoids, such as *Triportheus* and curimatids, used mainly when the larger preferred fishes are not available; (5) catfishes, including large pimelodids such as *Brachyplatystoma* and *Pseudoplatystoma* species, *Hypophthalmus* where these occur, and smaller callichthyids and loricariids in areas lacking larger species; (6) cichlids, especially *Cichla*, which are very important as subsistence and sport fish over much of the continent, with *Crenicichla* and smaller cichlids mainly in subsistence catches; (7) osteoglossids – *Arapaima gigas*, its numbers now reduced where it has been heavily fished, and the smaller *Osteoglossum*, also a preferred food fish; and (8) sciaenid *Plagioscion* and large clupeid *Pellona* species. Numerous other kinds are also eaten by subsistence fishermen.

In the Central Amazon basin, data are available on relative numbers of species and seasonal changes in catches in various types of gear from three main areas, as discussed above (p. 126) (see also Annibal, 1985). The Manaus market fishes (Table 6.1) came from a huge area (mapped by Petrere, 1978*b*): from over 700 km upriver to 600 km downriver from Manaus, as well as long distances up tributary rivers. Petrere (1982, 1983*b*, 1985) analysed the catches of the 11 most important groups of fish, treating eight river systems separately (Amazon downstream from Manaus, Solimões and Japurá upstream, and tributary Madeira, Purus, Jurúa, Jutaí, Negro and Branco rivers). This brought out clearly the differences in types of fish caught from different areas, also how strongly changes in environmental conditions, particularly fluctuations in water level on the floodplains, affect both the fishing strategies and composition of the catches. Catches brought to market are also affected by factors other than the relative abundance of species in the rivers; smaller boats operate nearer to Manaus and their catch is more diverse and sold quicker (as fresher). The bigger boats fishing in more remote areas tend to concentrate on tambaqui (*Colossoma macropomum*) which commands a higher price. A Manaus boat might be away for 40 days, fishing in 100 different places. Fishing methods were described by Petrere (1978*b*), Goulding (1981) and Smith (1981). At Manaus nearly half (48%) of the fish landed came from openwater seines (*arrastão, redinha, lampara*), 34% from gillnets (*malhadeira*), 9% from large shore seines (*arrastadeira, rede* *grande*), 3% from trident spears used at night with a lamp, 2% from hooked lines, and smaller amounts from harpoons and local methods used for particular species.

What is the best unit of effort to use to compare catches caught by such varied gear and involving so many species? Petrere (1978*a*) thought number of fishermen × number of fishing days, but his later analyses (Petrere, 1982, 1983*b*) showed that 95–98% of the variance could be related to the number of fishermen operating the gear. Catches per man-hour for different types of gear were given by Smith (1981).

Catches from Asian freshwaters

In the Indo-Pacific region, Hora & Pillay (1962) estimated that there were 2.5 million ha of freshwater, made up of natural lakes (59%), man-made lakes (19%) and fishponds (22%), and other areas where fish culture could be developed in wet ricefields and brackish waters. Since then many large man-made lakes have been created, many in the Mekong system (see Lagler, 1976).

Of the yields from Asian waters shown in Table 14.1, the nutrient-rich shallow lakes in Indonesia were especially productive; the low catches from some Indian reservoirs were attributed to concentrating on cyprinids poorly adapted to lakes. In peninsular India the monsoon-filled rivers are dammed at many points to hold floodwater through the dry season and many of these impoundments are stocked with carp fry. The estuarine reaches have important fisheries based on anadromous clupeids (*Hilsa*). The Indian carps most commonly stocked, *Catla catla, Cirrhina mrigala, Labeo rohita* and *L. calbasu*, spawn in the June–September monsoon in the flooded river shallows. They do not spawn in ponds, though they may spawn in some large reservoirs, and since 1957 spawning has been induced by injecting the fish with pituitary hypophysis hormone. The fry are very hardy and are transported long distances to stock dams. These species differ in food habits, which makes them suitable complementary species, though they all take varying combinations of algae, zooplankton, decomposing vegetable matter, detritus and organic waste. These carps grow fast – *L. calbasu* to 35 cm (450 g) in one year, *L. rohita* to 45 cm (675 g), *Catla catla* to 45 cm (900 g), *C. mrigala* to 60 cm (up to 2 kg) – and can grow very large, e.g. *Catla* to 1.8 m. Catfishes such as *Mystus* species are also important food fish from Indian rivers.

The Mekong River system, with the Grand Lac of Kampuchea flooded annually by water backed up the Tonle Sap and several large man-made lakes, has very valuable fisheries (Lagler, 1976). Catches from the Grand Lac have declined since surrounding forest was cleared (Dussart, 1974).

In the Lower Mekong, below the Tonle Sap, fishes are caught mainly when migrating downriver as the water falls in October to February (Blache & Goossens, 1954). Other species move from the sea into the estuary, where earth dams are constructed to retain these fishes in flooded areas for culture. More than 150 of the 800 species inhabiting the Mekong system make up the bulk of the Mekong catch, mainly cyprinids (species of *Cirrhinus, Thynnichthys, Cyclocheilichthys, Labeo, Puntius, Hampala, Pangasius*) and siluroids (such as *Kryptopterus, Mystus, Wallago*), with species of *Channa, Notopterus, Clupea* and *Ambassis*.

Standing stocks

The tropical waters carry many more fish species than comparable waters in the temperate zones, but are the standing stocks (ichthyomasses) higher? Is fish production per unit area greater than in the temperate zones?

Estimates of ichthyomasses from various tropical and temperate waters, compared in Table 14.2, are based where possible on samples collected by chemofishing or some method which catches all the fish present. But on marine reefs estimates generally have to be made by surveying the numbers of different types of fish in a quadrat or along a transect (observed while SCUBA swimming), then collecting a sample of fish (e.g. by rotenone poisoning or explosive) to obtain fresh weights from which total biomass is computed, and the margins of error are great. These reef samples do not represent production in these areas, as they probably include some fishes which have fed over a much wider area. Production, the rate at which fish flesh is produced, is not known as there is very little information on growth rates of reef fish.

Data are as yet only fragmentary but they show that estimates of biomass vary very much with the conditions at the time of sampling. This is very clear from the variations in biomass across a coral reef, for example at One Tree Island (Great Barrier Reef) where values ranged from 175 kg ha^{-1} on the windward upper reef slopes to 1950 kg ha^{-1} at the transition between reef and offreef floor (Goldman & Talbot, 1976). On coral reefs the structure of the reef affects the ichthyomass distribution very markedly. An artificial reef of concrete blocks (800 blocks spread over 50 m in 9 m of water) established for 2 yr in the Virgin Islands housed 55 fish species totalling 87 kg weight, 11 times the concentration of fishes in two nearby natural reefs (Randall, 1963). This artificial reef was colonized mainly by juveniles of species which fed in nearby eel-grass beds (haemulids of four species making up 41% of the ichthyomass, serranids 18%).

Table 14.2. *Comparison of ichthyomasses in some tropical and temperate waters (for further data from freshwaters see Welcomme, 1979)*

Location	Biomass (kg ha⁻¹)		Authority
	Mean	Range	
Marine reefs			
Red Sea fringing reef	380		Clark *et al.*, 1968[a]
Caribbean patch reef	450		Bardach (1959)
Virgin Is (two reefs)	1590		Randall (1963)
Hawaii Keahole Point (two)	1850		Brock, 1954[a]
One Tree I, GBR		175–1950	Goldman & Talbot (1976)
Freshwaters			
Africa			
L. Chad		30–5620	Loubens (1969)
L. George (Uganda)		60–900	Burgis *et al.* (1973)
L. Tanganyika (pelagic)		150–850	FAO (1982)
L. Malawi (pelagic)		86–94	FAO (1982)
Kafue floodplain (Zambia)			
openwater lagoon	337 highwater:	426 lowwater	Lagler *et al.* (1971)
vegetated lagoon	2682 highwater:	592 lowwater	
river channel	337 highwater:	204 lowwater	
grassmarsh	64 highwater:	dry	
Kafue R. channel (3 sites)		106–676	Kapetsky (1974)
Niger R. coffer dams (Nigeria)	60+		Motwani & Kanwai (1970)

Sokoto R. pools (Nigeria)			
sand bottom	785	691–1007	Holden (1963)
mud bottom	233	196–270	
mixed bottom	1012	585–1440	
Chari R. (W. Africa)			
river arm	5116		Loubens (1969)
river arm 2 months later	1600		
river arm following year	369		
South America			
Parana R. floodplain (Argentina)			
temporary lagoons (8)	1264	175–6500	Bonetto *et al.* (1969)
permanent lagoons (4)	918	550–1287	
Apure R. floodplain (Venezuela)	982		Mago (1970)
Magdalena R. floodplain (Colombia)			
openwater	55.7	0.23–251	Kapetsky (1976)
bays	78.9	20–2323	
Solimões Amazon (Manaus)	1600		Bayley (1982)
Asia[b]			
Gombak R. (Malaya)	180		Bishop (1973)
Streams (3) (Thailand)		81–186	Geisler *et al.* (1979)
Temperate freshwaters for comparison			
R. Thames (England)	659		IBP
Loch Leven (Scotland)	130		IBP
Horokiwi stream (New Zealand)	311		Allen (1951)

[a] Cited in Goldman & Talbot (1976).
[b] For Mekong habitats see Pantulu in Davies & Walker (1986).

In African freshwaters, on the Kafue floodplain (Zambia) where nine 0.25 ha areas were chemofished, biomasses from open water and the river channel (both 337 kg ha^{-1}) were much less than in a vegetated lagoon (2682 kg ha^{-1}) (Lagler et al., 1971). Furthermore, in this openwater lagoon the biomass was higher at low water (426 kg ha^{-1}) than at high water (337 kg ha^{-1}), though estimates for the whole vast area of floodplain suggested that the low-water ichthyomass was actually less, only about 60% of the high-water ichthyomass (Welcomme, 1979). In the Kafue River channel, Kapetsky (1974) obtained very different values at three sites, which he thought reflected fishing pressures in the three places. In the Sokoto River (Nigeria), pools with both sand and mud carried higher biomasses than sand- or mud-bottomed pools, the latter containing mainly small fishes (Holden, 1963). In Lake Chad, Loubens (1969) found ichthyomass estimates to vary from 30 to 5620 kg ha^{-1}. Even in the relatively uniform open waters of Lake George (Uganda) ichthyomasses ranged from 60 to 900 kg ha^{-1} (Burgis et al., 1973). In an arm of the Chari River (West Africa) the ichthyomass was greatly reduced 2 months after the initial samples (at the end of the flood) and was very small in the same place the following year, indicating the seasonal variation and great variation from year to year (Loubens, 1969).

In South America, on the Magdalena floodplain (Colombia), Kapetsky (1976) estimated biomasses between 0.23 and 251 kg ha^{-1} in the open water and between 20 and 2323 kg ha^{-1} in the bays; there was also a negative correlation between fish stock density and water level (see Welcomme, 1979). In the Middle Paraná near Santa Fé (Argentina), fish communities in eight temporary oxbow lakes studied at low water had a mean biomass of over 1200 kg ha^{-1} of the dry season area of the lake, but numbers of fish and biomasses varied very much from lake to lake (Bonetto et al., 1969). The biomass was over 900 kg ha^{-1} from the low-water area of four permanent lakes. These examples all stress the difficulties of trying to make comparisons between different locations. Only small areas can be sampled comprehensively and the extrapolation of such data to whole floodplains or lakes many thousands of hectares in extent multiplies any initial errors or effects due to sampling bias. Furthermore, the fishes move about a good deal and the presence of migrant fishes at the time of sampling will greatly affect results. Dry-season pools are refuges for fishes from a wide area, so the ichthyomass is not produced within the pool (as it would be in a fishpond); the flooded area which has contributed to the production of the fish is difficult to estimate and is likely to vary considerably from year to year.

Biological production

The difficulties of determining fish growth rates in tropical waters complicate estimates of the biological production and data are scanty. In two freshwater lagoons in Cuba, production figures based mainly on growth data for introduced North American species (bluegill, *Lepomis macrochirus*, and largemouth bass, *Micropterus salmoides*) were determined by Holčík (1970) as 220 and 276 kg ha^{-1}, 70–85% of the ichthyomass (Table 14.3). This production/biomass (P/B) ratio (0.68–0.85) was higher than some from temperate waters (Czechoslovakia) and Holčík attributed this to the fast growth rates, early maturity and short life cycles in tropical waters.

In Africa, the available production for each of 20 species of fishes in Lake Kariba was less than the biomass, but the estimated total production (i.e. including that of the fishes which did not survive to be caught at the end of the year) exceeded the biomass. On the Kafue floodplain, for three tilapia species studied by Kapetsky (1974), production, calculated from growth rate data, equalled biomass in *Oreochromis macrochir*, exceeded it (159%) in *Tilapia rendalli* and was only 75% of the biomass in the larger-growing *O. andersonii* (Table 14.3).

Coulter (1981) summarized information on standing stocks and production for the pelagic clupeids and their centropomid predators in Lake Tanganyika. These fish populations are unusual in that predator and prey biomasses are nearly equivalent, which can only be sustained if the prey are substantially more productive than their predators; the clupeids had a life span of 12–15 months, the centropomids 4–5 years. From growth data Coulter determined that the prey species had a mean annual production of 700 kg ha^{-1}, mean annual biomass of 230 kg ha^{-1} (P/B ratio 3.04) and an estimated clupeid yield of 350 kg ha^{-1} yr^{-1}.

In the equatorial 290 km^2 Lake George (Uganda) production of the algivorous tilapia *O. niloticus*, mainstay of the fishery, was estimated to be 5500 tonnes (fresh wt) yr^{-1} (equivalent to 189 kg ha^{-1} yr^{-1}), of which over half was cropped each year (2790 tonnes (fresh wt), equivalent to 96 kg ha^{-1} yr^{-1}). As the mean biomass of tilapia had been estimated as 3.7 g (fresh wt) m^{-2}, this would give a surprisingly high $P:B$ ratio of 5:1 (suggesting that the biomass may have been underestimated; the great variability of biomass, 60–900 kg ha^{-1}, in this lake was shown in Burgis *et al.* (1973)).

For Amazon fishes, from fortnightly samples collected from three stations over a 190 km stretch of Solimões whitewater near Manaus,

Table 14.3. *Ecological production, yields and turnover rates in some tropical fishes*

	Mean biomass (kg ha^{-1}) B	Production (kg ha^{-1} yr^{-1}) total A	available P	Yield (kg ha^{-1} yr^{-1}) total Y(A)	available Y(P)	A/B ratio (turnover)
Neotropics						
Mexican coastal lagoons (Warburton, 1979)	–	345		–	–	3.25 demersal \ 8.44 pelagic
Cuban lagoons (Holčík, 1970)						
L. Sabanilla	321	220		–	–	0.68
L. Luisa	325	276		–	–	0.85
Amazon, Solimões	1600	2800				1.75
Manaus (Bayley, 1982)						
Africa						
L. Kariba (Mahon & Balon, 1977)						
20 spp, of which	827	1224	720	400	202	1.48
9 'preferred' spp	383	298	227	193	143	0.78
6 'secondary' spp	343	548	333	71	31	1.60
5 'accompanying' small spp	100	377	159	135	28	3.77
Kafue Eastern floodplain (Kapetsky, 1974)						
Tilapia rendalli	125	198	110	18	8	1.59
Oreochromis macrochir	145	145	96	39	23	1.00
Oreochromis andersoni	157	119	92	23	15	0.75
L. Tanganyika (Coulter, 1981)						
pelagic prey spp	230	700		350		3.04
predators	37+					<1.0
L. George (Burgis & Dunn, 1978)		189		96		5.1
Temperate Waters (IBP data)						
R. Thames (England)	120	197				1.6
Loch Leven (Scotland)	130	170				1.3
Loch Leven (Scotland)	25	44				1.8
Horokiwi stream (New Zealand) (Allen 1951)	311	533				1.7

Bayley (1982) estimated fish biomasses at high and low water; these gave a mean annual biomass of ca 160 g m^{-2} (1600 kg ha^{-1}), with considerable fluctuations. Fish production was computed from estimated growth rates to be of the order of 280 g m^{-2} (2800 kg ha^{-1}). This indicated a very high P/B ratio (1.75) in these riverine fishes (in which fecundity and juvenile mortality are probably very high). Piscivores made up a high proportion, 35% of the total biomass, 25% of the production in the study area.

Production can, however, be quite high in relation to biomass in some temperate waters. IBP data collated in Le Cren & Lowe-McConnell (1980) indicated that P/B ratios ranged from 0.9 to 2.3 for some Soviet lakes, and from 1.2 to 3.7 for some small streams in England. In Loch Leven (Scotland) fish production was estimated to be 130% of biomass. In Allen's (1951) classic studies of trout production in New Zealand streams, he concluded that production greatly exceeded biomass (P/B = 1.7), though Le Cren considered Allen's production estimates too high.

Thus, although tropical studies on biological production of fish are in their infancy, it seems clear that standing stocks and production can be considerably higher than those usual in temperate waters. The transfer efficiencies to produce these fish stocks were considered in an earlier chapter (p. 276). The proportions cropped by man (yields), already considered, depend very much on the efficiency of fishing methods.

Types of fish community and responses to exploitation

In earlier chapters we have seen how tropical fish communities range from those in which fish populations respond to changes in environmental conditions (particularly variations in nutrient supply) seasonally and from year to year, to communities in which populations remain very stable throughout the year and from year to year. The associated attributes of these extreme types of community were summarized in Table 11.1. The former, characteristically found on floodplains, in upwelling areas of lakes or sea and in estuaries of great rivers, are in many ways more like communities in the temperate zones, also subject to great seasonal variation, than are the very diverse communities found where environmental conditions fluctuate little and have remained stable long enough for such communities to evolve, aided by structural cover (reef and rock) providing shelter from predation.

These tropical fish communities demonstrate clearly the general characteristics of 'pioneer' and 'mature' communities identified by Margalef (1968). Those subject to seasonal variation, for example in the pelagic zones, behave in many ways like pioneer communities in that they tend to be populated by relatively few species with marked dominants, fishes with short life cycles ('r-strategists') whose numbers fluctuate markedly with the nutrient supply, fast-growing fishes with a high production/biomass ratio, mobile fishes with relatively simple behaviour, lacking territoriality and complex breeding behaviour. In the more mature communities, life cycles tend to be longer and learning processes more important. Reef fishes tend to have longer life cycles than pelagic ones, many undergoing sex change, and their stores of learned information must be far greater (see for example Helfman & Schultz, 1984; Diamant & Shpigel, 1985). Many (most) reef fish are territorial, homing to the same places each night, often over regular routes; their social relationships include specific breeding displays, parental care of the young in many species and complex social and interspecific relationships at many levels, even interphylum mutualism with crustaceans, echinoderms or coelenterates. In the pelagic zone, trophic specialization is generally for grabbing food as low as possible in the trophic web (phytoplankton or detritus) or to feed on fishes which do this. In reef and rock faunas, adaptive radiations have led to trophic specializations to use the many diverse food sources. In the tropics the availability of plant food throughout the year enables many fishes to be direct herbivores.

For conservation it is very important to realize that these two extreme types of fish community are likely to react very differently to fishing pressures and other perturbations such as pollution. Communities in which r-selection predominates, with their very fast population turnover and already subject to natural fluctuations in numbers, are likely to be able to recover relatively rapidly if the perturbations are not too drastic or longlasting. The short life cycles characteristically found in tropical waters mean, however, that relatively few age groups are represented in catches and responses to fishing pressure will be much more immediate than in temperate seas. The very diverse communities, on the other hand, seemingly so 'stable', are likely to be much more easily damaged and to take longer to recover. Stability has been essential for their evolution but they are likely to be very easily disrupted when any unusual perturbation to which they have not become adjusted during their evolution is encountered. Once disrupted, the very complex web of interrelationships will recover very slowly, or not at all if irrevocable damage has been done.

Life history characteristics and responses to exploitation at different fishing levels have been examined by Regier (1977), Garrod & Knights (1979) and Adams (1980), who also concluded that the K-selected species would be highly sensitive to overfishing and that once depleted, stocks would take a long time to recover.

The greatest care is needed, therefore, to avoid damage to the very complex ecosystems, such as the colourful world of coral reefs, and perhaps even more to the very diverse communities of cichlid fishes in the African Great Lakes, which lack the recolonizing pelagic larvae of coral reef fishes. The present catastrophic decline in Lake Victoria's cichlid populations stresses the dangers posed by introduced species to the indigenous fish. Fishes in riverine ecosystems are likely to be more resilient, since these are already subject to environmental fluctuations and the fishes are less stenotopic. Many of the tropical forest river fishes are, however, very dependent on forest products for food, and clearing the forests is having profound effects on these fish faunas.

From studies cited in this book, it is clear that natural productivity varies a great deal, limited by availability of nutrients at the base of the food webs. This is remedied in fish culture by additions of nutrients and by feeding the fish; in natural waters there is not much we can do about it except learn to manage the fish populations to keep catches within the limits of sustained production. Tropical fish research is still in its infancy and we need long-term approaches to its problems. We need to know what controls are necessary to maintain standing stocks to give optimal sustained yields. In the sea, fish populations are less likely to be fished to extinction than they can be in circumscribed freshwaters, even in rivers (for example the demise of *Labeo altivelis* populations fished on their spawning runs up the Luapula from Lake Mweru (de Kimpe, 1964)). As any water body can only produce a limited amount of fish flesh a year, whether fishery management should aim at a few larger 'preferred' species or more numerous smaller species (as discussed for the Amazon floodplain by Bayley (1981)) should be part of local economic planning.

In a few places (notably Sri Lanka which lacked any suitable lacustrine species and in Madagascar), stocking with exotic species such as tilapia has boosted production. But the early promise of such fisheries has not always lasted (for example, in the Sepik River, Papua New Guinea). In numerous other cases introductions have led to problems, ranging from introduced species becoming pests (as did snappers and groupers stocked for sport fishing around some Hawaiian islands), to the extermination of local species (as in Lake Lanao, Philippines and Lake Titicaca in the Andes). As we have seen, in Gatun Lake (Panama) and in Lake Victoria, both huge lakes, introductions (whether accidental or deliberate) have changed the whole fish faunas. The demise of the cyprinid species flock in

Lake Lanao in less than 25 years after extraneous species were introduced has shown how the products of millennia of evolution can be so quickly destroyed.

Introductions made under pressure from sport fishing interests are often particularly harmful, as introduced species are generally (1) piscivorous, and (2) large species, unlikely to lead to increased production of fish for food as production is lost at each step of the food chain that produces them. Introductions of non-piscivorous species can also radically affect indigenous species in unforeseen ways, as introduced tilapias did in Lake Victoria.

By analogy with tropical rainforests, tropical conditions generally have been thought to be very 'stable'. Studies cited in this book have shown that conditions in tropical waters are in many cases very variable, seasonally and from year to year, their fish communities very mobile, their species composition changing dynamically. Man-made lake studies have shown how quickly riverine fish faunas can become modified into lacustrine ones. Even in the less-seasonal lacustrine environments, major changes have occurred during the course of this work: much of Lake Chad has dried up due to long-term climatic events, and fish communities in Gatun Lake, Panama, and in Lake Victoria have changed radically due to biotic pressures from introduced species.

We have a great responsibility to see that these uniquely complex and colourful fish faunas, especially those in the African Great Lakes and on coral reefs, are preserved, not only as sources of food and recreation for future generations, but also as objects of wonder and material which can contribute greatly to the scientific understanding of the evolutionary processes at work on our planet.

APPENDIX: FISH FAMILIES IN TROPICAL WATERS

Fish families encountered in tropical communities mentioned in the text, indicating size of family by approximate numbers of genera and species (not all tropical) after Nelson (1984), and freshwater families (*) found in Neotropical (N), Ethiopian (E) or Asian (A) regions

		No. genera	No. species	*
Class CHONDRICHTHYES				
Subclass ELASMOBRANCHII				
Rhincodontidae	whale shark	1	1	
Carcharhinidae	requiem sharks	24	91	
Sphyrnidae	hammerheads	1	9	
Pristidae	sawfishes	1	6	
Torpedinidae	electric rays	10	38	
Rhinobatidae	guitarfishes	9	48	
Dasyatidae	stingrays	10	90	
Potamotrygonidae	freshwater stingrays	2	14	N
Myliobatidae	eagle rays	5	27	
Mobulidae	manta rays	2	10	

(continued)

Fish families in tropical waters – *continued*

		No. genera	No. species	*
Class *OSTEICHTHYES*	Bony fishes			
Subclass *DIPNEUSTI*	lungfishes			
Ceratodontidae	Australian lungfish	1	1	
Lepidosirenidae	S. American lungfish	1	1	N
Protopteridae	African lungfishes	1	4	E
Subclass *BRACHIOPTERYGII*	Tassel-finned fishes			
Polypteridae	bichirs	2	11	E
Subclass *ACTINOPTERYGII*	Ray-finned fishes			
Subdivision *TELEOSTEI*				
Infradivision *Osteoglossomorpha*				
Order *Osteoglossiformes*				
Osteoglossidae	bony tongues	4	6	N,E,A
Pantodontidae	butterflyfishes	1	1	E
Notopteridae	featherbacks	3	6	E,A
Mormyridae	elephantfishes	16	190	E
Gymnarchidae		1	1	E
Infradivision *Elopomorpha*				
Order *Elopiformes*				
Elopidae	tenpounders	1	5	
Megalopidae	tarpon	1	2	
Albulidae	bonefishes	2	4	
Order *Anguilliformes*				
Anguillidae	freshwater eels	1	16	E,A
Moringuidae	worm eels	2	10	
Muraenidae	moray eels	12	110	
Nemichthyidae	snipe eels	3	9	
Ophichthyidae	snake eels	49	236	
Congridae	garden eels	4	14	

Infradivision Clupeomorpha				
Order Clupeiformes				
Denticipitidae	denticle herring	1	1	E
Clupeidae	herring/sardines	50	190	
Engraulidae	anchovies	16	139	
Chirocentridae	wolf-herrings	1	1	
Infradivision Euteleostei				
Superorder *OSTARIOPHYSI*				
Series *Anotophysi*				
Order *Gonorynchiformes*				
Chanidae	milkfish	1	1	
Kneriidae		4	24	E
Phractolaemidae		1	1	E
Series *Otophysi*				
Order *Cypriniformes*				
Cyprinidae	minnow/carps	194	2070	E,A
Homalopteridae	hillstream loaches	27	110	A
Cobitidae	loaches	21	175	A
Order *Characiformes*				
Citharinidae	characins	2	7	E
Distichodontidae		15	69	E
Hemiodontidae		9	50	N
Curimatidae		30	138	N
Anostomidae		10	105	N
Erythrinidae	trahiras	3	5	N
Lebiasinidae		2	11	N
Gasteropelecidae	hatchetfishes	3	9	N
Ctenoluciidae	pike-characids	2	4	N
Hepsetidae		1	1	E
Characidae	characins	166	841	N,E

(continued)

Fish families in tropical waters – *continued*

		No. genera	No. species	*
Order *Siluriformes*	catfishes			
Bagridae	bagrids	27	205	E,A
Siluridae	sheatfish	15	70	A
Schilbeidae	schilbeids	20	60	E,A
Pangasiidae	pangasiids	8	25	A
Amphilidae	loach catfishes	7	47	E
Sisoridae	sisorid catfishes	20	65	A
Clariidae	airbreathers	13	100	E,A
Heteropneustidae	airsac catfishes	1	2	A
Malapteruridae	electric catfishes	1	2	E
Akysidae	stream catfishes	3	8	A
Amblycipidae	torrent catfishes	2	5	A
Chacidae	squarehead catfishes	1	2	A
Ariidae	seacatfishes	20	120	N,E,A
Plotosidae	eeltail catfishes	8	30	E,A
Mochokidae	squeakers	10	150	E
Doradidae	thorny catfishes	37	80	N
Auchenipteridae		19	60	N
Pimelodidae	longwhiskered catfishes	56	290	N
Ageneiosidae	barbelless catfishes	2	25	N
Helogenidae		2	4	N
Cetopsidae	whale catfishes	4	12	N
Hypophthalmidae	loweye catfishes	1	2	N
Aspredinidae	banjo catfishes	8	25	N
Trichomycteridae	parasitic catfishes	27	175	N
Callichthyidae	armoured catfishes	8	110	N
Loricariidae	suckermouth armoured	70	600	N
Astroblepidae	andean catfishes	1	35	N

Taxon	Common name			
Order Gymnotiformes				
Sternopygidae		5	11	N
Rhamphichthyidae		2	2	N
Hypopomidae		4	12	N
Apteronotidae		10	25	N
Gymnotidae	knifefishes	1	3	N
Electrophoridae	electric eel	1	1	N
Superorder PROTACANTHOPTERYGII				
Order Aulopiformes				
Synodontidae	lizardfishes	5	39	
Superorder PARACANTHOPTERYGII				
Order Ophidiiformes				
Carapidae	pearlfishes	6	27	
Order Batrachoidiformes				
Batrachoididae	toadfishes	19	64	
Order Lophiiformes				
Antennariidae	frogfishes	15	60	
Ogcocephalidae	batfishes	9	57	
Order Gobiesociformes				
Gobiesocidae	clingfishes	35	110	
Superorder ACANTHOPTERYGII				
Order Cyprinodontiformes				
Exocoetidae	flyingfishes	8	48	
Hemiramphidae	halfbeaks	12	80	
Belonidae	needlefishes	9	32	
Scomberesocidae	sauries	4	4	
Oryziatidae	ricefishes	1	7	
Cyprinodontiformes (*sensu* Parenti, 1981)				
Aplocheilidae	rivulines	15	210	N,E,A
Cyprinodontidae	killifishes	29	268	N,E,A
Poeciliidae	live-bearers	22	150	N
Anablepidae	foureyes	1	3	N

(*continued*)

Fish families in tropical waters – *continued*

		No. genera	No. species	*
Order *Atheriniformes*				
Atherinidae	silversides	29	160	
Melanotaeniidae	rainbowfishes	8	50	
Series *PERCOMORPHA*				
Order *Beryciformes*				
Holocentridae	squirrelfishes	8	61	
Order *Syngnathiformes*				
Aulostomidae	trumpetfishes	1	3	
Fistulariidae	cornetfishes	1	4	
Centriscidae	shrimpfishes	2	4	
Syngnathidae	pipefishes/seahorses	55	230	
Order *Dactylopteriformes*				
Dactylopteridae	flying gurnard	4	4	
Order *Synbranchiformes*				
Synbranchidae	swamp eels	4	15	N,E,A
Order *Scorpaeniformes*				
Scorpaenidae	scorpionfishes	60	310	
Triglidae	searobins	14	86	
Platycephalidae	flatheads	22	64	
Order *Perciformes*				
Centropomidae	snooks	6	35	
Serranidae	seabasses	35	370	
Grammistidae	soapfishes	10	24	
Teraponidae	grunters	15	39	
Kuhliidae	aholeholes	4	17	
Priacanthidae	bigeyes	3	18	
Apogonidae	cardinalfishes	26	192	
Pomatomidae	bluefishes	2	3	
Rachycentridae	cobia	1	1	

Family	Common name			
Echeneidae	remoras	7	8	
Carangidae	jacks/pompanos	25	140	
Coryphaenidae	dolphinfishes	1	2	
Leiognathidae	pony, slipmouth	3	21	
Bramidae	pomfrets	6	18	
Lutjanidae	snappers	17	185	
Caesionidae	fusiliers	4	30	
Lobotidae	tripletails	2	6	
Gerridae	mojarras	7	40	
Haemulidae	grunts	17	175	
Sparidae	porgies	29	100	
Lethrinidae	emperors	4	25	
Nemipteridae	threadfin breams	3	35	
Sciaenidae	croakers/drums	50	210	
Mullidae	goatfishes	3	55	
Pempheridae	sweepers	2	25	
Toxotidae	archerfishes	1	6	
Kyphosidae	seachubs	17	45	
Ephippidae	spadefishes	5	17	
Scatophagidae	scats	2	4	
Chaetodontidae	butterflyfishes	10	114	
Pomacanthidae	angelfishes	7	74	
Nandidae	leaffishes	7	10	
Cichlidae	cichlids	84	680	N,E,A
Embiotocidae	surfperch	13	23	
Pomacentridae	damselfishes	25	235	N,E,A
Cirrhitidae	hawkfishes	9	32	
Mugilidae	mullets	13	95	
Sphyraenidae	barracudas	1	18	
Polynemidae	threadfins	7	32	
Labridae	wrasses	57	500	

(continued)

Fish families in tropical waters – *continued*

		No. genera	No. species	*
Scaridae	parrotfishes	98	312	
Chaenopsidae	pike-blennies	10	56	
Blenniidae	combtooth blennies	53	301	
Callionymidae	dragonets	10	130	
Eleotridae	sleepers	40	150	
Gobiidae	gobies	200	1500	
Acanthuridae	surgeonfishes	10	77	
Siganidae	rabbitfishes	2	25	
Trichiuridae	cutlassfishes	9	17	
Scombridae	mackerels/tunas	15	48	
Xiphiidae	swordfish	1	1	
Istiophoridae	billfishes	3	10	
Nomeidae	driftfishes	3	15	
Stromateidae	butterfishes	3	13	
Anabantidae	climbing gouramies	3	40	E,A
Belontiidae	gouramies	11	28	A
Helostomatidae	kissing gourami	1	1	A
Osphronemidae	giant gourami	1	1	A
Channidae	snakeheads	2	12	E,A
Mastacembelidae	spiny eels	4	60	E,A
Order *Pleuronectiformes*				
Psettodidae	flatfishes	1	2	
Cynoglossidae	tonguesoles	3	103	
Soleidae	soles	31	117	
Order *Tetraodontiformes*				
Balistidae	leatherjackets	42	135	
Ostraciidae	boxfishes	13	30	
Tetraodontidae	puffers	16	118	
Diodontidae	porcupinefishes	2	15	

REFERENCES

Ackermann, W. C., White, G. F. & Worthington, E. B. (1973) *Man-made Lakes: Their Problems and Environmental Effects*. Geophysical Monograph, no. 17. Washington, D.C.: American Geophysical Union

Adams, P. B. (1980) Life history patterns in marine fishes and their consequences for fisheries management. *Fishery Bulletin of the U.S. Fish & Wildlife Service*, **78**, 1–12

Albaret, J. J. (1979*a*) Reproduction et fecundité des poissons d'eau douce de Côte d'Ivoire. *Revue d' Hydrobiologie Tropicale*, **15**, 347–71

Albaret, J. J. (1979*b*) Revue des recherches entreprises sur la fecondité des poissons d'eau douce africains. In *Première Réunion Limnologie Africaine, Nairobi, 1979* Paris: ORSTOM

Albrecht, H. (1963) Colour signals as dynamic means of demarcating territory. In *Signals in the Animal World*, ed. D. Burkhardt, W. Schleidt & H. Altner, pp. 115–16. London: Allen & Unwin

Albrecht, H. (1969) Behavior of four species of Atlantic damselfishes from Colombia, S. America. *Zeitschrift für Tierpsychologie*, **26**, 226–76

Aldenhoven, J. M. (1983) Sex change on a coral reef and the social mechanisms that produce it. In *Abstracts, 18th International Ethological Conference, Brisbane, 1983*, p. 3 Brisbane: University of Queensland Press

Alevison, W. S. (1976) Mixed schooling and its possible significance in a tropical western Atlantic parrotfish and surgeonfish. *Copeia*, **4**, 796–8

Ali, M. T. (1984) Fish and fisheries of Lake Nubia, Sudan. *Hydrobiologia*, **110**, 305–14

Allen, G. R. (1972) *Anemone Fishes*. Neptune City, N.J.: TFH Publications Inc. (2nd edn 1975)

Allen, K. R. (1951) The Horokiwi stream: a study of a trout population. *Fisheries Bulletin, New Zealand Marine Department*, **10**, 1–238

Almeida, R. G. (1980) Aspectos taxonomicos e habitos alimentares de tres especies de *Triportheus* (Pisces: Characoidei, Characidae), do lago do Castanho, Amazonas. Master's thesis, INPA, Manaus, Brazil

Anderson, G. R. V., Ehrlich, A. H., Ehrlich, P. R., Roughgarden, J. D., Russell, B. C. & Talbot, F. H. (1981) The community structure of coral reef fishes. *American Naturalist*, **117**, 476–95

Angermeier, P. I. & Karr, J. R. (1983) Fish communities along environmental gradients in a system of tropical streams. *Environmental Biology of Fishes*, **9**(2), 39–57

Annibal, S. R. P. (1985) Pêche et hydrologie en Amazonie Centrale. *Verhandlungen der Internationalen Vereinigung für theoretische und angewandte Limnologie*, **22**, 2692–7

Anon. (1980) The biology of damselfishes. Symposium, ASIH, June 1978, Tempe, Arizona. *Bulletin of Marine Science, Special Issue*, no. 30. (Foreword by A. R. Emery & R. E. Thresher)

Axelrod, H. R. & Burgess, W. E. (1976) *African Cichlids of Lakes Malawi and Tanganyika*. 5th edn. Neptune City, N.J.: TFH

Ayala, A. F. (1970) Competition, coexistence and evolution. In *Essays in Evolution and Genetics in Honour of Th Dobzansky*, ed. M. K. Hecht & W. C. Steere, pp. 121–58. Amsterdam: North Holland Pub. Co.

Azevedo, P. de & Gomes, A. L. (1943) Contribução ao estudo da biologia da Traira, *Hoplias malabaricus*. *Boletim de Industria Animal, São Paulo, n.s.*, **5**, 15–64

Bagenal, T. (ed.) (1978) *Methods for Assessment of Fish Production in Freshwaters*. IBP Handbook no. 3, 3rd edn. (1968, 1971 edns edited by W. H. Ricker) Oxford: Blackwell Scientific Publications

Bailey, R. G., Churchfield, S., Petr, T. & Pimm, R. (1978) The ecology of the fishes in Nyumba ya Mungu reservoir, Tanzania. *Biological Journal of the Linnean Society of London*, **10**, 109–37

Balirwa, J. S. & Bugenyi, F. W. B. (1980) Notes on the fisheries of the R. Nzoia, Kenya. *Biological Conservation*, **18**, 53–8

Balon, E. K. (1971) Age and growth of *Hydrocynus vittatus* Cast. in L. Kariba Sinazongwe area. *Fisheries Research Bulletin Zambia*, **5**, 89–118

Balon, E. K. (1974) Fish production of a tropical ecosystem. In *Lake Kariba, a Man-made tropical Ecosystem in Central Africa*, ed. E. K. Balon & A. G. Coche, Monographiae Biologicae, no. 24, Part II. pp. 253–676. The Hague: Dr W. Junk

Balon, E. K. (1977) Early ontogeny of *Labeotropheus* Ahl, 1927 (Mbuna, Cichlidae, L. Malawi), with a discussion on advanced protective styles in fish reproduction and development. *Environmental Biology of Fishes*, **2**, 147–76

Balon, E. K. (1984) Patterns in the evolution of reproductive styles in fishes. In *Fish Reproduction*, ed. G. W. Potts & R. J. Wootton, pp. 35–53. London: Academic Press

Balon, E. K. & Coche, A. G. (1974). *Lake Kariba, a Man-made Tropical Ecosystem in Central Africa*. Monographiae Biologicae, no. 24. The Hague: Dr W. Junk

Balon, E. K. & Stewart, D. J. (1983) Fish assemblages in a river with unusual gradient (Luongo, Africa – Zaire system), reflections on river zonation and description of another new species. *Environmental Biology of Fishes*, **9**, 225–52

Banister, K. E. & Bailey, R. G. (1979) Fishes collected by the Zaire River Expedition, 1974–75. *Zoological Journal of the Linnean Society of London*, **66**, 205–49

Banister, K. E. & Clarke, M. A. (1980) A revision of large *Barbus* (Pisces, Cyprinidae) of L. Malawi with a reconstruction of the history of the southern African Rift Valley lakes. *Journal of Natural History, London*, **14**, 483–542

Bardach, J. E. (1958) On the movements of certain Bermuda reef fishes. *Ecology*, **39**, 139–46

Bardach, J. E. (1959) The summer standing crop of fish on a shallow Bermuda reef. *Limnology & Oceanography*, **4**, 77–85

Bardach, J. E. & Todd, J. H. (1970) Chemical communication in fish. In *Communication by Chemical Signals*, ed J. W. Johnston, D. G. Moulton & A. Turk, pp. 205–40. New York: A. C. Meridith Corporation

Bardach, J. E., Winn, H. E. & Menzel, D. W. (1959) The role of the senses in the feeding of the nocturnal reef predators *Gymnothorax moringa* and *G. vicinus*. *Copeia*, **2**, 133–9

Barel, C. N. D., Dorit, R., Greenwood, P. H., Fryer, G. *et al*. (1985) Destruction of fisheries in Africa's lakes. *Nature, London*, **315**, 19–20

Barlow, G. W. (1974*a*) Contrasts in social behavior between Central American cichlid fishes and coral-reef surgeon fishes. *American Zoologist*, **14**, 9–34

Barlow, G. W. (1974*b*) Extraspecific imposition of social groupings among surgeon fishes (Pisces: Acanthuridae). *Journal of Zoology, London*, **174**, 333–40

Barlow, G. W. (1981) Patterns of parental investment, dispersal and size among coral-reef fishes. *Environmental Biology of Fishes*, **6**, 65–85

Barlow, G. W. (1983) The benefits of being gold: behavioral consequences of polychromatism in the midas cichlid, *Cichlasoma citrinellum*. *Environmental Biology of Fishes*, **8**, 235–47

Barnett, C. (1982) The chemosensory responses of young cichlid fish to parents and predators. *Animal Behaviour*, **30**, 35–42

Barthem, R. B. (1985) Ocorrência, distribuição e biologia dos peixes da Baía de Marajó, Estuário Amazônico. *Boletim do Museu Paraense Emilio Goeldi Zoologia*, **2**, 49–69

Bauer, J. A., Jr & Bauer, S. E. (1981) Reproductive biology of pygmy angelfishes of the genus *Centropyge* (Pomacanthidae). *Bulletin of Marine Science*, **31**, 495–513

Bayley, P. B. (1973) Studies on the migratory characin *Prochilodus platensis* Holmberg, 1889. *Journal of Fish Biology*, **5**, 25–40

Bayley, P. B. (1977) Changes in species composition of the yields and catch per unit effort during the development of the fishery at L. Turkana, Kenya. *Archiv für Hydrobiologie*, **79**, 111–32

Bayley, P. B. (1981) Fish yield from the Amazon in Brazil: comparison with African river yields and management possibilities. *Transactions of the American Fisheries Society*, **110**, 351–9

Bayley, P. B. (1982) Central Amazon fish populations: biomass, production and some dynamic characteristics. Thesis for Degree of Doctor of Philosophy, Dalhousie University, Canada, 308 pp.

Baylis, J. R. (1981) The evolution of parental care in fishes, with reference to Darwin's rule of male sexual selection. *Environmental Biology of Fishes*, **6**, 223–51

Beadle, L. C. (1981) *The Inland Waters of Tropical Africa. An Introduction to Tropical Limnology*, 2nd edn. London: Longman

Beauchamp, R. S. A. (1964) The rift valley lakes of Africa. *Verhandlungen der Internationalen Vereinigung für theoretische und angewandte Limnologie*, **15**, 91–9

Beddington, J. R. & May, R. M. (1982) The harvesting of interacting species in a natural ecosystem. *Scientific American*, **247**(5), 42–9

Beeckman, W. & de Bont, A. F. (1985) Characteristics of the Nam Ngum reservoir ecosystem as deduced from the food of the most important species. *Verhandlungen der Internationalen Vereinigung für theoretische und angewandte Limnologie*, **22**, 2643–9

Bell-Cross, G. (1965) Movement of fish across the Congo–Zambezi watershed in the Mwinilunga District of Northern Rhodesia. In *Proceedings, Central African Medical Congress, Lusaka, 1963*, pp. 415–24. Oxford: Pergamon Press

Benda, R. S. (1981) A comparison of bottom trawl catch rates in the Kenyan waters of L. Victoria. *Journal of Fish Biology*, **18**, 609–13

Benech, V. (1974) Données sur la croissance de *Citharinus citharus* dans la bassin tchadien. *Cahiers d'ORSTOM, série Hydrobiologie*, **8**, 23–33

Benech, V. (1975) Croissance, mortalité et production de *Brachysynodontis batensoda* (Pisces, Mochocidae) dans l'archipel sud-est du lac Tchad. *Cahiers d'ORSTOM série Hydrobiologie*, **9**, 91–103

Bergstrand, E. & Cordone, A. J. (1971) Exploratory bottom trawling in L. Victoria. *African Journal of Tropical Hydrobiology & Fisheries*, **1**, 13–23

Berns, S., Chave, E. H. & Peters, H. M. (1978) On the biology of *Tilapia squamipinnis* (Gunther) from L. Malawi (Teleostei, Cichlidae). *Archiv für Hydrobiologie*, **84**, 218–46

Berra, T. M., Moore, R. & Reynolds, L. F. (1975) The freshwater fishes of the Laloki River system of New Guinea. *Copeia*, **2**, 316–26

Bishop, J. E. (1973) *Limnology of a small Malayan River Sungai Gombak*. Monographiae Biologicae, no. 22. The Hague: Dr W. Junk

Blache, J. & Goossens, J. (1954) Monographie piscicole d'une zone de pêche au Cambodge. *Cybium, Bulletin de l'Association des Amis du Laboratoire des Pêches Coloniales, Paris*, **8**, 1–49

Blache, J., Miton, F., Stauch, A., Iltis, A. & Loubens, G. (1964) Les poissons du bassin du Tchad et du bassin adjacent du Mayo Kebbi. *Mémoires ORSTOM, Paris*, pp. 1–483

Blackburn, M. (1965) Oceanography and the ecology of tunas. *Oceanography & Marine Biology*, **3**, 299–322

Black-Cleworth, P. (1970) The role of electrical discharges in the non-reproductive social behaviour of *Gymnotus carapo* (Gymnotidae, Pisces). *Animal Behaviour Monographs*, **3**(1), 1–77

Blake, B. F. (1977*a*) L. Kainji, Nigeria: a summary of the changes within the fish population since the impoundment of the Niger in 1968. *Hydrobiologia*, **53**, 131–7

Blake, B. F. (1977*b*) The effect of the impoundment on L. Kainji, Nigeria, on the indigenous species of mormyrid fishes. *Freshwater Biology*, **7**, 37–42

Blake, B. F. (1977*c*) Aspects of the reproductive biology of *Hippopotamyrus pictus* from L. Kainji, with notes on four other mormyrid species. *Journal of Fish Biology*, **11**, 437–45

Blaxter, J. H. S. & Hunter, J. R. (1982) The biology of clupeoid fishes. *Advances in Marine Biology*, **20**, 1–223

Blumer, L. S. (1982) A bibliography and categorization of bony fishes exhibiting parental care. *Zoological Journal of the Linnean Society of London*, **75**, 1–22

Bohnsack, J. A. (1983) Resilience of reef communities in the Florida Keys following a January 1977 hypothermal fish kill. *Environmental Biology of Fishes*, **9**, 41–53

Bonetto, A. A. & Castello, H. P. (1985) *Pesca y piscicultura en aguas continentales de America Latina*. Secretaria General de la Organización de los Estados Americanos, Programa Regional de Desarrollo Cientifico y Tecnológico, Washington, D.C. Monografia no. 31, 118 p.

Bonetto, A., Cordiviola de Yuan, E., Pignalberi, C. & Oliveros, O. (1969) Ciclos hidrológicos del rio Paraná y las poblaciones de peces contenidas en las cuencas temporarias de su valle de inundación. *Physis, B. Aires*, **29**(78), 213–23

Bonetto, A., Dioni, W. & Pignalberi, C. (1969) Limnological investigations on biotic communities in the Middle Paraná River valley. *Verhandlungen Internationalen Vereinigung für Theoretische und Angewandte Limnologie*, **17**, 1035–50

Bonetto, A. & Pignalberi, C. (1964) Nuevos aportes al conocimiento de las migracion es de los peces en los rios mesopotamicos de la Republica Argentina. *Communicacions Instituto Nacional Limnologia, Santo Tome*, **1**, 1–14

Bonetto, A. A., Pignalberi, C., Cordiviola de Yuan, E. & Oliveros, O. (1971) Informaciones complentarias sobre migraciones de peces en la Cuenca del Plata. *Physis, B. Aires*, **30**(81), 505–20

Bowen, S. H. (1979) A nutritional constraint in detritivory by fishes: the stunted population of *Sarotherodon mossambicus* in L. Sibaya, S. Africa. *Ecological Monographs*, **49**, 17–31

Bowen, S. H. (1983) Detritivory in neotropical fish communities. *Environmental Biology of Fishes*, **9**, 137–44

Bowmaker, A. P. (1969) Contribution to knowledge of the biology of *Alestes macrophthalmus* Gthr (Characidae). *Hydrobiologia*, **33**, 302–41

Bowmaker, A. P. (1973) Potamodromesis in the Mwenda River, L. Kariba. In *Manmade Lakes: Their Problems and Environmental Effects*, ed. W. C. Ackermann *et al.*, pp. 159–64. Geophysical Monograph no. 17. Washington: American Geophysical Union

Bowmaker, A. P., Jackson, P. B. N. & Jubb, R. A. (1978) Freshwater fishes. In *Biogeography and Ecology of Southern Africa*, ed. M. J. A. Werger, pp. 1181–230. The Hague: Dr W. Junk

Braum, E. (1983) Beobachtungen über eine reversible Lippenextension und ihre Rolle bei der Notatmung von *Brycon* sp. und *Colossoma macropomum*. *Amazoniana*, **7**, 355–74

Breder, C. M. & Rosen, D. E. (1966) *Modes of Reproduction in Fishes*. New York: American Museum of Natural History

Brichard, P. (1978) *Fishes of Lake Tanganyika*. Neptune City, N.J.: TFH Publications Inc.

Brichard, P. (1979) Unusual brooding behaviors in L. Tanganyika cichlids. *Buntbarsche Bulletin*, **74**, 10–12

Briggs, J. C. (1960) Fishes of worldwide (circumtropical) distribution. *Copeia*, **3**, 171–80

Briggs, J. C. (1966) Zoogeography and evolution. *Evolution, Lancaster, Pa*, **20**, 282–9

Briggs, J. C. (1967) Relationships of the tropical shelf regions. *Studies in Tropical Oceanography, Miami*, **5**, 847 pp.

Brosset, A. (1982) Le peuplement de Cyprinodonts du bassin de l'Ivindo, Gabon. *(La) Terre et la Vie*, **36**, 233–92

Bruton, M. N. (1979) The breeding biology and early development of *Clarias gariepinus* (Pisces: Clariidae) in L. Sibaya, S. Africa, with a review of breeding in *Clarias*. *Transactions of the Zoological Society of London*, **35**, 1–45

Burchard, J. V., Jr (1965) Family structure in the dwarf cichlid *Apistogramma trifasciatum*. *Zeitschrift für Tierpsychologie*, **22**, 150–62

Burgis, M. J., Darlington, J. P. E., Dunn, I. G., Gwahaba, G. G. & McGowan, L. M. (1973) The biomass and distribution of organisms in L. George, Uganda. *Proceedings of the Royal Society of London, B*, **184**, 271–98

Burgis, M. J. & Dunn, I. G. (1978) Production in three contrasting ecosystems. In *Ecology of Freshwater Fish Production*, ed. S. D. Gerking, pp. 137–58. Oxford: Blackwell

Cadwalladr, D. A. (1965) Notes of the breeding biology and ecology of *Labeo victorianus* Blgr (Cyprinidae) of L. Victoria. *Revue de Zoologie et de Botanique Africaines*, **72**, 109–34

Carmouze, J. P., Durand, J. R. & Levêque, C. (1983) *Lake Chad. Ecology and Productivity of a Shallow Tropical Ecosystem*. Monographiae Biologicae no. 53. The Hague: Dr W. Junk

Carter, G. S. (1935) Respiratory adaptations of the fishes of the forest waters with descriptions of the accessory respiratory organs of *Electophorus electricus* and *Plecostomus plecostomus*. *Journal of the Linnean Society (Zoology)*, **39**, 219–33.

Carter, G. S. & Beadle, L. C. (1930) Notes on the habits and development of *Lepidosiren paradoxa*. *Journal of the Linnean Society of London (Zoology)*, **27**, 197–203.

Carter, G. S. & Beadle, L. C. (1931) The fauna of the swamps of the Paraguayan Chaco in relation to its environment, II. Respiratory adaptations in the fishes. *Journal of the Linnean Society of London (Zoology)*, **37**, 327–68

Carvalho, F. M. (1980) Composição quimica e reprodução do mapara (*Hypophthalmus edentatus* Spix, 1829) do lago do Castanho, Amazonas (Siluriformes, Hypophthalmidae). *Acta Amazonica*, **10**(2), 379–89

Chapman, D. W. & Well, P. (1978a) Growth and mortality of *Stolothrissa tanganicae*. *Transactions of the American Fisheries Society*, **107**, 26–35

Chapman, D. W. & Well, P. (1978b) Observations on the biology of *Luciolates stappersii* in L. Tanganyika (Tanzania). *Transactions of the American Fisheries Society*, **107**, 567–73

Chartock, M. A. (1983) The role of *Acanthurus guttatus* in cycling algal production to detritus. *Biotropica*, **15**, 117–21

Chave, E. H. & Eckert, D. B. (1974) Ecological aspects of the distributions of fishes at Fanning Is. *Pacific Science*, **28**, 297–317

Chevey, P. & Le Poulain, F. (1940) La pêche dans les eaux douces du Cambodge. *Mémoires de l'Institut Oceanographique de l'Indochine, Hanoi*, **5**, 1–193

Chichocki, F. P. (1976) Cladistic history of cichlid fishes and reproductive strategies of the American cichlid genera *Acarichthys, Biotodoma* and *Geophagus*. PhD Thesis, University of Michigan, vol. 2, 710 pp.

Coates, D. (1985) Fish yield estimates for the Sepik River, Papua New Guinea, a large floodplain system east of "Wallace's line". *Journal of Fish Biology*, **27**, 431–43.

Coe, M. (1966) The biology of *Tilapia grahami* Blgr. in L. Magadi, Kenya. *Acta Tropica*, **23**, 146–77

Cohen, D. M. (1970) How many recent fishes are there? *Proceedings of the California Academy of Sciences, 4th series*, **38**, 341–6

Collette, B. B. & Earle, S. A. (eds) (1972) Results of the Tektite program: ecology of coral reef fishes. *Science Bulletin Natural History Museum, Los Angeles County*, **14**, 1–180

Collette, B. B. & Naven, C. E. (1983) *FAO Species Catalogue*. vol. 2. *Scombrids of the world: An annotated and illustrated Catalogue of Tunas, Mackerels, Bonitos and related species known to date*. FAO Fisheries Synopses no. 125. Rome: FAO

Collette, B. B. & Talbot, F. H. (1972) Activity patterns of coral reef fishes with emphasis on nocturnal–diurnal changeover. *Science Bulletin Natural History Museum, Los Angeles County*, **14**, 98–124

Connell, J. H. (1978) Diversity in tropical rain forests and coral reefs. *Science, New York*, **199**, 1302–10

Corbet, P. S. (1961) The food of non-cichlid fishes in the L. Victoria basin, with remarks on their evolution and adaptation to lacustrine conditions. *Proceedings of the Zoological Society of London*, **136**, 1–101

Cordivioli de Yuan, E. & Pignalberi, C. (1981) Fish populations in the Paraná River. 2. *Santa Fe and Corrientes areas. Hydrobiologia*, **77**, 261–72

Coulter, G. W. (1963) Hydrobiological changes in relation to biological production in southern L. Tanganyika. *Limnology & Oceanography*, **8**, 463–77

Coulter, G. W. (1966) Hydrobiological processes and the deepwater fish community in L. Tanganyika. Ph.D. Thesis, Queens University, Belfast

Coulter, G. W. (1968) The deep benthic fishes at the south end of L. Tanganyika with special reference to distribution and feeding in *Bathybates* species, *Hemibates stenosoma* and *Chrysichthys* species. *Fisheries Research Bulletin, Zambia* 4 (1965–66), 33–8

Coulter, G. W. (1970) Population changes within a group of fish species in L. Tanganyika following their exploitation. *Journal of Fish Biology*, **2**, 329–53

Coulter, G. W. (1976) The biology of *Lates* species (Nile Perch) in L. Tanganyika, and the status of the pelagic fishery for *Lates* species and *Luciolates stappersii* (Blgr). *Journal of Fish Biology*, **9**, 235–59

Coulter, G. W. (1981) Biomass, production and potential yield of the L. Tanganyika pelagic fish community. *Transactions of the American Fisheries Society*, **110**, 325–35

Courtenay, W. R. & Stauffer, J. R. (eds) (1984) *Distribution, Biology and Management of Exotic Fishes*. Baltimore: Johns Hopkins University Press

Crapon de Caprona, M. D. & Fritzsch, B. (1984) Interspecific fertile hybrids of haplochromine cichlids (Teleostei) and their possible importance for speciation. *Netherlands Journal of Zoology*, **34**, 503–38

Cridland, C. C. (1961) The reproduction of *Tilapia esculenta* under artificial conditions. *Hydrobiologia*, **18**, 177–84

Cushing, D. H. (1971) Survey of resources of the Indian Ocean and Indonesian area. *Indian Ocean Fishery Commission (IOFC) FAO/DEV/71/2,FAO, Rome*, 1–12

Dafni, J. & Diamant, A. (1984) School-orientated mimicry, a new type of mimicry in fishes. *Marine Ecology, Progress Series*, **20**, 45–50

Daget, J. (1952) Biologie et croissance des espèces du genre *Alestes*. *Bulletin de l'Institut Français d'Afrique Noire*, **14**, 191–225

Daget, J. (1954) Les poissons du Niger Superieur. *Mémoires de l'Institut Français d'Afrique Noire, IFAN-DAKAR*, no. 36, 1–391

Daget, J. (1957) Données récentes sur la biologie des poissons dans le delta central du Niger. *Hydrobiologia*, **9**, 321–47

Daget, J. (1962) Le genre *Citharinus*. *Revue de Zoologie et Botanique Africaines*, **66**, 81–106

Daget, J. & Durand, J.-R. (1981) Poissons. In *Flore et Faune aquatique de l'Afrique Sahelo-Soudienne*, vol. 2, ed. J.-R. Durand & C. Levêque, pp. 688–771. Paris: ORSTOM

Daget, J., Gosse, J. P. & Thys van den Audenaerde, D. F. E. (eds) (1984) *Check-list of the Freshwater Fishes of Africa*. (CLOFFA) vol. 1. Paris: ORSTOM-MRAC

Daget, J., Planquette, N. & Planquette, P. (1973) Premières données sur la dynamique des peuplements de poissons du Bandama (Côte d'Ivoire). *Bulletin de Museum National d'Histoire Naturelle*, 3e ser., no. 151, *Ecologie générale*, **7**, 129–42

Dahlgren, B. T. (1979) The effects of population density on fecundity and fertility in the guppy, *Poecilia reticulata*. *Journal of Fish Biology*, **15**, 71–91

Dansoko, D., Breman, H. & Daget, J. (1976) Influence de la secheresse sur les populations d'*Hydrocynus* dans le delta Central du Niger. *Cahiers ORSTOM, sér. Hydrobiologie*, **10**(2), 71–6

Davies, B. R. & Walker, K. F. (1986) *The Ecology of River Systems*. Monographiae Biologicae, **60**, Dordrecht: Dr W. Junk

Davis, T. L. O. (1982) Maturity and sexuality in barramundi, *Lates calcarifer* (Bloch), in the Northern Territory and south-eastern Gulf of Carpentaria. *Australian Journal of Marine and Freshwater Research*, **33**, 529–45

Davis, T. L. O. (1984) A population of sexually precocious barramundi, *Lates calcarifer*, in the Gulf of Carpentaria, Australia. *Copeia*, **1**, 144–9

Davis, T. L. O. (1985) The food of barramundi, *Lates calcarifer* (Bloch) in coastal waters of Van Dieman Gulf and the Gulf of Carpentaria, Australia. *Journal of Fish Biology*, **26**, 669–82

Day, J. H., Blaber, S. J. M. & Wallace, J. H. (1981) Estuarine fishes. In *Estuarine Ecology*, ed. J. H. Day, pp. 197–221. Rotterdam: A. A. Balkema

Diamant, A. & Sphigel, M. (1985) Interspecific feeding associations of groupers (Serranidae) with octopuses and moray eels in the Gulf of Eilat (Aqaba). *Environmental Biology of Fishes*, **13**, 153–9

Dobzhansky, Th. (1950) Evolution in the tropics. *American Scientist*, **38**, 209–21

Doherty, P. J., Williams, D. McB. & Sale, P. F. (1985) The adaptive significance of larval dispersal in coral reef fishes. *Environmental Biology of Fishes*, **12**, 81–90

Dominey, W. J. (1984) Effects of sexual selection and life history on speciation: species flocks in African cichlids and Hawaiian *Drosophila*. In *Evolution of Fish Species Flocks*, ed. A. A. Echelle & I. Kornfield, pp. 231–49. Orono: University of Maine at Orono Press

Dooley, J. K. (1972) Fishes associated with pelagic Sargassum complex, with a discussion of the *Sargassum* community. *Contributions in Marine Science, University of Texas*, **16**, 1–32

Dorn, E. (1983) Uber die Atmungsorgane einiger luftatmender Amazonasfische. *Amazoniana*, **7**, 375–95

Dudley, R. G. (1974) Growth of *Tilapia* of the Kafue floodplain, Zambia: predicted effects of the Kafue Gorge dam. *Transactions of the American Fisheries Society*, **103**, 281–91

Dudley, R. G. (1979) Changes in growth and size distribution of *Sarotherodon macrochir* and *S. andersoni* from the Kafue floodplain, Zambia, since construction of the Kafue dam. *Journal of Fish Biology*, **14**, 205–23

Dumont, H. J. (1986) The Tanganyika sardine in L. Kiru: another ecodisaster in Africa? *Environmental Conservation*, **13**, 143–8

Dunn, I. G., Burgis, M. J., Ganf, G. G., McGowan, L. M. & Viner, A. B. (1969) Lake George, Uganda: a limnological survey. *Verhandlungen Internationalen Vereinigung für theoretische und angewandte Limnologie*, **17**, 284–8

Durand, J. R. (1978) Biologie et dynamique des populations d'*Alestes baremose* (Pisces Characidae) du bassin Tchadien. *Travaux et Documents de l'ORSTOM*, no. 98, 332 pp.

Durand, J. R. & Levêque, C. (1981) *Flore et Faune aquatiques de l'Afrique Sahelo-Soudanienne*, 2 vols. Paris: ORSTOM

Dussart, B. H. (1974) Biology of inland waters in humid tropical Asia. In *Natural Resources of Humid Tropical Asia*, pp. 331–53. Natural Resources Research no. 12. Paris: UNESCO.

Eccles, D. H. (1974) An outline of the physical limnology of L. Malawi (L. Nyasa). *Limnology and Oceanography*, **19**, 730–42

Eccles, D. H. & Lewis, D. S. C. (1976) A revision of the genus *Docimodus* Blgr (Cichlidae), a group of fishes with unusual feeding habits from L. Malawi. *Zoological Journal, Linnean Society of London*, **58**, 165–72

Eccles, D. H. & Lewis, D. S. C. (1981) Midwater spawning in *Haplochromis chrysonotus* (Blgr) (Cichlidae) in L. Malawi. *Environmental Biology of Fishes*, **6**, 201–2

Echelle, A. A. & Kornfield, I. (1984) *Evolution of Fish Species Flocks*. Orono: University of Maine at Orono Press

Edwards, R. R. C. (1985) Growth rates of Lutjanidae (snappers) in tropical Australian waters. *Journal of Fish Biology*, **26**, 1–4

Egler, W. A. & Schwassmann, H. O. (1964) Limnological studies in the Amazon estuary. *Verhandlungen Internationalen Vereinigung für theoretische und angewandte Limnologie*, **15**, 1059–66

Ehrlich, P. R. (1975) The population biology of coral reef fishes. *Annual Review of Ecology & Systematics*, **6**, 211–47

Ehrlich, P. R. & Erhlich, A. H. (1973) Co-evolution: heterotypic schooling in Caribbean reef fishes. *American Naturalist*, **107**, 157–60

Ehrlich, P. R., Talbot, F. H., Russell, B. C. & Anderson, G. V. R. (1977) The behaviour of chaetodontid fishes with special reference to Lorenz's 'poster coloration' hypothesis. *Journal of Zoology, London*, **183**, 213–28

Eibl-Eibesfeldt, I. (1965), (translated by G. Vevers) *Land of a Thousand Atolls. A study of marine life in the Maldive & Nicobar Islands*. London: MacGibbon & Kee

Eigenmann, C. H. (1912) The freshwater fish fauna of British Guiana, including a study of the ecological grouping of species. *Memoirs of the Carnegie Museum*, **5**, 1–578

Ellis, C. M. A. (1978) Biology of *Luciolates stappersi* in L. Tanganyika (Burundi). *Transactions of the American Fisheries Society*, **107**, 557–66

Emery, A. R. (1974) Comparative ecology and functional osteology of 14 species of damselfish (Pisces: Pomacentridae) at Alligator reef, Florida Keys. *Bulletin of Marine Science*, **23**, 649–770

Emery, A. R. (1978) The basis of fish community structure: marine and freshwater comparisons. *Environmental Biology of Fishes*, **3**, 33–47

Erdman, D. S. (1977) Spawning patterns of fish from the northwestern Caribbean. In *Symposium on Progress of Marine Research Caribbean and adjacent Regions, Caracas*, ed. H. B. Stewart, pp. 145–69. FAO Fish Report no. 200. Rome: FAO

Fagade, S. O. & Olaniyan, C. I. O. (1973) The food and feeding relationships of the fishes in the Lagos Lagoon. *Journal of Fish Biology*, **5**, 205–25

FAO (1973) Proceedings of meeting on fisheries management. *Journal of the Fisheries Research Board of Canada*, **30**(12), 1936–2217

FAO (1974–84) *Species Identification Sheets for Fishery Purposes: Eastern Indian Ocean (Fishing Area 57) & Western Central Pacific (FA 71)*, 4 vols, 1974 (ed. W. Fischer & P. J. P. Whitehead); *Western Central Atlantic (FA 31)*, 6 vols, 1978 (ed. W. Fischer); *Eastern Central Atlantic (FA 34 & part of 47)*, 6 vols, 1981; (ed. W. Fischer, G. Bianchi & W. B. Scott); *Western Indian Ocean (FA 51)*, 5 vols, 1984 (ed. W. Fischer & G. Bianchi). Rome: FAO

FAO (1975) Report of the consultation on fisheries problems in the Sahelian zone, Bamako, Mali, 1974. *FAO CIFA/OP*, no 4, Annex 6, pp. 18–22

FAO (1978) *Actas des Simposio sobre Acuicultura en America Latina, Montevideo, Uruguay, 1974*. FAO Informes de Pesca, no. 159. 3 vols. Rome: FAO

FAO (1979) Review of the state of world fishery resources. *FAO Fisheries Circular*, no. 710. (Rev. 1). Rome: FAO

FAO (1981) Review of the state of world fishery resources. *FAO Fisheries Circular*, no. 710. (Rev. 2). Rome: FAO

FAO (1982) Fishery expansion project, Malawi. Biological studies on the pelagic ecosystem of L. Malawi. *FAO, FI:DP/MLW/75/019, Technical Report* no. 1, 182 pp.

FAO (1985) Development and management of the fisheries of L. Victoria. *FAO Fisheries Report*, no. 335 (FIPIR 335(en)), 145 pp.

Fernald, R. D. & Hirata, N. R. (1977) Field study of *Haplochromis burtoni*; habitats and co-habitant. *Environmental Biology of Fishes*, **2**, 299–308

Fernando, C. H. (1984) Reservoirs and lakes of SE Asia (Oriental Region). In *Lakes and Reservoirs*, ed. F. Taub, pp. 411–46. Netherlands: Elsevier

Fernando, C. H. & de Silva, S. S. (1984) Man-made lakes: ancient heritage and modern biological resource. In *Ecology and Biogeography in Sri Lanka*, ed. C. H. Fernando, pp. 431–51. Monographiae Biologicae no. 57. The Hague: Dr W. Junk

Ferreira, E. J. G. (1981) Alimentação dos adultos de doze espécies de ciclídeos (Perciformes, Cichlidae) do Rio Negro, Brasil. Master's Thesis, INPA, Manaus.

Ferreira, E. J. G. (1984) The fish fauna of the Curuá-Una reservoir Santarem, Para. II. Food and feeding habits of the main species. *Amazoniana*, **9**, 1–76

Fink, S. V. & Fink, W. L. (1981) Interrelationships of the ostariophysan fishes (Teleostei). *Zoological Journal of the Linnean Society*, **72**, 297–353

Fink, W. L. & Fink, S. V. (1979) Central Amazonia and its fishes. *Comparative Biochemistry & Physiology*, **62A**, 13–29

Fischelson, L. (1977) Sociobiology of feeding behaviour of coral fish along the coral reef of the Gulf of Eilat (= Gulf of Aqaba), Red Sea. *Israel Journal of Zoology*, **26**, 114–34

Fischelson, L., Popper, D. & Avidor, A. (1974) Biosociology and ecology of pomacentrid fishes around the Sinai Peninsular (northern Red Sea). *Journal of Fish Biology*, **6**, 119–33

Fittkau, E.-J. (1967) On the ecology of Amazonian rain forest streams. *Atas do Simposio sobre a Biota Amazonica, Rio de Janeiro*, **3** (*Limnology*), 97–108

Fittkau, E.-J. (1973) Crocodiles and the nutrient metabolism of Amazonian waters. *Amazoniana*, **4**, 103–33

Fontenele, O. (1950) Contibuçao para o conhecimento de biologia des Tucunares (Cichlidae) en cativiero *Revista Brasileira de Biologia*, **10**, 503–19

Fricke, H. W. (1975) The role of behaviour in marine symbiotic animals. *Symposia of the Society of Experimental Biology*, no. 29, 581–94

Fricke, H. W. (1977), Community structure, social organization and ecological requirements of coral reef fish (Pomacentridae). *Helgoländer wissenschaftliche Meeresuntersuchungen*, **30**, 412–26

Fricke, H. W. (1980) Control of different mating systems in a coral reef fish by one environmental factor. *Animal Behaviour*, **28**, 561–9

Frost, W. E. (1955) Observations on the biology of eels (*Anguilla* spp) of Kenya Colony, E. Africa. *Fishery Publications Colonial Office, London*, **6**, 1–28

Fryer, G. (1959) The trophic interrelationships and ecology of some littoral communities of L. Nyasa with especial reference to the fishes, and a discussion of the evolution of a group of rock-frequenting Cichlidae. *Proceedings of the Zoological Society of London*, **132**, 153–281

Fryer, G. (1961) Observations on the biology of the cichlid fish *Tilapia variabilis* in the northern waters of L. Victoria (E. Africa). *Revue de Zoologie et de Botanique Africaines*, **64**, 1–33

Fryer, G. (1972) Conservation of the Great Lakes of East Africa: a lesson and a warning. *Biological Conservation*, **4**, 256–62

Fryer, G. (1977) Evolution of species flocks of cichlid fishes in African lakes. *Zeitschrift für zoologische Systematik und Evolutionsforschung*, **15**, 141–65

Fryer, G. (1984) The conservation and rational exploitation of the biota of Africa's Great Lakes. In *Convention of Threatened Natural Habitats*, ed. A. V. Hall, p. 135–54. S. African National Scientific Programme Report no 92. Pretoria: CSIR

Fryer, G., Greenwood, P. H. & Peake, J. F. (1983) Punctuated equilibria, morphological stasis and the palaentological documentation of speciation: a biological appraisal of a case history in an African lake. *Biological Journal of the Linnean Society*, **20**, 195–205

Fryer, G., Greenwood, P. H. & Peake, J. F. (1985) The demonstration of speciation in fossil molluscs and living fishes. *Biological Journal of the Linnean Society*, **26**, 325–36

Fryer, G. & Iles, T. D. (1969) Alternative routes to evolutionary success as exhibited by African cichlid fishes of the genus *Tilapia* and the species flocks of the Great Lakes. *Evolution (Lancaster, Pa)*, **23**, 359–69

Fryer, G. & Iles, T. D. (1972) *The Cichlid Fishes of the Great Lakes of Africa*. Edinburgh: Oliver & Boyd

Furtado, J. I. & Mori, S. (eds) (1982) *Tasek Bera: The Ecology of a Freshwater Swamp*. Monographiae Biologicae no. 47. The Hague: Dr W. Junk

Ganapati, S. V. & Sreenivasan, A. (1972) Energy flow in aquatic ecosystems in India. In *Productivity Problems in Freshwaters*, ed. Z. Kajak & A. Hilbricht-Ilkowska, pp. 457–75. Warsaw: PWN Polish Scientific Publishers

Garrod, D. J. & Horwood, J. W. (1984) Reproductive strategies and the response to exploitation. In *Fish Reproduction*, ed. G. W. Potts & R. J. Wootton, pp. 367–84. London: Academic Press

Garrod, D. J. & Knights, B. J. (1979) Fish stocks: their life-history characteristics and response to exploitation. *Symposia of the Zoological Society of London*, **44**, 361–82

Geisler, R. (1969) Investigations about free oxygen, biological oxygen demand and oxygen consumption of fishes in a tropical 'blackwater' (Rio Negro). *Archiv für Hydrobiologie*, **66**, 307–25

Geisler, R. Schmidt, G. W. & Sookvibul, S. (1979) Diversity and biomass of fishes in three typical streams in Thailand. *Internationale Revue der gesamten Hydrobiologie*, **64**, 673–97

Gery, J. (1964) Poissons characoides nouveaux ou non signales de l'Ilha do Bananal, Bresil. *Vie et Milieu, Paris*, supplement 17, 447–71

Gery, J. (1969) The freshwater fishes of South America. In *Biogeography and Ecology in South America*, ed. E.-J. Fittkau *et al.*, pp. 828–48. Monographiae Biologicae no. 9. The Hague: Dr W. Junk

Gery, J. (1977) *Characoids of the World*. Neptune City, NJ: TFH Publications Inc.

Gery, J. (1984) The fishes of Amazonia. In *The Amazon: Limnology and Landscape Ecology of a Mighty Tropical River and its Basin*, ed. H. Sioli, pp. 353–70. Dordrecht: Dr W. Junk

Gittleman, J. L. (1981) The phylogeny of parental care in fishes. *Animal Behaviour*, **29**, 936–41

Gladfelter, W. B. (1979) Twilight migrations and foraging activities of the copper sweeper *Pempheris schomburgki* (Pempheridae). *Marine Biology (Berlin)*, **50**, 109–19

Gladfelter, W. B. & Gladfelter, E. H. (1978) Fish community structure as a function of habitat structure on West Indian patch reefs. *Revista de Biologia Tropical*, **26** (supplement 1), 65–84

Gladfelter, W. B. & Johnson, W. S. (1983) Feeding niche separation in a guild of tropical reef fishes (Holocentridae). *Ecology*, **64**, 552–63

Gladfelter, W. B., Ogden, J. C. & Gladfelter, E. H. (1980) Similarity and diversity among coral reef fish communities: a comparison between tropical western Atlantic (Virgin Is) and tropical Central Pacific (Marshall Is) patch reefs. *Ecology*, **61**, 1156–68

Glasser, J. W. (1979) The role of predation in shaping and maintaining the structure of communities. *American Naturalist*, **113**, 631–41

Glucksman, J., West, G. & Berra, T. M. (1976) The introduced fishes of Papua New Guinea with special reference to *Tilapia mossambica*. *Biological Conservation*, **9**, 37–44

Godoy, M. P. de (1959) Age, growth, sexual maturity, behaviour, migration, tagging and transplantation of the Curimbata (*Prochilodus scrofa* Stdr, 1881) of the Mogi Guassu River, S. Paulo State, Brasil. *Anais de Academia Brasileira de Ciencias*, **31**, 447–77

Godoy, M. P. de (1967) Dez anos de observacoes sobre periodicidade migratoria de peixes do Rio Mogi Guassu. *Revista Brasileira de Biologia*, **27**, 1–12

Godoy, M. P. de (1975) *Peixes do Brasil, subordem Characoidei, bacia do Rio Mogi Guassu*. Piracicaba, São Paulo: Editoria Franciscam (4 vols) (In Portuguese)

Goldman, B. & Talbot, F. H. (1976) Aspects of the ecology of coral reef fishes. In *Biology & Geology of Coral Reefs*, vol. 3, *Biology 2*, ed. O. A. Jones & R. Endean, pp. 125–54. New York: Academic Press

Gorman, D. T. & Karr, J. R. (1978) Habitat structure and stream fish communities. *Ecology*, **59**, 507–15

Gosline, W. A. (1965) Vertical zonation of inshore fishes in the upper water layers of the Hawaiian Is. *Ecology*, **46**, 823–31

Gosse, J.-P. (1963) Le milieu aquatique et l'écologie des poissons dans la région de Yangambi. *Annales Musée Royal de l'Afrique Centrale, Tervuren*, sér. 8, *Sci Zool.*, no. 116, 113–270

Goulding, M. (1980) *The Fishes and the Forest: Explorations in Amazonian Natural History*. Los Angeles: University of California Press

Goulding, M. (1981) *Man and Fisheries on an Amazon Frontier*. Developments in Hydrobiology, no. 4. The Hague: Dr W. Junk

Goulding, M. & Carvalho, M. L. (1982) Life history and management of the tambaqui (*Colossoma macropomum*: Characidae): an important Amazonian food fish. *Revista Brasileira de Zoologia, S. Paulo*, **1**(2), 107–33

Goulding, M. & Carvalho, M. L. (1984) Ecology of Amazonian needlefishes (Belonidae). *Revista Brasileira de Zoologia, S. Paulo*, **2**(3), 99–111

Goulding, M. & Ferreira, E. J. G. (1984) Shrimp-eating fishes and a case of prey-switching in Amazon rivers. *Revista Brasileira de Zoologia, S. Paulo*, **2**(3), 85–97

Green, J. (1977) Haematology and habits in catfish of the genus *Synodontis*. *Journal of Zoology*, **182**, 39–50

Greenfield, D. W. & Greenfield, T. A. (1982) Habitat and resource partitioning between two species of *Acanthemblemaria* (Pisces: Chaeropsidae) with comments on the chaos hypothesis. *Smithsonian Contributions to Marine Science*, no. 12, 499–507

Greenwood, P. H. (1955) Reproduction in the catfish *Clarias mossambicus* Peters. *Nature, London*, **176**, 516

Greenwood, P. H. (1958) Reproduction in the E. African lung-fish *Protopterus aethiopicus*. *Proceedings of the Zoological Society of London*, **130**, 547–67

Greenwood, P. H. (1965) The cichlid fishes of L. Nabugabo, Uganda. *Bulletin of the British Museum (Natural History), Zoology*, **12**, 315–57

Greenwood, P. H. (1974) Cichlid fishes of L. Victoria: the biology and evolution of a species flock. *Bulletin of the British Museum (Natural History), Zoology*, supplement 6, 1–134

Greenwood, P. H. (1975). Classification. In *A History of Fishes*, 3rd edn, ed. J. R. Norman, pp. 378–96. London: Ernest Benn

Greenwood, P. H. (1976a) A review of the family Centropomidae. *Bulletin of the British Museum (Natural History), Zoology*, **29**, 1–81

Greenwood, P. H. (1976b) Fish fauna of the Nile. In *The Nile, Biology of an Ancient River*, ed. J. Rzoska, pp. 127–41. Monographiae Biologicae no. 29. The Hague: Dr W. Junk

Greenwood, P. H. (1979) Towards a phyletic classification of the 'genus' *Haplochromis* and related taxa. *Bulletin of the British Museum (Natural History), Zoology*, **35**, 265–322

Greenwood, P. H. (1980) *Ibid.* Part II. The species of Lakes Victoria, Nabugabo, Edward, George and Kivu. *Bulletin of the British Museum (Natural History), Zoology*, **39**, 1–101

Greenwood, P. H. (1981) *The Haplochromine Fishes of the East African Lakes. Collected papers on their taxonomy, biology and evolution (with an introduction and species index).* Munich: Kraus International Publications. (Reprints of numerous earlier papers published in *Bulletin of the British Museum (Natural History), Zoology.*)

Greenwood, P. H. (1983*a*) The *Ophthalmotilapia* assemblage of cichlid fishes reconsidered. *Bulletin of the British Museum (Natural History), Zoology*, **44**, 249–90

Greenwood, P. H. (1983*b*) On *Macropleurodus, Chilotilapia* (Teleostei: Cichlidae) and the interrelationships of African cichlid species flocks. *Bulletin of the British Museum (Natural History), Zoology*, **45**, 209–31

Greenwood, P. H. (1983*c*) The zoogeography of African freshwater fishes: bioaccountancy or biogeography? In *Evolution, Time and Space: The Emergence of the Biosphere*, ed. R. W. Sims, J. H. Price & P. E. S. Whalley, pp. 179–99. London: Academic Press

Greenwood, P. H. (1984) African cichlids and evolutionary theories. In *Evolution of Fish Species Flocks*, ed. A. A. Echelle & I. Kornfield, pp. 141–54. Orono: University of Maine at Orono Press

Greenwood, P. H. & Lund, J. W. G. (eds) (1973) A discussion on the biology of an equatorial lake: Lake George, Uganda. *Proceedings of the Royal Society of London, B*, **184**, 227–346

Grossman, G. D., Moyle, P. B. & Whitaker, J. O. (1982) Stochasticity in structural and functional characteristics of an Indiana stream fish assemblage. A test of community theory. *American Naturalist*, **120**, 423–54

Grove, A. T. (ed.) (1985) *The Niger and its Neighbours.* Netherlands: Balkema

Gulland, J. A. (1970) Food chain studies and some problems in world fisheries. In *Marine Food Chains*, ed. J. H. Steele, pp. 458–67. Edinburgh: Oliver & Boyd

Gulland, J. A. (1971) *The Fish Resources of the Ocean.* Byfleet, Hants, UK: FAO & Fishery News (Books) Ltd

Gulland, J. A. (1978) Fishery management: new strategies for new conditions. *Transactions of the American Fisheries Society*, **107**, 1–11

Gulland, J. A. (1983) *Fish Stock Assessment: A Manual of Basic Methods.* Chichester: John Wiley & Sons

Gundermann, N. & Popper, D. M. (1984) Notes on the Indo-Pacific mangal fishes and on mangrove related fisheries. In *Hydrobiology of the Mangal*, ed. D. Por & I. Dor, pp. 201–6. The Hague: Dr W. Junk

Gwahaba, J. J. (1973) Effects of fishing on the *Tilapia nilotica* populations of L. George, Uganda, over the past 20 years. *East African Wildlife Journal*, **11**, 317–28

Haines, A. K. (1983) Fish fauna and ecology. In *The Purari – Tropical Environment of a High Rainfall River Basin*, ed. T. Petr, pp. 367–84. Monographiae Biologicae no. 51. The Hague: Dr W. Junk

Halstead, B. W. (1953) Some general considerations of the problems of poisonous fishes and ichthyosarcotoxism. *Copeia*, **1**, 31–3

Harmelin-Vivien, M. L. & Bouchon, C. (1976) Feeding behavior of some carnivorous fishes (Serranidae & Scorpaenidae) from Tulear (Madagascar). *Marine Biology (Berlin)*, **37**, 329–40

Hecky, R. E., Fee, E. J., Kling, H. J. & Rudd, J. W. M. (1981) Relationships between primary production and fish production in L. Tanganyika. *Transactions of the American Fisheries Society*, **110**, 336–45

Heilgenberg, W. & Bastian, J. (1980) Species specificity of electric organ discharges in sympatric gymnotoid fishes of the Rio Negro. *Acta Biologica Venezuelica*, **10**, 187–203

Helfman, G. S. (1978) Patterns of community structure in fishes: summary and overview. *Environmental Biology of Fishes*, **3**, 129–48

Helfman, G. S. & Schultz, E. T. (1984) Social transmission of behavioral traditions in a coral fish. *Animal Behaviour*, **32**, 379–84

Henderson, H. F. & Welcomme, R. L. (1974) The relationship of yield to morpho-edaphic index and numbers of fishermen in African inland fisheries. *CIFA Occasional Paper*, no. 1, 19 pp. Rome: FAO

Hepher, B. (1967) Some biological aspects of warmwater fish pond management. In *The Biological Basis of Freshwater Fish Production*, ed. S. K. Gerking, pp. 417–28. Oxford: Blackwell Scientific Publications

Hepher, B. & Pruginin, Y. (1982) Tilapia culture in ponds under controlled conditions. In *The Biology and Culture of Tilapias*, ed. R. S. V. Pullin & R. H. Lowe-McConnell, pp. 185–203. Manila, Philippines: ICLARM

Hiatt, R. W. & Strasburg, D. W. (1960) Ecological relationships of the fish fauna on coral reefs of the Marshall Is. *Ecological Monographs*, **30**, 65–127

Hickling, C. F. (1960) The Malacca *Tilapia* hybrids. *Journal of Genetics*, **57**, 1–10

Hobson, E. S. (1965) Diurnal–nocturnal activity of some inshore fishes in the Gulf of California. *Copeia*, **3**, 291–302

Hobson, E. S. (1968) Predatory behaviour of some shore fishes in the Gulf of California. *Research Reports, U.S. Fish & Wildlife Service*, **73**, 1–92

Hobson, E. S. (1972) Activity of Hawaiian reef fishes during the evening and morning transitions between daylight and darkness. *Fishery Bulletin, U.S. Fish & Wildlife Service*, **70**, 715–40

Hobson, E. S. (1973) Diel feeding migrations in tropical reef fishes. *Helgoländer wissenschaftliche Meeresuntersuchungen*, **24**, 361–70

Hobson, E. S. (1974) Feeding relationships of teleost fishes on coral reefs in Kona, Hawaii. *Fishery Bulletin, U.S. Fish & Wildlife Service*, **72**, 915–1031

Hobson, E. S. (1975) Feeding patterns among tropical reef fishes. *American Science*, **63**, 382–92

Hoffman, S. G. (1983) Sex related foraging behaviour in sequentially hermaphroditic hogfishes (*Bodianus* sp). *Ecology*, **64**, 798–808

Holčík, J. (1970) Standing crop, abundance, production and some ecological aspects of fish populations in some inland waters of Cuba. *Věstník Ceskoslovenské Zoologické Spolecnosti (Praha)*, **34**, 184–201

Holden, M. J. (1963) The populations of fish in dry season pools of the R. Sokoto. *Fishery Publications of the Colonial Office, London*, **19**, 1–58

Holzberg, S. (1978) A field and laboratory study of the behaviour and ecology of *Pseudotropheus zebra* (Blgr), an endemic cichlid of L. Malawi (Pisces: Cichlidae). *Zeitschrift für zoologische Systematik und Evolutionsforschung*, **16**, 171–87

Hoogerhoud, R. J. C. (1984) A taxonomic reconsideration of the haplochromine genera *Gaurochromis* Greenwood 1980 and *Labrochromis* Regan (1970). *Netherlands Journal of Zoology*, **34**, 539–65

Hoogerhoud, R. J. C., Witte, F. & Barel, C. D. N. (1983) The ecological differentiation of two closely resembling *Haplochromis* species from L. Victoria (*H. Iris* and *H. hiatus*): Cichlidae. *Netherlands Journal of Zoology*, **33**, 283–305

Hopkins, C. D. (1974a) Electric communication in the reproductive behaviour of *Sternopygus macrurus* (Gymnotoidei). *Zeitschrift für Tierpsychologie*, **35**, 518–35

Hopkins, C. D. (1974b) Electric communication in fish. *American Scientist*, **62**, 426–37

Hopson, A. J. (1972) *A Study of the Nile Perch in L. Chad*. Overseas Research Publication, no. 19. London: HMSO

Hopson, A. J. (ed.) (1982) *Lake Turkana. A report on the findings of the Lake Turkana Project 1972–75*. London: Overseas Development Administration

Hora, S. L. (1930) Ecology, bionomics and evolution of the torrential fauna, with special reference to the organs of attachment. *Philosophical Transactions of the Royal Society, B*, **218**, 171–82

Hora, S. L. & Pillay, T. V. R. (1962) *Handbook of Fish Culture in the Indo-Pacific Region.* FAO Fisheries Biology, Technical Publication no. 14, 204 pp. Rome: FAO

Hori, M. (1983) Feeding ecology of 13 species of *Lamprologus* (Cichlidae) coexisting at a rocky shore of L. Tanganyika. *Physiology & Ecology, Japan*, **20**, 129–49

Hori, M., Yamaoka, K. & Takamura, K. (1983) Abundance and micro-distribution of cichlid fishes on a rocky shore in L. Tanganyika. *African Study Monographs, Kyoto*, **3**, 25–38

Howes, G. J. (1980) The anatomy, phylogeny and classification of bariliine cyprinid fishes. *Bulletin of the British Museum (Natural History), Zoology*, **37**, 129–98

Howes, G. J. (1983) Problems in catfish anatomy and phylogeny exemplified by the Neotropical Hypophthalmidae (Siluroidei). *Bulletin of the British Museum (Natural History), Zoology*, **45**, 1–39

Howes, G. J. (1984) A review of the anatomy, phylogeny and biogeography of the African neoboline cyprinid fishes. *Bulletin of the British Museum (Natural History), Zoology*, **47**, 151–85

Howes, G. J. (1985) The phylogenetic relationships of the electric catfish family Malapteruridae. *Journal of Natural History*, **19**, 37–67

Ihering, R. von (1928) *Da Viva dos Peixes.* São Paulo: Comp. Melhoramentos

Iles, T. D. (1960) A group of zooplankton feeders of the genus *Haplochromis* (Cichlidae). *Annals & Magazine of Natural History*, **17**, 257–80

Iles, T. D. (1971) Ecological aspects of growth in African cichlid fishes. *Journal du Conseil Permanent International pour l'Exploration de la Mer*, **33**, 362–84

Iles, T. D. (1973) Dwarfing or stunting in the genus *Tilapia* (Cichlidae): a possibly unique recruiting mechanism. *Rapport et Proces-Verbaux des Réunions du Conseil Permanent International pour l'Exploration de la Mer*, **164**, 247–54

Inger, R. F. (1955) Ecological notes on the fish fauna of a coastal drainage of N. Borneo. *Fieldiana: Zoology, Chicago*, **37**, 47–90

Ita, E. O. (1978) Analysis of fish distribution in Kainji lake, Nigeria. *Hydrobiologia*, **58**, 233–44

Itzkowitz, M. (1977) Social dynamics of mixed-species groups of Jamaican reef fishes. *Behavior, Ecology, Sociobiology*, **2**, 361–84

Ivlev, V. S. (1961) *Experimental Ecology of the Feeding of Fishes.* New Haven, CT: Yale University Press

Jackson, P. B. N. (1959) Revision of the clariid catfishes of Nyasaland, with descriptions of a new genus and seven new species. *Proceedings of the Zoological Society of London*, **132**, 109–28

Jackson, P. B. N., Iles, T. D., Harding, D. & Fryer, G. (1963) *Report on the Survey of Northern Lake Nyasa 1954–55.* Zomba, Malawi: Government Printer

Jayaram, K. C. (1974) Ecology and distribution of freshwater fishes, amphibia and reptiles. In *Ecology & Biogeography in India*, ed. M. S. Mani, pp. 517–84. Monographiae Biologicae no. 23. The Hague: Dr W. Junk

Jeffries, M. J. & Lawton, J. H. (1984) Enemy-free space and the structure of ecological communities. *Biological Journal of the Linnean Society*, **23**, 269–86

Jhingran, V. G. (1975) *Fish and Fisheries of India.* Delhi: Hindustan Publishing Corporation

Johannes, R. E. (1978) Reproductive strategies of coastal marine fishes in the tropics. *Environmental Biology of Fishes*, **3**, 141–60

Johnson, D. S. (1967) Distributional patterns in Malayan freshwater fish. *Ecology*, **48**, 722–30

Jones, R. (1982) Ecosystems, food chains and fish yields. In *Theory and Management of Tropical Fisheries*, ed. D. Pauly & G. J. Murphy, pp. 195–239. Manila, Philippines: ICLARM

Jubb, R. A. (1977) Comments on Victoria Falls as a physical barrier for downstream dispersal of fishes. *Copeia*, **1**, 198–9

Junk, W. J. (1973) Investigations on the ecology and production biology of the floating meadows (*Paspalo–Echinochloetum*) on the Middle Amazon. Part 2: The aquatic fauna in the root zone of the floating vegetation. *Amazoniana*, **4**, 9–102

Junk, W. J. (1984) Ecology, fisheries and fish culture in Amazoniana. In *The Amazon. Limnology and Landscape Ecology of a Mighty Tropical River and its Basin*, ed. H. Sioli, pp. 443–76. Dordrecht: Dr W. Junk

Junk, W. J., Soares, G. M. & Carvalho, F. M. (1983) Distribution of fish species in a lake of the Amazon river floodplain near Manaus (Lago Camaleão), with special reference to extreme oxygen conditions. *Amazoniana*, **7**, 397–431

Kajak, Z. & Hillbricht-Ilkowska, A. (eds) (1972) *Productivity Problems of Freshwaters.* Warsaw: PWN Polish Scientific Publishers (for IBP–UNESCO)

Kapetsky, J. M. (1974) The Kafue river floodplain: an example of preimpoundment potential for fish production. In *Lake Kariba: A Man-made Tropical Ecosystem in Central Africa*, ed. E. K. Balon & A. G. Coche, pp. 497–523. The Hague: W. Junk

Kapetsky, J. M. (1976) Fish populations of the floodplain lakes of the Magdalena River. Bogata, INDERENA-FAO, 30 pp. (mimeo)

Kawanabe, H. (1981) Territorial behaviour of *Tropheus moorei* (Osteichthyes: Cichlidae) with a preliminary consideration on the territorial forms in animals. *African Study Monographs, Kyoto*, **1**, 101–8

Kawanabe, H. (ed.) (1983) *Ecological and Limnological Study on L. Tanganyika and its Adjacent Regions, vol. 2*. Kyoto, Japan: Kyoto University

Keenleyside, M. H. A. (1955) Some aspects of schooling behaviour of fish. *Behaviour*, **8**, 183–248

Keenleyside, M. H. A. (1979) *Diversity and Adaptation in Fish Behaviour.* Zoophysiology and Ecology, no. 11. Berlin: Springer-Verlag

Keenleyside, M. H. A. (1981) Parental care patterns of fishes. *American Naturalist*, **117**, 1019–22

Kelly, C. D. & Hourigan, T. F. (1983) The function of conspicuous coloration in chaetodontid fishes: a new hypothesis. *Animal Behaviour*, **31**, 615–17

Kimpe, P. de (1964) Contribution à l'étude hydrobiologique du Luapula-Moero. *Annales du Musée Royal de l'Afrique Centrale, Tervuren*, ser. 8, *Science Zoologiques*, **128**, 1–238

Kipling, C. (1984) A study of perch (*Perca fluviatilis*) and pike (*Esox lucius*) in Windermere from 1941 to 1982. *Journal du Conseil Permanent International pour l'Exploration de la Mer*, **41**, 259–67

Kirschbaum, F. (1984) Reproduction of weakly electric teleosts: just another example of convergent development? *Environmental Biology of Fishes*, **10**, 3–14

Knöppel, H.-A. (1970) Food of Central Amazonian fishes. *Amazoniana*, **2**, 257–352

Kocher, T. D. & McKaye, K. R. (1983) Defence of heterospecific cichlids by *Cyrtocara moorii* in L. Malawi, Africa. *Copeia*, **2**, 544–7

Kock, R. L. (1982) Patterns of abundance and variation in reef fishes near an artificial reef at Guam. *Environmental Biology of Fishes*, **7**, 121–36

Kornfield, I. L. (1978) Evidence for rapid speciation in African cichlid fishes. *Experientia* (*Basel*), **34**, 335–6

Kornfield, I. & Carpenter, K. E. (1984) Cyprinids of L. Lanao, Philippines: taxonomic validity, evolutionary rates and speciation scenarios. In *Evolution of Species Flocks*, ed. A. A. Echelle & I. Kornfield, pp. 69–84. Orono: University of Maine at Orono Press

Kornfield, I. L., Smith, D. C., Gagnon, R. B. & Taylor, J. N. (1982) The cichlid fish of Cuatro Cienegas, Mexico: direct evidence of conspecificity among distinct morphs. *Evolution*, **36**, 658–64

Kornfield, I. L. & Taylor, J. N. (1983) A new species of polymorphic fish, *Cichlasoma minckleyi* from Cuatro Cienegas, Mexico. *Proceedings of the Biological Society of Washington*, **96**, 253–69

Kramer, D. L. (1978) Reproductive seasonality in the fishes of a tropical stream. *Ecology*, 5, 976–85

Kramer, D. L. (1983) The evolutionary ecology of the respiratory mode in fishes: an analysis based on the costs of breathing. *Environmental Biology of Fishes*, 9, 67–80

Kramer, D. L., Lindsey, C. C., Moodie, G. E. E. & Stevens, E. D. (1978) The fishes and the aquatic environment of the Central Amazon basin, with particular reference to respiratory patterns. *Canadian Journal of Zoology*, 56, 717–29

Kristensen, I. (1970) Competition in three cyprinodont fish species in the Netherlands Antilles. *Uitgaven van de Natuurwetenschappelijke Studiekring voor Suriname*, 32(119), 82–101

Kullander, S. O. (1983) *A revision of the South American cichlid genus* Cichlasoma (*Teleostei: Cichlidae*). Stockholm: Naturhistorika Rijksmuseet

Lack, D. (1954) *The Natural Regulation of Animal Numbers*. Oxford: Clarendon Press

Lagler, K. (ed.) (1976) *Fisheries and Integrated Mekong River Basin Development*. Terminal Report of the Mekong Basinwide Fishery Studies. Field Investigations, Appendix, vol. 1; 2 Executive Volume. School of Natural Resources, University of Michigan, Ann Arbor, USA.

Lagler, K. F., Kapetsky, J. M. & Stewart, D. J. (1971) *The fisheries of the Kafue flats, Zambia, in relation to the Kafue Gorge dam*. University of Michigan Technical Report. Rome: FAO. no. FI: SF/ZAM 11. Technical Report, 1, 1–161

Lassig, B. R. (1983) The effects of a cyclonic storm on coral reef fish assemblages. *Environmental Biology of Fishes*, 9, 55–63

Latif, A. F. A. (1984) Lake Nasser – the new man-made lake in Egypt. In *Lakes and Rivers*, ed. F. B. Taub, pp. 385–410. Amsterdam: Elsevier

Lauder, G. V. & Liem, K. F. (1983) The evolution and interrelationships of the Actinopterygian fishes. *Bulletin of the Museum of Comparative Zoology (Harvard)*, 150, 95–197

Lauzanne, L. (1976) Régimes alimentaires et relations trophiques des poissons du lac Tchad. *Cahiers ORSTOM, sér. Hydrobiologie*, 10, 267–310

Le Cren, E. D. & Lowe-McConnell, R. H. (1980) *The Functioning of Freshwater Ecosystems*. IBP no. 22. Cambridge: Cambridge University Press

Lek, S. & Lek, S. (1977) Ecologie et biologie de *Micralestes acutidens* (Peters, 1852) du bassin du lac Tchad. *Cahiers ORSTOM, sér. Hydrobiologie*, 11, 255–68

Lek, S. & Lek, S. (1978) Etude de quelques espèces de petits Mormyridae du bassin du lac Tchad. I. Observations sur la répartition et l'écologie. *Cahiers ORSTOM, sér. Hydrobiologie*, 12, 225–36

Lelek, A. (1973) Sequence of changes in fish populations of the new tropical man-made lake, Kainji, Nigeria. *Archiv für Hydrobiologie*, 71, 381–420

Levêque, C. (1979) Biological productivity of L. Chad. In *Première Réunion Limnologie Africaine, Nairobi, 1979*, pp. 1–30. Paris: ORSTOM

Levêque, C., Bruton, M. & Ssentongo, G. (in press). *Biology and Ecology of African Freshwater Fishes*, Paris: ORSTOM

Levine, J. S. & MacNichol, E. F. (1982) Color vision in fishes. *Scientific American*, 246(2), 108–17

Lewis, D. S. S. (1974a) The effects of the formation of L. Kainji (Nigeria) upon the indigenous fish population. *Hydrobiologia*, 45, 281–301

Lewis, D. S. C. (1974b) The food and feeding habits of *Hydrocynus forskahlii* Cuv. and *H. brevis* Gthr in L. Kainji, Nigeria. *Journal of Fish Biology*, 6, 349–63

Lewis D. S. C. (1981) Preliminary comparisons between the ecology of haplochromine cichlid fishes of L. Victoria and L. Malawi. *Netherlands Journal of Zoology*, 31, 746–61

Lewis, D. S. C. (1982) Problems of species definition in L. Malawi cichlid fishes (Pisces: Cichlidae). *J. L. B. Smith Institute of Ichthyology, Special Publication*, no. 23, 1–5

Liem, K. F. (1974) Evolutionary strategies and morphological innovations in cichlid pharyngeal jaws. *Systematic Zoology*, **22**, 425–41

Liem, K. F. (1978) Modulatory multiplicity in the functional repertoire of the feeding mechanism in cichlid fishes. 1. Piscivores. *Journal of Morphology*, **158**, 323–60

Liem, K. F. (1980) Adaptive significance of intra- and interspecific differences in the feeding repertoires of cichlid fishes. *American Zoologist*, **20**, 295–314

Liem, K. F. & Kaufman, L. S. (1984) Intraspecific macroevolution: functional biology of the polymorphic cichlid species *Cichlasoma minckleyi*. In *Evolution of Species Flocks*, ed. A. A. Echelle & I. Kornfield, pp. 203–15. Orono: University of Maine at Orono Press

Liem, K. F. & Osse, J. W. M. (1975) Biological versatility, evolution, and food resource exploitation in African cichlid fishes. *American Zoologist*, **15**, 427–54

Liem, K. F. & Stewart, D. F. (1976) Evolution of the scale-eating cichlid fishes of L. Tanganyika: a generic revision with description of new species. *Bulletin of the Museum of Comparative Zoology Harvard*, **147**, 319–50

Limbaugh, C. (1961) Cleaning symbiosis. *Scientific American*, **205**(2), 42–9

Limberger, D. (1983) Pairs and harems in a cichlid fish, *Lamprologus brichardi*. *Zeitschrift für Tierpsychologie*, **62**, 115–44

Lissmann, H. W. (1958) On the function and evolution of electric organs in fish. *Journal of Experimental Biology*, **35**, 156–91

Lissmann, H. W. (1961) Ecological studies on gymnotids. In *Bioelectrogenesis*, ed. C. Chagas & A. Paes de Carvalho, pp. 215–26. Amsterdam: Elsevier

Lissmann, H. W. (1963) Electric location by fishes. *Scientific American*, **208**, 50–9

Lobel, P. S. (1978) Diel, lunar and seasonal periodicity in the reproductive behavior of the pomacanthid fish *Centropyge potteri* and some reef fishes in Hawaii. *Pacific Science*, **32**(2), 193–207

Lobel, P. S. (1980) Herbivory by damselfishes and their role in coral reef community ecology. *Bulletin of Marine Science*, **30**, 273–89

Lobel, P. S. & Robinson, A. R. (1983) Reef fishes at sea: ocean currents and the advection of larvae. In *The Ecology of Deep and Shallow Coral Reefs*, ed. M. L. Reaka, pp. 29–38. NOAA's Undersea Research Program, vol. 1. Rockville, MD

Lobel, P. S. & Robinson, A. R. (1986) Transport and entrapment of fish larvae by ocean mesoscale eddies and currents in Hawaiian waters. *Deep-Sea Resarch*, **33**, 483–500

Longhurst, A. R. (1957) The food of the demersal fish of a West African estuary. *Journal of Animal Ecology*, **26**, 369–87

Longhurst, A. R. (1960) A summary of the food of West African demersal fish. *Bulletin de l'Institut Français Afrique Noire*, **22**, sér. A., no. 1, 276–82

Longhurst, A. R. (1963) Bionomics of fisheries resources of the Eastern Tropical Atlantic. *Fishery Publications Colonial Office, London*, no. 20, 1–66

Longhurst, A. R. (1965) A survey of the fish resources of the eastern Gulf of Guinea. *Journal du Conseil Permanent International pour l'Exploration de la Mer*, **29**, 302–34

Longhurst, A. R. (1969) Species assemblages in tropical demersal fisheries. *Proceedings, Symposium on Oceanography & Fisheries Research in the Tropical Atlantic*, pp. 147–68. UNESCO, SC N.S. 67/D 60/AF

Longhurst, A. R. (1971) The clupeid resources of tropical seas. *Oceanography & Marine Biology*, **9**, 349–85

Low, R. M. (1971) Interspecific territoriality in a pomacentrid reef fish, *Pomacentrus flavicauda*. *Ecology*, **52**, 648–54

Lowe, R. H. (1952) Report on the *Tilapia* and other fish and fisheries of L. Nyasa. *Fishery Publications, Colonial Office, London*, **1**(2), 1–126

Lowe, R. H. (1953) Notes on the ecology and evolution of Nyasa fishes of the genus *Tilapia*, with a description of *T. saka* Lowe. *Proceedings of the Zoological Society of London*, **22**, 1035–41

Lowe, R. H. (McConnell) & Longhurst, A. R. (1961) Trawl fishing in the tropical Atlantic. *Nature, London*, **192**, 620–3

Lowe-McConnell, R. H. (1958) Observations on the biology of *Tilapia nilotica* L. in East African waters. *Revue de Zoologie et de Botanique Africaines*, **57**, 129–70

Lowe-McConnell, R. H. (1959) Breeding behaviour patterns and ecological differences between *Tilapia* species and their significance for evolution within the genus *Tilapia*. *Proceedings of the Zoological Society of London*, **132**, 1–30

Lowe-McConnell, R. H. (1962) The fishes of the British Guiana continental shelf, Atlantic coast of South America, with notes on their natural history. *Journal of the Linnean Society (Zoology)*, **44**, 669–700

Lowe-McConnell, R. H. (1964) The fishes of the Rupununi savanna district of British Guiana, Pt 1. Groupings of fish species and effects of the seasonal cycles on the fish. *Journal of the Linnean Society (Zoology)*, **45**, 103–44

Lowe-McConnell, R. H. (1966) The sciaenid fishes of British Guiana. *Bulletin of Marine Science*, **16**, 20–57

Lowe-McConnell, R. H. (1969a) Speciation in tropical freshwater fishes. In *Speciation in Tropical Environments*, ed. R. H. Lowe-McConnell, pp. 51–75. London: Academic Press. (also in *Biological Journal of the Linnean Society*, **1**.)

Lowe-McConnell, R. H. (1969b) The cichlid fishes of Guyana, S. America, with notes on their ecology and breeding behaviour. *Zoological Journal of the Linnean Society*, **48**, 255–302

Lowe-McConnell, R. H. (1975) *Fish Communities in Tropical Freshwaters: Their Distribution, Ecology and Evolution*. London: Longman

Lowe-McConnell, R. H. (1977) *Ecology of Fishes in Tropical Waters*. Studies in Biology, no. 76. London: Edward Arnold

Lowe-McConnell, R. H. (1979) Ecological aspects of seasonality in fishes of tropical waters. *Symposia of the Zoological Society of London*, no. *44*, 219–41

Lowe-McConnell, R. H. (1982) Tilapias in fish communities. In *The Biology and Culture of Tilapias*, ed. R. S. V. Pullin & R. H. Lowe-McConnell, pp. 83–113. ICLARM Conference Proceedings 7. Manila, Philippines: International Center for Living Aquatic Resources

Lowe-McConnell, R. H. (1984) The status of studies on South American freshwater food fishes. In *Evolutionary Ecology of Neotropical Freshwater Fishes*, ed. T. M. Zaret, pp. 139–56. The Hague: Dr W. Junk

Lowe-McConnell, R. H. (1985) The biology of the river systems with particular reference to the fishes. In *The Niger and its Neighbours*, ed. A. T. Grove, pp. 101–40. Netherlands: Balkema

Lowe-McConnell, R. H. & Howes, G. J. (1981) Pisces. In *Aquatic Biota of Tropical South America, Part 2: Anarthropoda*, ed. S. H. Hurlbert, G. Rodriguez & N. D. Santos, pp. 218–29. San Diego: San Diego State University Press

Luckhurst, B. E. & Luckhurst, K. (1977) Recruitment patterns of coral reef fishes on the fringing reef of Curaçao, Netherlands Antilles. *Canadian Journal of Zoology*, **55**, 681–9

Luckhurst, B. E. & Luckhurst, K. (1978a) Diurnal space utilization in coral reef fish communities. *Marine Biology (Berlin)*, **49**, 325–32

Luckhurst, B. E. & Luckhurst, K. (1978b) Analysis of the influence of substrate variables on coral reef fish communities. *Marine Biology (Berlin)*, **49**, 317–23

Lüling, K. H. (1958) Zur Lebenweise von *Synbranchus marmoratus*. *Bonner Zoologische Beiträge*, **1**(9), 68–94

Lüling, K. H. (1962) Zur Okologie von *Pterolebias peruensis* Myers, 1954, (Pisces, Cyprinodontidae) am See Quista Cocha (Amazonia peruana). *Bonner zoologische Beiträge*, **4**, 353–9

Lüling, K. H. (1963) Die Quisto Cocha und ihre haufigen Fische (Amazonia peruana). *Beiträge zur Neotropischen Fauna*, **3**, 34–56

Lüling, K. H. (1964) Zur Biologie und Okologie von *Arapaima gigas (Pisces, Osteoglossidae). Zeitschrift für Morphologie und Okologie der Tiere*, **54**, 436–530

Lüling, K. F. (1971*a*) Okologische Beobachtungen und Untersuchungen am Biotop des *Rivulus beniensis* (Cyprinodontidae). *Beitrage Neotropischen Fauna*, **6**, 163–93

Lüling, K. F. (1971*b*) *Aequidens vittata* (Heck.) und andere Fische des Rio Huallaga im Ubergangsbereich zur Hylaea. *Zoologische Beitrage*, **17**, 193–226

Lüling, K. F. (1975) Ichthyologische und gewasserkuntliche Beobachtungen und untersuchungen an der Yarina Cocha, in der Umbgebung von Pucallpa und am Rio Pacaya (mittlerer und unterer Ucayali, Ostperu). *Zoologische Beitrage, Neue Folge*, **21**, 29–96

Lundberg, J. G. & Stager, J. C. (1985) Microgeographical diversity in the Neotropical knifefish *Eigenmannia macrops* (Gymnotiformes). *Environmental Biology of Fishes*, **13**, 173–81

McKaye, K. R. (1977) Competition for breeding sites between the cichlid fishes of L. Jiloa, Nicaragua. *Ecology*, **58**, 291–302

McKaye, K. R. (1980) Seasonality in habitat selection by the gold color morph of *Cichlasoma citrinellum* and its relevance to sympatric speciation in the family Cichlidae. *Environmental Biology of Fishes*, **5**, 75–8

McKaye, K. R. (1981) Field observation on death feigning: a unique hunting behavior by the predatory cichlid, *Haplochromis livingstoni*, of L. Malawi. *Environmental Biology of Fishes*, **6**, 361–5

McKaye, K. R. (1983) Ecology and breeding behaviour of a cichlid fish, *Cyrtocara eucinostomus*, on a large lek in L. Malawi, Africa. *Environmental Biology of Fishes*, **8**, 81–96

McKaye, K. R. (1984) Behavioural aspects of cichlid reproductive strategies: patterns of territoriality and brood defence in Central American substratum spawners and African mouth brooders. In *Fish Reproduction: Strategies and Tactics*, ed. G. W. Potts & R. J. Wootton, pp. 245–73. London: Academic Press

McKaye, K. R. & Gray, W. N. (1984) Extrinsic barriers to gene flow in rock-dwelling cichlids of Lake Malawi: macrohabitat heterogeneity and reef colonization. In *Evolution of Fish Species Flocks*, ed. A. A. Echelle & I. Kornfield, pp. 169–83. Orono: University of Maine at Orono Press

McKaye, K. R. & Kocher, T. (1983) Head ramming behaviour by three paedophagous cichlids in L. Malawi, Africa. *Animal Behaviour*, **31**, 206–10

McKaye, K. R, Kocher, T., Reinthal, P., Harrison, P. & Kornfield, I. (1984) Genetic evidence for allopatric and sympatric differentiation among color morphs of a L. Malawi cichlid fish. *Evolution, Lancaster, Pa*, **38**, 215–19

McKaye, K. R., Kocher, T., Reinthal, P. & Kornfield, I. (1982) A sympatric sibling species complex of *Petrotilapia* Trewavas from L. Malawi analysed by enzyme electrophoresis (Pisces, Cichlidae). *Zoological Journal of the Linnean Society*, **76**, 91–6

McKaye, K. R. & McKaye, N. M. (1977) Communal care and kidnapping of young by parental cichlids. *Evolution, Lancaster, Pa*, **31**, 674–81

McKaye, K. R. & Marsh, A. (1983) Food switching by two specialized algae-scraping cichlid fishes in L. Malawi, Africa. *Oecologia (Berlin)*, **56**, 245–8

McKaye, K. R. & Oliver, M. F. (1980) Geometry of a selfish school: defence of cichlid young by a bagrid catfish in L. Malawi. *Animal Behaviour*, **28**, 1287

MacNae, W. (1968) A general account of the flora and fauna of mangrove swamps and forests in the Indo-Pacific region. *Advances in Marine Biology*, **6**, 73–270

Mago, F. M. L. (1970) Estudios preliminares sobre la ecologia de los peces de los llanos de Venezuela. *Acta Biologica Venezuelica*, **7**, 71–102

Mahon, R. & Balon, E. K. (1977) Fish production in L. Kariba, reconsidered. *Environmental Biology of Fishes*, **1**, 215–18

Man, H. S. H. & Hodgkiss, I. J. (1977*a*) Studies on the ichthyofauna in Plover Cove

Reservoir Hong Kong: I. Sequence of fish population changes. *Journal of Fish Biology*, 10, 493–503

Man, H. S. H. & Hodgkiss, I. J. (1977b) Studies on the ichthyofauna in Plover Cove Reservoir, Hong Kong: feeding and food relations. *Journal of Fish Biology*, 11, 1–13

Mani, M. S. (ed.) (1974). *Ecology and Biogeography in India*. Monographiae Biologicae no. 23. The Hague: Dr W. Junk

Mann, R. K., Mills, C. A. & Crisp, D. T. (1984) Geographical variation in the life-history tactics of some species of freshwater fish. In *Fish Reproduction: Strategies and Tactics*, ed. G. W. Potts & R. J. Wootton, pp. 171–86. London: Academic Press

Margalef, R. (1968) *Perspectives in Ecological Theory*. Chicago: University of Chicago Press

Marlier, G. (1967) Ecological studies on some lakes in the Amazon valley. *Amazoniana*, 1, 91–115

Marlier, G. (1968) Les poissons du lac Rondo et leur alimentaires trophiques du lac Redondo; les poissons du Rio Preto da Eva. *Cadernos Amazonia (INPA, Manaus)*, 11, 21–57

Marlier, G. (1973) Limnology of the Congo and Amazon rivers. In *Tropical Forest Ecosystems in Africa and South America: A Comparative Review*, ed. B. J. Meggers, E. S. Eyensu & W. D. Duckworth, pp. 223–38. Washington, D.C.: Smithsonian Institution Press

Marsh A. C. (1983) A taxonomic study of the fish genus *Petrotilapia* (Pisces: Cichlidae) from L. Malawi, Pt 1. *Ichthyological Bulletin of Rhodes University*, 48, 1–14

Marsh, A. C. & Ribbink, A. J. (1981) A comparison of the abilities of three species of *Petrotilapia* (Cichlidae, L. Malawi) to penetrate deep water. *Environmental Biology of Fishes*, 6, 367–9

Marsh, A. C. & Ribbink, A. J. (1985) Feeding site specialization in three sympatric species of *Petrotilapia* from L. Malawi. *Biological Journal of the Linnean Society*, 25, 331–8

Marsh, A. C. & Ribbink, A. J. (1986) Feeding schools among L. Malawi cichlid fishes. *Environmental Biology of Fishes*, 15, 75–9

Marsh, A. C., Ribbink, A. J. & Marsh, B. A. (1981) Sibling species complexes in sympatric populations of *Petrotilapia* Trewavas (Cichlidae, L. Malawi). *Zoological Journal of the Linnean Society*, 71, 253–64

Marsh, B. A., Marsh, A. C. & Ribbink, A. J. (1986) Reproductive seasonality in a group of rock-frequenting cichlid fishes in L. Malawi. *Journal of Zoology*, 208, 9–20

Marten, G. G. (1979) Predator removal: effect on fisheries yields in L. Victoria (East Africa). *Science, New York*, 203, 646–8

Marten, G. G. & Polovina, J. J. (1982) A comparative study of fish yields from various tropical ecosystems. In *Theory & Management of Tropical Fisheries*, ed. D. Pauly & G. I. Murphy, pp. 255–8. ICLARM Conference Proceedings, no. 9. Manila, Philippines: ICLARM

Matthes, H. (1961) *Boulengerochromis microlepis*, a L. Tanganyika fish of economical importance. *Bulletin of Aquatic Biology*, 3(24), 1–15

Matthes, H. (1964) Les poissons du lac Tumba et de la region d'Ikela. Etude systematique et écologique. *Annales Musée Royal de l'Afrique Centrale, Sciences Zoologiques*, 126, 1–204

Mayland, H. J. (1982) *Der Malawi-See und seine Fische*. Hannover: Landbuch

Mayland, H. J. (1984) *Mittelamerika Cichliden und Lebendgebärende*. Hannover: Landbuch

Maynard Smith, J. (1966) Sympatric speciation. *American Naturalist*, 100, 637–50

MBA (Marine Biological Association of India) (1964) *Proceedings of the Symposium on Scombroid Fishes, 1962, Mandapam Camp, India*

Melak, J. M. (1976) Primary production and fish yields in tropical lakes. *Transactions of the American Fisheries Society*, 105, 575–80

Melzak, M. (1981) Chemical warfare on the coral reef. *New Scientist*, **89**, 733–5

Merona, B. de (1981) Zonation icthylogique du bassin du Bandama (Côte d'Ivoire). *Revue Hydrobiologie Tropicale*, **14**, 63–75

Merona, B. de (1983) Modèle d'estimation rapide de la croissance des poissons. Application aux poissons d'eau douce d'Afrique. *Revue Hydrobiologie Tropicale*, **16**, 103–13

Merona, B. de (1985) Les peuplements de poissons et la pêche dans le bas Tocantins (Amazonie Bresilienne) avant la fermature du barrage de Tucurui. *Verhandlungen Internationalen Vereinigung für theoretische und angewandte Limnologie*, **22**, 2698–703

Miller, R. R. (1966) Geographical distribution of Central American freshwater fishes. *Copeia*, **4**, 773–802

Miyadi, D. (1960) Perspectives of experimental research on social interference among fishes. In *Perspectives in Marine Biology*, ed. A. A. Buzzati-Traverso, pp. 469–79. University of California Press

Mizuno, N. & Furtado, J. I. (1982) Ecological notes on fishes. In *Tasek Bera. The Ecology of a Freshwater Swamp*, ed. J. I. Furtado & S. Mori, pp. 321–54. Monographiae Biologicae, no. 47. The Hague: Dr W. Junk

Moe, M. R., Jr (1969) Biology of the Red Grouper *Epinephalus morio* (Val.) from the Eastern Gulf of Mexico. *Professional Paper Series, Florida State Board Conservation Marine Laboratory*, no. 10, 1–95

Mohsin, A. K. M. & Ambak, M. A. (1983) *Freshwater Fishes of Peninsular Malaysia*. Pertanian, Malaysia: Penerbit Universiti

Moodie, G. E. E. & Power, M. (1982) The reproductive biology of an armoured catfish, *Loricaria uracantha*, from Central America. *Environmental Biology of Fishes*, **7**, 143–8

Moore, R. (1979) Natural sex inversion in the giant perch (*Lates calcarifer*). *Australian Journal of Marine & Freshwater Research*, **30**, 803–13

Moore, R. (1982) Spawning and early life history of barramundi, *Lates calcarifer* (Bloch), in Papua New Guinea. *Australian Journal of Marine & Freshwater Research*, **33**, 647–61

Moore, R. & Reynolds, L. F. (1982) Migration patterns of barramundi, *Lates calcarifer* (Bloch), in Papua New Guinea, *Australian Journal of Marine & Freshwater Research*, **33**, 671–82

Moriarty, D. J. W., Darlington, J. P. E., Dunn, I. G., Moriarty, C. M. & Tevlin, M. P. (1973) Feeding and grazing in L. George, Uganda. *Proceedings of the Royal Society, London, B*, **184**, 299–319

Motwani, M. P. & Kanwai, Y. (1970) Fish and fisheries of the coffer-dammed right channel of the R. Niger at Kainji. In *Kainji Lake Studies*, vol. 1, *Ecology*, ed. S. A. Visser, pp. 27–48. Ibadan, Nigeria: Nigerian Institute of Social & Economic Research, Ibadan University Press

Moyer, J. T., Thresher, R. E., & Colin, P. L. (1983) Courtship, spawning and inferred social organization of American angelfishes (Genera *Pomacanthus, Holocentrus & Centropyge*: Pomacanthidae). *Environmental Biology of Fishes*, **9**, 25–39

Moyle, P. B. & Senanayake, F. R. (1984) Resource partitioning among the fishes of rainforest streams in Sri Lanka. *Journal of Zoology, London*, **202**, 195–223

Munro, J. L., Gaut, V. C., Thompson, R. & Reeson, R. H. (1973) The spawning seasons of Caribbean reef fishes. *Journal of Fish Biology*, **5**, 69–84

Murdoch, W. W. (1969) Switching in general predators: experiments on predator specificity and stability of prey populations. *Ecological Monographs*, **39**, 335–54

Myrberg, A. A. (1966) Parental recognition of young in cichlid fishes. *Animal Behaviour*, **14**, 565–71

Myrberg, A. A., Jr (1972) Ethology of the bicolor damselfish *Eupomacentrus partitus* (Pisces: Pomacentridae): a comparative analysis of laboratory and field behaviour. *Animal Behaviour Monographs*, **5**(3), 197–283

Nagoshi, M. (1983) Distribution, abundance and parental care of the genus *Lamprologus* (Cichlidae) in L. Tanganyika. *African Study Monographs, Kyoto*, **3**, 39–47

Nagoshi, M. (1985) Growth and survival in larval stage of the genus *Lamprologus* (Cichlidae) in L. Tanganyika. *Verhandlungen der Internationalen Vereiningung für theoretische und angewandte Limnologie*, **22**, 2663–70

Nakamura, H. (1969) *Tuna Distribution and Migration*. London: Fishing News (Books) Ltd

Nakamura, I. (1985) *FAO Species Catalogue*. vol. 5. *Billfishes of the World: An annotated and illustrated Catalogue of Marlins, Sailfishes, Spearfishes and Sword Fishes known to date*. FAO Fisheries Synopsis, no. 125. Rome: FAO

Nelson, J. S. (1984) *Fishes of the World*, 2nd edn. New York: John Wiley

Nelson, K. (1964) Temporal patterning and courtship behaviour in the glandulocaudine fishes. *Behaviour*, **24**, 90–146

Neudecker, S. & Lobel, P. S. (1982) Mating systems of chaetodontid and pomacanthid fishes at St Croix. *Zeitschrift für Tierpsychologie*, **59**, 299–318

Nordlie, F. G. & Kelso, D. O. (1975) Trophic relationships in a tropical estuary. *Revista de Biologia Tropical*, **23**, 77–99

Northcote, T. G., Arcifa, M. S. & Froehlich, O. (1985) Effects of impoundment and drawdown on the fish community of a South American river. *Verhandlungen der internationalen Vereinigung für theoretische und angewandte Limnologie*, **22**, 2704–11

Novoa, D. F. & Ramos, F. (1978) *Las pesquerias comerciales del Rio Orinoco*. Caracas: Corporacion Venezolana de Guayana. 161 pp. (In Spanish)

Obeng, L. (ed.) (1968) *Man-made Lakes: The Accra Symposium*. Accra: Ghana University Press

Odum, E. P. (1969) The strategy of ecosystem development. *Science, New York*, **164**, 262–70

Odum, H. T. & Odum, E. P. (1955) Trophic structure and productivity of a windward coral reef community on Eniwetok Atoll. *Ecological Monographs*, **25**, 1–291

Ogden, J. C. & Ehrlich, P. R. (1977) The behavior of heterotypic resting schools of juvenile grunts (Pomadasyidae). *Marine Biology (Berlin)*, **42**, 273–80

Ogden, J. C. & Lobel, P. S. (1978) The role of herbivorous fishes and urchins in coral reef communities. *Environmental Biology of Fishes*, **3**, 49–63

Oglesby, R. T. (1982) The morphoedaphic index symposium – overview and observations. *Transactions of the American Fisheries Society*, **111**, 171–5

Okedi, J. (1969) Observations on the breeding and growth of certain mormyrid fishes of the L. Victoria basin. *Revue de Zoologie et Botanique Africaines*, **79**, 34–64

Okedi, J. (1970) A study of the fecundity of some mormyrid fishes from L. Victoria. *East African Agriculture and Forestry Journal*, **35**, 436–42

Okedi, J. (1971) The food and feeding habits of the small mormyrid fishes of L. Victoria. *African Journal of Tropical Hydrobiology & Fisheries*, **1**, 1–12

Okemwa, E. N. (1984) Potential fishery of Nile Perch *Lates niloticus* Linne (Pisces, Centropomidae) in Nyanza Gulf of L. Victoria, E. Africa. *Hydrobiologia*, **108**, 121–6

Olatunde, A. A. (1977) The distribution, abundance and trends in the establishment of the family Schilbeidae (Siluriformes) in L. Kainji, Nigeria. *Hydrobiologia*, **56**, 69–80

Oliveira, A. M. E. de (1972) Peixes estuarios do Nordeste Oriental Brasileiro. *Archivos de Ciencias do Mar, Brazil*, **12**, 35–41

Oppenheimer, J. R. (1970) Mouth breeding in fishes. *Animal Behaviour*, **18**, 493–503

Orians, G. H. (1975) Diversity, stability and maturity in natural ecosystems. In *Unifying Concepts in Ecology*, ed. W. H. van Dobben & R. H. Lowe-McConnell, pp. 139–50. The Hague: Dr W. Junk

Ormond, R. F. G. (1980) Aggressive mimicry and other interspecific feeding associations among Red Sea coral reef predators. *Journal of Zoology*, **191**, 247–62

ORSTOM (1983) *Hydrobiological Activities in the Tropical Zone*. (Bibliography) Paris: Office de la Recherche Scientifique et Technique Outre-Mer

Paine, R. T. (1966) Food web complexity and species diversity. *American Naturalist*, **100**, 65–75

Paine, R. T. (1969) A note on trophic complexity and community stability. *American Naturalist*, **103**, 91–3

Paiva, M. P. (1973) Recursos pesqueiros e a pesca na Bacia do Rio Parnaiba (Brasil). *Boletim Cearense de Agronomica, Fortaleza, Ceara*, **14**, 49–82

Paiva, M. P. (1982) *Grandes Represas do Brasil*. Brasilia: Editerra

Parenti, L. (1981) A phylogenetic and biogeographical analysis of cyprinodontiform fishes (Teleostei, Atherinomorpha). *Bulletin American Museum of Natural History*, **168** (art 4). 335–557

Parenti, L. (1984a) Killifish classification. *The Aquarist*, June, 20–2

Parenti, L. (1984b) A taxonomic revision of the Andean killifish genus *Orestias* (Cyprinodontiformes, Cyprinodontidae). *Bulletin of the American Museum of Natural History*, **178**, 107–214

Parenti, L. (1984c) Biogeography of the Andean killifish genus *Orestias* with comments on the species flock concept. In *Evolution of Fish Species Flocks*, ed. A. A. Echelle & I. Kornfield, pp. 85–92. Orono, Maine: University of Maine Press

Partridge, B. L. (1982) The structure and function of fish schools. *Scientific American*, **246**(6), 90–9

Patrick, R. (1964) A discussion of the result of the Catherwood Expedition to the Peruvian headwaters of the Amazon. *Verhandlungen der internationalen Vereinigung für theoretische und angewandte Limnologie*, **15**, 1084–90

Paugy, D. (1978) Ecologie et biologie des *Alestes baremoze* (Pisces, Characidae) des rivières de Côte d'Ivoire. *Cahiers ORSTOM, sér. Hydrobiologie*, **12**, 245–75

Paugy, D. (1979–80a) Ecologie et biologie des *Alestes imberi* (Pisces, Characidae) des rivières de Côte d'Ivoire: comparaison meristique avec *A. nigricauda. Cahiers ORSTOM, sér. Hydrobiologie*, **13**, 129–41

Paugy, D. (1979–80b) Ecologie et biologie des *Alestes nurse* des rivières de Côte d'Ivoire. *Cahiers ORSTOM. Hydrobiologie*, **13**, 143–59

Pauly, D. (1981) On the interrelationships between natural mortality, growth parameters, and mean environmental temperature in 175 fish stocks. *Journal du Conseil Permanent International pour l'Exploration de la Mer*, **39**, 175–92

Pauly, D. (1983) Some simple methods for the assessment of tropical fish stocks. *FAO (Rome) Fisheries Technical Papers*, no. 234, 52 pp. (FAO(FIRM/T234)(En)

Pauly, D. & Murphy, G. J. (eds) (1982) *Theory and Management of Tropical Fisheries*. ICLARM Conference Proceedings no. 9. Manila, Philippines: ICLARM

Perrone, M., Jr (1978) Mate size and breeding success in a monogamous cichlid fish. *Environmental Biology of Fishes*, **3**, 193–201

Perrone, M., Jr & Zaret, T. M. (1979) Parental care patterns of fishes. *American Naturalist*, **113**, 351–61

Peters, H. M. (1963) Eizahl, Eigenwicht und Gelegeentwicklung in der Gattung *Tilapia* (Cichlidae, Teleostei). *Internationale Revue der gesamten Hydrobiologie*, **48**, 547–76

Petr, T. (1968) Distribution, abundance and food of commercial fish in the Black Volta and Volta man-made lake in Ghana during its first period of filling (1964–1966). I. Mormyridae. *Hydrobiologia*, **32**, 417–48

Petr, T. (1974) Distribution, abundance of food of commercial fish in the Black Volta and the Volta man-made lake in Ghana during the filling period (1964–68). II. Characidae. *Hydrobiologia*, **45**, 303–37

Petr, T. (ed.) (1983) *The Purari – Tropical Environment of a High Rainfall River Basin*. Monographiae Biologicae, no. 51. The Hague: Dr W. Junk

Petr, T. (1984) Indigenous fish and stocking of lakes and reservoirs on tropical islands of the Indo-Pacific. *FAO (Rome) Fisheries Reports*, no. 312. (FIRI/R 312)

Petrere, M., Jr (1978a) [Fishing and fishing pressure in the State of Amazonas, I. Per capita fishing pressure.] *Acta Amazonica*, **8**, 439–54. (In Portuguese)

Petrere, M., Jr (1978b) [Fishing and fishing pressure in the State of Amazonas. II. Locations, capture methods and landing statistics.] *Acta Amazonica*, **8**, suppl. 2, 1–54 (In Portuguese)

Petrere, M., Jr (1982) Ecology of the fisheries in the River Amazon and its tributaries in the Amazonas State (Brazil). Ph.D. thesis, University of East Anglia, UK

Petrere, M., Jr (1983a) Yield per recruit of the tambaqui (*Colossoma macropomum* Cuvier) in the Amazonas State, Brazil. *Journal of Fish Biology*, **22**, 133–44

Petrere, M., Jr (1983b) Relationships among catches, fishing effort and river morphology for eight rivers in Amazonas State (Brazil), during 1976–1978. *Amazoniana*, **8**, 281–96

Petrere, M., Jr (1985a) A pesca comercial no Rio Solimões–Amazonas e suis afluentes: analise dos informos do pescado desembarcado no Mercado Municipal de Manaus (1976–78). *Ciencia e Cultura*, **37**, 1987–99

Petrere, M., Jr (1985b) Migraciones de peces de agua dulce en America Latina; algunos comentarios. *COPESCAL Documento Ocasional*, (1), 17 p. Rome: FAO

Phillips, P. C. (1981) Diversity and fish community structure in a Central American mangrove embayment. *Revista de Biologia Tropical*, **29**, 227–36

Pianka, E. R. (1970) On r and K selection. *American Naturalist*, **100**, 593–7

Pillay, T. V. R. (1958) Biology of the Hilsa, *Hilsa ilisha* (Ham.) of the R. Hooghly. *Indian Journal of Fisheries*, **5**, 201–57

Pillay, T. V. R. (1967) Estuarine fisheries in West Africa. In *Estuaries*, ed. G. Lauff, pp. 639–46. Washington: American Association for the Advancement of Science, Publication no. 83

Poll, M. (1953) *Résultats, Scientifique Explorations Hydrobiologique de Lac Tanganyika (1946–7). Poissons non-Cichlidae*, 3(5A), pp. 1–251. Bruxelles: Institut Royal des Sciences Naturelles de Belgique

Poll, M. (1956) *Résultats, Scientifique Explorations Hydrobiologique de Lac Tanganyika (1946–7). Poissons Cichlidae*, 3(5B), pp. 1–619. Buxelles: Institut Royal des Sciences Naturelles de Belgique

Poll, M. (1957) Les genres des poissons d'eau douce de l'Afrique. *Annales du Musée Royal du Congo Belge, Tervuren, Sciences Zoologiques*, **54**, 1–191

Poll, M. (1959a) Recherches sur la fauna ichthyologique de la région du Stanley Pool. *Annales Musée Royal du Congo Belge, Tervuren, Sciences Zoologiques*, **71**, 75–174

Poll, M. (1959b) Aspects nouveau de la faune ichthyologique du Congo Belge. *Bulletin de la Société Zoologique de France*, **84**, 259–71

Poll, M. (1973) Nombre et distribution géographique des poissons d'eau douce africains. *Bulletin du Museum National d'Histoire Naturelle, Paris*, 3 sér., no. 150, *Ecologie générale*, **6**, 113–28

Poll, M. (1976) Poissons. *Exploration du Parc National de l'Upemba. Mission G. F. de Witte*, fasc. 73, 1–127. (Fondation pour favorisés les recherches scientifiques en Afrique.)

Poll, M. (1980) Ethologie comparée des poissons fluviatiles et lacustres africains. *Bulletin de l'Academie Royal de Belgique (Classe Sciences)*, 5e sér., **66**, 78–97

Por, F. D. & Dor, I. (1984) *Hydrobiology of the Mangal: The Ecosystem of the Mangrove Forests*. Dordrecht: Dr W. Junk

Potts, G. W. (1973) The ethology of *Labroides dimidiatus* (C & V) (Labridae, Pisces) in Aldabra. *Animal Behaviour*, **21**, 250–91

Potts, G. W. (1980) The predatory behaviour of *Caranx melampygus* in the channel environment of Aldabra Atoll (Indian Ocean). *Journal of Zoology, London*, **192**, 323–50

Potts, G. W. & Wootton, R. J. (1984) *Fish Reproduction: Strategies & Tactics*. London: Academic Press

Power, M. E. (1983) Grazing responses of tropical freshwater fishes to different scales of variation in their food. *Environmental Biology of Fishes*, **9**, 25–37

Power, M. (1984*a*) Habitat quality and the distribution of algal-grazing catfish in a Panamanian stream. *Journal of Animal Ecology*, **53**, 357–74

Power, M. (1984*b*) Depth distribution of armoured catfish: predator-induced resource avoidance? *Ecology*, **65**, 523–8

Prowze, G. A. (1972) Some observations on primary and fish production in experimental ponds at Malacca, Malaysia. In *Productivity Problems of Freshwaters*, ed. Z. Kajak & A. Hillbricht-Ilkowska, pp. 555–61

Pullin, R. S. V. & Lowe-McConnell, R. H. (1982) *The Biology and Culture of Tilapias*. ICLARM Conference Proceedings no. 7. Manila, Philippines: International Center for Living Aquatic Resources Management

Rahel, F. J., Lyons, J. D. & Cochran, P. A. (1984) Stochastic or deterministic regulation of assemblage structure? It may depend on how the assemblage is defined. *American Naturalist*, **124**, 583–9

Randall, J. E. (1961) A contribution to the biology of the convict surgeon fish of the Hawaiian Is, *Acanthurus triostegus sandivicensis*. *Pacific Science*, **15**, 215–72

Randall, J. E. (1963) An analysis of the fish populations of artificial and natural reefs in the Virgin Is. *Caribbean Journal of Science*, **3**, 31–47

Randall, J. E. (1967) Food habits of reef fishes of the West Indies. *Studies in Tropical Oceanography*, **5**, 665–847

Randall, J. E. & Brock, V. E. (1960) Observations on the ecology of epinephaline and lutjanid fishes of the Society Is, with emphasis on food habits. *Transactions of the American Fisheries Society*, **89**, 9–16

Randall, J. E. & Randall, H. A. (1960) Examples of mimicry and protective resemblance in tropical marine fishes. *Bulletin of Marine Science of the Gulf & Caribbean*, **10**, 444–80

Randall, J. E. & Randall, H. A. (1963) The spawning and early development of the Atlantic parrotfish, *Sparisoma rubripinne*, with notes on other scarid and labrid fishes. *Zoologica, New York*, **48**, 49–60

Rasa, O. A. E. (1969) Territoriality of the establishment of dominance by means of visual clues in *Pomacentrus jenkinsi* (Pomacentridae). *Zeitschrift für Tierpsychologie*, **26**, 825–45

Regier, H. A. (1977) Fish communities and aquatic ecosystems. In *Fish Population Dynamics*, ed. J. A. Gulland, pp. 134–55. Chichester: John Wiley & Sons

Regier, H. A. & Cowell, E. G. (1972) Applications of ecosystem theory, succession, diversity, stability, stress and conservation. *Biological Conservation*, **4**, 83–8

Reid, G. M. (1985) *A Revision of African species of* Labeo *(Pisces: Cyprinidae)*. Theses Zoologicae, vol. 6. Braunschweig: Von J. Kramer

Reid, G. M. & Sydenham, H. (1979) A check list of Lower Benue river fishes and an ichthyological review of the Benue River (W. Africa). *Journal of Natural History*, **13**, 41–67

Reid, S. B. (1983) La biologia de los bagres rayados *Pseudoplatystoma fasciatum* y *P. tigrinum* en la cuenca del Rio Apure, Venezuela. *Revista UNELLEZ de Ciencia y Technologia*, **1**, 13–41

Reid, S. B. (1986) Cryptic adaptations of small catfishes *Sorubim lima* (Pimelodidae) in Venezuela. *Biotropica*, **18**, 86–8

Reinboth, R. (ed.) (1975) *Intersexuality in the Animal Kingdom*. Heidelberg: Springer-Verlag

Reynolds, J. D. (1970) Biology of the small pelagic fishes in the new Volta lake in Ghana, Pt I. The lake and the fish: feeding habits. *Hydrobiologia*, **35**, 568–603

Reynolds, J. D. (1974) Biology of the small pelagic fishes in the new Volta Lake in Ghana, Pt III. Sex and reproduction. *Hydrobiologia*, **45**, 489–508

Reynolds, L. F. & Moore, R. (1982) Growth rates of barramundi, *Lates calcarifer* (Bloch), in Papua New Guinea. *Australian Journal of Marine & Freshwater Research*, **33**, 663–70

Reznick, D. & Endler, J. A. (1982) The impact of predation on life history evolution in Trinidadian guppies (*Poecilia reticulata*). *Evolution*, **36**, 160–77

Ribbink, A. J. (1977) Cuckoo among L. Malawi cichlid fish. *Nature, London*, **267**, 243–4

Ribbink, A. J. (1984*a*) The feeding behaviour of a cleaner, scale and skin and fin eater of L. Malawi (*Docimodus evelynae*, Pisces, Cichlidae). *Netherlands Journal of Zoology*, **34**, 182–96

Ribbink, A. J. (1984*b*) Is the species flock concept tenable? In *Evolution of Fish Species Flocks*, ed. A. A. Echelle & I. Kornfield, pp. 21–5. Orono: University of Maine Press

Ribbink, A. J. & Hill, B. J. (1979) Depth equilibrium by two predatory cichlid fish from L. Malawi. *Journal of Fish Biology*, **14**, 507–10

Ribbink, A. J. & Lewis, D. S. C. (1982) *Melanochromis crabro* sp. nov.: a cichlid fish from L. Malawi which feeds on ectoparasites and catfish eggs. *Netherlands Journal of Zoology*, **32**, 72–87

Ribbink, A. J., Marsh, A. C., & Marsh, B. A. (1981) Nest-building and communal care of young by *Tilapia rendalli* Dumeril in L. Malawi. *Environmental Biology of Fishes*, **6**, 219–22

Ribbink, A. J., Marsh, A. C., Marsh, B. & Sharp, B. J. (1980) Parental behaviour and mixed broods among cichlid fish of L. Malawi. *South African Journal of Zoology*, **15**, 1–6

Ribbink, A. J., Marsh, A. C., Marsh, B. A. & Sharp, B. J. (1983*a*) The zoogeography, ecology and taxonomy of the genus *Labeotropheus* Ahl 1927 (Pisces: Cichlidae) of L. Malawi. *Zoological Journal of the Linnean Society*, **79**, 223–43

Ribbink, A. J., Marsh, B. A., Marsh, A. C. & Sharp, B. J. (1983*b*) A preliminary survey of the cichlid fishes of rocky habitats in L. Malawi. *South African Journal of Zoology*, **18**(3), 160 pp

Ribeiro, M. C. L. de B. (1983) As migracões dos jaraquis (Pisces, Prochilodontidae) no Rio Negro, Amazonas, Brasil. Master's thesis, INPA, Manaus, Brazil

Ringuelet, R. A., Aramburu, R. H. & Alonso de Aramburu, A. (1967) *Los Peces Argentinos de Agua Dulce*. Buenos Aires: Comision de Investigacion Cientifica

Rinne, J. N. & Wanjala, A. B. (1983) Maturity, fecundity and breeding seasons of the major catfishes in L. Victoria, E. Africa. *Journal of Fish Biology*, **23**, 357–63

Roberts, T. R. (1972) Ecology of fishes in the Amazon and Congo basins. *Bulletin of the Museum of Comparative Zoology, Harvard*, **143**, 117–47

Roberts, T. R. (1975) Geographical distribution of African freshwater fishes. *Zoological Journal of the Linnean Society*, **57**, 249–319

Roberts, T. R. (1978) An ichthyological survey of the Fly River in Papua New Guinea with descriptions of new species. *Smithsonian Contributions to Zoology*, **281**, 1–72

Roberts, T. R. & Stewart, D. J. (1976) An ecological and systematic survey of fishes in the rapids of the lower Zaire or Congo River. *Bulletin of the Museum of Comparative Zoology, Harvard*, **147**, 239–317

Robertson, R. (1972) Social control of sex reversal in a coral-reef fish. *Science, New York*, **177**, 1007–9

Robertson, R. (1973) Sex change under the waves. *New Scientist*, **58**, 538–40

Robertson, R. (1983) On the spawning behaviour and spawning cycles of 8 surgeon fishes (Acanthuridae) from the Indo-Pacific. *Environmental Biology of Fishes*, **9**, 193–223

Robertson, R. & Justines, G. (1982) Protogynous hermaphroditism and gonochorism in four Caribbean reef gobies. *Environmental Biology of Fishes*, **7**, 137–42

Robertson, D. R. & Lassig, B. (1980) Spatial distribution patterns and coexistence of a group of territorial damselfishes from the Great Barrier reef. *Bulletin of Marine Science*, **30**, 187–203

Robertson, D. R., Polunin, N. V. C. & Leighton, K. (1979) The behavioural ecology of three Indian Ocean surgeon fishes (*Acanthurus lineatus, A. leucostictus & Zebrasoma scopas*): their feediing strategies, and social and mating systems. *Environmental Biology of Fishes*, **4**, 125–70

Robertson, D. R., Reinboth, R. & Bruce, R. (1982) Gonochorism, protogynous sex-change and spawning in three sparisomatine parrotfishes from the western Indian Ocean. *Bulletin of Marine Science*, **32**, 868–79

Robertson, D. R., Sweatman, H. P. A., Fletcher, E. & Cleland, M. G. (1976) Schooling as a mechanism for circumventing the territoriality of competitors. *Ecology*, **57**, 1208–20

Robertson, D. R. & Warner, R. R. (1978) Sexual patterns in the labroid fishes of the Western Caribbean. 2. The parrot fishes. *Smithsonian Contributions to Zoology*, **255**, 1–26

Robison, B. H. (1972) Distribution of the midwater fishes in the Gulf of California. *Copeia*, **3**, 448–61

Rosa, H., Jr (ed.) (1963) *Proceedings, World Scientific Meeting on Tunas and Related Species*. Rome: FAO Fish Report no. 6

Rosen, D. E. (1978) Vicariant patterns and historical explanation in biogeography. *Systematic Zoology*, **27**, 159–88

Rosendahl, B. R. & Livingstone, D. A. (1983) Rift lakes of East Africa. New seismic data and implications for future research. *Episodes*, **1**, 14–19

Rothschild, B. J. & Suda, A. (1977) Population dynamics of tuna. In *Fish Population Dynamics*, ed. J. A. Gulland, pp. 309–34. Chichester: John Wiley

Russell, B. C., Allen, G. R. & Lubbock, H. R. (1976) New cases of mimicry in marine fishes. *Journal of Zoology, London*, **180**, 407–23

Ryder, R. A. (1982) The morpho-edaphic index – use, abuse and fundamental concepts. *Transactions of the American Fisheries Society*, **111**, 154–64

Rzoska, J. (1978) *On the Nature of Rivers with Case Stories of Nile, Zaire and Amazon*. The Hague: Dr W. Junk

Sage, R. D. & Selander, R. K. (1975) Trophic radiation through polymorphism in cichlid fishes. *Proceedings of the National Academy of Sciences of the United States*, **72**, 4669–73

Sagua, V. O. (1978) The effect of Kainji dam, Nigeria, upon fish production in the R. Niger below the dam at Faku. In *Symposium on River and Floodplain Fisheries in Africa, Bujumbura Burundi, November 1977. Review and Experience Papers*, ed. R. L. Welcomme, pp. 210–24. Rome: FAO. CIFA Technical Paper no. 5

Sagua, V. O. (1979) Observations on the food and feeding habits of the African electric catfish *Malapterurus electricus*. *Journal of Fish Biology*, **16**, 61–70

Saigal, B. N. (1964) Studies on the fishery and biology of the commercial catfishes of the Ganga R. system. II. Maturity, spawning and food of *Mystus* (*Osteobagrus*) *aor*. *Indian Journal of Fisheries*, **11**, 1–44

Sale, P. F. (1977) Maintenance of high diversity in coral reef fish communities. *American Naturalist*, **111**, 337–59

Sale, P. F. (1978a) Reef fishes and other vertebrates: a comparison of social structures. In *Contrasts in Behaviour*, ed. E. S. Reece & F. J. Lighter, pp. 313–46. New York: John Wiley

Sale, P. F. (1978b) Coexistence of coral reef fishes – a lottery for living space. *Environmental Biology of Fishes*, **3**, 85–102

Sale, P. F. (1979) Recruitment, loss and coexistence in a guild of territorial coral reef fishes. *Oecologia (Berlin)*, **42**, 159–77

Sale, P. F. (1980a) Assemblages of fish on patch reefs – predictable or unpredictable? *Environmental Biology of Fishes*, **5**, 243–9

Sale, P. F. (1980b) The ecology of fishes on coral reefs. *Oceanography & Marine Biology*, **18**, 367–421

Sale, P. F., Doherty, P. J. & Douglas, W. A. (1980) Juvenile recruitment strategies and the coexistence of territorial pomacentrid fishes. *Bulletin of Marine Science*, **30**, 147–58

Sale, P. F. & Dybdahl, R. (1975) Determinants of community structure for coral reef fishes in an experimental habitat. *Ecology*, **56**, 1345–55

Santos, G. M. (1982) Caracterizacao, habitos alimentares e reproductivos de quatro especies de 'aracus' e consideracoes ecologicas sobre o gupo no lago Janauaca-Am (Osteichthyes, Characoidei, Anostomidae). *Acta Amazonica*, **12**, 713–39

Santos, U. de M. (1973) Beobachtungen über Wasserbewegungen, chemische Schichtung und Fischwanderungen in Varzea-seen am mittleren Solimoes (Amazonas). *Oecologia (Berlin)*, **13**, 239–46

Sato, Tetsu (1986) A brood parasitic catfish of mouthbrooding cichlid fishes in L. Tanganyika. *Nature*, **323**, 58–9

Saul, W. G. (1975) An ecological study of fishes at a site in upper Amazonian Ecuador. *Proceedings of the Academy of Natural Sciences of Philadelphia*, **127**, 93–134

Sazima, I. (1983) Scale-eating in characoids and other fishes. *Environmental Biology of Fishes*, **9**, 87–101

Scheel, J. (1968) *Rivulins of the Old World*. Neptune City, N.J.: TFH Publications

Schroder, J. H. (1980) Morphological and behavioural differences between the BB/OB and B/W colour morphs of *Pseudotropheus zebra* Blgr (Pisces; Cichlidae). *Zeitschrift für zoologische Systematik und Evolutionsforschung*, **18**, 69–76

Schut, J. de Silva, S. S. & Kortmulder, K. (1984) Habitat, associations and competition of eight *Barbus* (= *Puntius*) species (Cyprinidae) indigenous to Sri Lanka. *Netherlands Journal of Zoology*, **34**, 159–81

Schwarz, A. L. (1985) The behaviour of fishes in their acoustic environment. *Environmental Biology of Fishes*, **13**, 3–15

Schwassmann, H. O. (1976) Ecology and taxonomic status of different geographic populations of *Gymnorhamphthichthys hypostomus* Ellis (Pisces, Cypriniformes, Gymnotoidei). *Biotropica*, **8**, 25–40

Schwassmann, H. O. (1978) Times of annual spawning and reproductive strategies in Amazonian fishes. In *Rhythmic Activity of Fishes*, ed. J. E. Thorpe, pp. 187–200. London: Academic Press

Shapiro, D. Y. (1981) Size, maturation and the social control of sex reversal in the coral reef fish *Anthias squamipinnis*. *Journal of Zoology, London*, **193**, 105–28

Shapiro, D. Y. (1984) Sex reversal and sociodemographic processes in coral reef fishes. In *Fish Reproduction*, ed. G. W. Potts & R. J. Wootton, pp. 103–18. London: Academic Press

Shpigel, M. (1982) Niche overlap among two species of coral dwelling fishes of the genus *Dascyllus* (Pomacentridae). *Environmental Biology of Fishes*, **7**, 65–8

Shomura, R. S. & Williams, F. (eds) (1974) *Proceedings of the International Billfish Symposium, Kailua-Kona, Hawaii, 1972*. NOAA technical Report, NMFS SSRF, no. 675. Washington, D.C.: US Department of Commerce, National Marine Fisheries Service

Shulman, M. J., (1985) Coral reef fish assemblages: intra and interspecific competition for shelter sites. *Environmental Biology of Fishes*, **13**, 81–92

Simpson, A. C. (1982) A reviews of the database of tropical multispecies stocks in the SE Asian region. In *Theory & Management of Tropical Fisheries*, ed. Pauly & Murphy, pp. 5–31. ICLARM Conference Proceedings no. 9. Manila, Philippines: ICLARM

Simpson, B. R. C. (1979) The phenology of annual killifishes. *Symposia of the Zoological Society of London*, no. 44, 243–61

Simpson, J. G. (1971) The present status of exploitation and investigation of the clupeoid resources of Venezuela. Symposium on Investigations & Resources of the Caribbean and adjacent Regions, *FAO Fish Report*, no. 71–2, 263–78

Sioli, H. (1964) General features of the limnology of Amazonia. *Verhandlungen der internationalen Vereinigung für theoretische und angewandte Limnologie*, **15**, 1053–8

Sioli, H. (ed.) (1984) *The Amazon: Limnology and Landscape Ecology of a Mighty River and its Basin*. Monographiae Biologicae. Dordrecht: Dr W. Junk

Smith, C. L. (1972) A spawning aggregation of Nassau groupers, *Epinephalus striatus* (Bloch). *Transactions of the American Fisheries Society*, **101**, 257–61

Smith, C. L. (1978) Coral reef fish communities: a compromise view. *Environmental Biology of Fishes*, **3**, 109–28

Smith, C. L. & Tyler, J. C. (1972) Space resource sharing in a coral reef community. *Science Bulletin of the Natural History Museum, Los Angeles County*, **14**, 125–78

Smith, C. L. & Tyler, J. C. (1975) Succession and stability in fish communities of dome-shaped patch reefs in the West Indies. *American Museum Novitates*, no. 2572, 1–18

Smith, C. L., Tyler, J. C. & Feinberg, M. N. (1981) Population ecology and biology of the pearlfish (*Carapus bermudensis*) in the lagoon at Bimini, Bahamas. *Bulletin of Marine Science*, **31**, 876–902

Smith, H. M. (1945) The freshwater fishes of Siam or Thailand. *Bulletin of the U.S. National Museum*, **188**, 1–622

Smith, N. J. H. (1981) *Man, Fishes, and the Amazon.* New York: Columbia University Press

Soares, M. G. M. (1979) Aspectos ecologicos (alimentação e reprodução) dos peixes do igarape do Porto, Arpuana, MT. *Acta Amazonica*, **9**, 325–52

Southwood, T. R. E. (1977) Habitat, the templet for ecological strategies. *Journal of Animal Ecology*, **46**, 337–65

Springer, V. G. (1982) Pacific plate biogeography, with special reference to shore fishes. *Smithsonian Contributions to Zoology*, no. 367, 1–182

Springer, V. G. & McErlean, A. J. (1962) A study of the behavior of some tagged South Florida coral reef fishes. *American Midland Naturalist*, **67**, 386–97

Sreenivasan, A. (1972) Energy transformations through primary productivity and fish production in some tropical freshwater impoundments and ponds. In *Productivity Problems of Freshwaters*, ed. Z. Kajak & A. Hillbricht-Ilkowska, pp. 505–14. Warsaw: PWN Polish Scientific Publishers

Starck, W. A., II (1971) Investigations on the Gray Snapper *Lutjanus griseus*. *Studies in Tropical Oceanography, Miami*, no. 10, 1–150

Stearns, S. C. (1976) Life history tactics: a review of the ideas. *Quarterly Review of Biology*, **51**, 3–47

Stearns, S. C. & Crandall, R. E. (1984) Plasticity for age and size at sexual maturity: a life-history response to unavoidable stress. In *Fish Reproduction*, ed. G. W. Potts & R. J. Wootton, pp. 13–33. London: Academic Press

Steele, J. H., Rasmusson, E. M., Glantz, M. H. & Arntz, W. E. (1984) The 1982–83 El Niño. *Oceanus, Woods Hole, Mass.*, **27**(2), 1–39

Stephenson, W. & Searles, R. B. (1960) Experimental studies on the ecology of intertidal environments at Heron I. 1. Exclusion of fish from Beach Rock. *Australian Journal of Marine and Freshwater Research*, **11**, 241–67

Stiassny, M. L. J. (1980) Phylogenetic versus convergent relationship between piscivorous cichlid fishes from Lakes Malawi and Tanganyika. *Bulletin of the British Museum (Natural History), Zoology*, **40**, 67–101

Suda, A. (1973) Development of fisheries for non-conventional species. (In FAO Technical Conference on Fishery Management and Development.) *Journal of the Fisheries Research Board of Canada*, **30**, 2121–58

Sund, P. N., Blackburn, M. & Williams, F. (1981) Tunas and their environment in the Pacific Ocean: a review. *Oceanography & Marine Biology*, **19**, 443–512

Sydenham, D. H. J. (1977) The qualitative composition and longitudinal zonation of the fish fauna of the R. Ogun, Western Nigeria. *Revue de Zoologie Africaine*, **91**, 974–96

Taborsky, M. (1984) Broodcare helpers in the cichlid fish *Lamprologus brichardi*: their costs and benefits. *Animal Behaviour*, **32**, 1236–52

Taborsky, M. & Limberger, D. (1981) Helpers in fish. *Behavior, Ecology and Sociobiology*, **8**, 143–5

Takamura, K. (1984) Interspecific relationship between two aufwuchs eaters *Petrochromis polydon* and *Tropheus moorei* (Pisces: Cichlidae) of L. Tanganyika, with a discussion on the evolution and functions of a symbiotic relationship. *Physiology & Ecology, Kyoto*, **20**, 59–69

Taki, Y. (1978) An analytical study of the fish fauna of the Mekong Basin as a biological production system in nature. *Research Institute of Evolutionary Biology, Tokyo, Special Publication*, no. 1, 1–74

Talbot, F. H. (1965) A description of the coral structure of Tutia Reef (Tanganyika Territory, E. Africa) and its fish fauna. *Proceedings of the Zoological Society of London*, **145**, 431–70

Talbot, F. H., Russell, B. C. & Anderson, G. V. R. (1978) Coral reef fish communities: unstable high diversity systems? *Ecological Monographs*, **48**, 425–40

Teas, H. J. (1983) *The Biology and Ecology of Mangroves*. The Hague: Dr W. Junk

Thomas, J. D. (1966) On the biology of the catfish *Clarias senegalensis* in a man-made lake in the Ghananian savanna with particular reference to its feeding habits. *Journal of Zoology, London*, **148**, 476–514

Thompson, R. & Munro, J. L. (1978) Aspects of the biology and ecology of Caribbean reef fishes: Serranidae (hinds and groupers). *Journal of Fish Biology*, **12**, 115–46

Thresher, R. E. (1984) *Reproduction in Reef Fishes*. Neptune, N.J.: TFH Publications

Todd, J. H. (1971) The chemical language of fishes. *Scientific American*, **224**, 98–108

Toews, D. R. & Griffith, J. S. (1979) Empirical estimates of potential fish yield for the Lake Bangweulu system Zambia, C. Africa. *Transactions of the American Fisheries Society*, **108**, 241–52

Townshend, T. J. & Wootton, R. J. (1984) Effects of food supply on the reproduction of the convict cichlid, *Cichlasoma nigrofasciatum*. *Journal of Fish Biology*, **24**, 91–104

Travers, R. (1984) A review of the Mastacembeloidei, a suborder of synbranchiform teleost fishes, Pt II. Phylogenetic analysis. *Bulletin of the British Museum (Natural History)*, *Zoology*, **47**, 83–150

Trewavas, E. (1982) Generic grouping of Tilapiini used in aquaculture. *Aquaculture*, **27**, 79–81

Trewavas, E. (1983) *Tilapiine fishes of the genera Sarotherodon, Oreochromis and Danakilia*. London: British Museum (Natural History)

Trewavas, E., Green, J. & Corbet, S. A. (1972) Ecological studies on crater lakes in West Cameroon: fishes of Barombi Mbo. *Journal of Zoology, London*, **167**, 41–95

Trivers, R. L. (1972) Parental investment and sexual selection. In *Sexual Selection and the Descent of Man*, ed. B. Campbell, pp. 136–76. London: Heinemann

Turner, J. L. (1977*a*) Changes in the size structure of cichlid populations of L. Malawi resulting from bottom trawling. *Journal of the Fisheries Research Board of Canada*, **34**, 232–8

Turner, J. L. (1977*b*) Some effects of demersal trawling in L. Malawi (L. Nyasa) from 1968 to 1974. *Journal of Fish Biology*, **10**, 261–73

Turner, J. L. (1978) Status of various multi-species fisheries of Lakes Victoria, Tanganyika and Malawi based on catch and effort data. In *Symposium on River and Floodplain Fisheries in Africa, Bujumbura, Burundi, November 1977*, ed. R. L. Welcomme, pp. 4–15. CIFA Technical Papert no. 5. Rome: FAO

Tweddle, D. (1983) Breeding behaviour of the mpasa, *Opsaridium microlepis* (Gthr) (Pisces: Cyprinidae) in L. Malawi. *Journal of the Limnological Society of South Africa*, **9**, 23–8

Vaas, K. F. (1952) Fisheries in the lake district along the R. Kapuas in West Borneo. *Proceedings of the Indo-Pacific Fisheries Council, Section II*, (10), 1–10

Vaas, K. F. & Sachlan, M. (1952) Notes on fisheries exploitation of the artificial lake Tjibury in West Java. *Contributions, Agricultural Research Station, Bogor*, no. 128, 1–22

Vaas, K. F., Sachlan, M. & Wiraatmadja, G. (1953) On the ecology and fisheries of some inland waters along the Rivers Ogan & Komering in SE Sumatra. *Contributions, Inland Fisheries Research Station, Bogor*, no. 3, 1–32

Van Dobben, W. H. & Lowe-McConnell, R. H. (eds) (1975) *Unifying Concepts in Ecology.* Report of the Plenary Sessions of the First International Congress of Ecology. The Hague: Dr W. Junk

van Oijen, M. J. P. (1982) Ecological differentiation among the piscivorous haplochromine cichlids of L. Victoria (East Africa). *Netherlands Journal of Zoology*, **32**, 336–63

van Oijen, M. J. P., Witte, F. & Witte-Maas, E. L. M. (1981) An introduction to the ecological and taxonomic investigations on the haplochromine cichlids from the Mwanza Gulf of L. Victoria. *Netherlands Journal of Zoology*, **31**, 149–74

Vannote, R. L., Minshall, G. W., Cummins, K. W., Sedell, J. R. & Cushing, C. E. (1980) The river continuum concept. *Canadian Journal of Fisheries & Aquatic Science*, **37**, 130–7

Vari, R. (1978) The Terapon perches (Teraponidae): a cladistic analysis and taxonomic revision. *Bulletin of the American Museum of Natural History*, **159**(5), 175–340

Verbeke, J. (1959) Le régime alimentaire des poissons des lacs Edouard et Albert (Congo Belge). *Résultats Scientifiques Exploration Hydrobiologiques, Lacs Kivu, Edouard et Albert (1952–54)*, **3**(3), 1–6

Waldner, R. E. & Robertson, D. R. (1980) Patterns of habitat partitioning by eight species of territorial Caribbean damselfishes (Pisces: Pomacentridae). *Bulletin of Marine Science*, **30**, 171–86

Warburton, K. (1979) Growth and production of some important species of fish in a Mexican coastal lagoon. *Journal of Fish Biology*, **14**, 449–64

Ward, J. A. & Samarakoon, J. I. (1981) Reproductive tactics of the Asian cichlids of the genus *Etroplus* in Sri Lanka. *Environmental Biology of Fishes*, **6**, 95–103

Ward, J. V. & Stanford, J. A. (1983) The intermediate-disturbance hypothesis: an explanation for biotic diversity patterns in lotic ecosystems. In *Dynamics of Lotic Ecosystems*, ed. T. D. Fontaine & S. M. Beydell, pp. 347–356 Ann Arbor, MI: Ann Arbor Science Publishers

Ward, J. A. & Wyman, R. L. (1977) Ethology and ecology of cichlid fishes of the genus *Etroplus* in Sri Lanka: preliminary findings. *Environmental Biology of Fishes*, **2**, 137–45

Ware, D. M. (1984) Fitness of reproductive strategies in teleost fishes. In *Fish Reproduction*, ed. G. W. Potts & R. J. Wootton, pp. 349–66. London: Academic Press

Warner, R. R. (1975) The adaptive significance of sequential hermaphroditism in animals. *American Naturalist*, **109**, 61–82

Warner, R. R. (1978) The evolution of hermaphroditism and unisexuality in aquatic and terrestrial vertebrates. In *Contrasts in Behaviour*, ed. R. G. Reese & F. J. Lighter, pp. 77–101. New York: John Wiley

Warner, R. R. & Hoffman, S. G. (1980) Population density and the economics of territorial defence in a coral reef fish. *Ecology*, **61**, 772–80

Warner, R. R. & Robertson, D. R. (1978) Sexual patterns in the labroid fishes of the Western Caribbean. I. The wrasses (Labridae). *Smithsonian Contributions to Zoology*, no. 254, 1–27

Warner, R. R., Robertson, D. R. & Leigh, E. G., Jr (1975) Sex change and sexual selection. *Science, New York*, **190**, 633–8

Watson, D. J. & Balon, E. K. (1984) Ecomorphological analysis of fish taxocenes in rainforest streams in northern Borneo. *Journals of Fish Biology*, **25**, 371–84

Weinstein, M. P. & Heck, K. L., Jr (1979) Ichthyofauna of seagrass meadows along the Caribbean Coast of Panama and in the Gulf of Mexico: composition, structure and community ecology. *Marine Biology (Berlin)*, **50**, 97–107

Weitzman, S. H. & Weitzman, M. (1982) Biogeography and evolutionary diversification in neotropical freshwater fishes, with comments on the refuge theory. In *Biological Diversification in the Tropics*, ed. G. T. Prance, pp. 403–22. New York: Colombia University Press

Welcomme, R. L. (1964) The habitats and habitat preferences of the young of the L. Victoria *Tilapia* (Cichlidae). *Revue de Zoologie et Botanique Africaines*, **70**, 1–28

Welcomme, R. L. (1966) Recent changes in the stocks of *Tilapia* in L. Victoria. *Nature, London*, **212**, 52–4

Welcomme, R. L. (1967*a*) Observations on the biology of the introduced species of *Tilapia* in L. Victoria. *Revue de Zoologie et Botanique Africaines*, **76**, 249–79

Welcomme, R. L. (1967*b*) The relationship between fecundity and fertility in the mouthbrooding cichlid fish *Tilapia leucosticta. Journal of Zoology, London*, **151**, 453–68

Welcomme, R. L. (1969) The biology and ecology of the fishes of a small tropical stream. *Journal of Zoology, London*, **158**, 485–529

Welcomme, R. L. (1970) Studies of the effects of abnormally high water levels on the ecology of fish in certain shallow regions of L. Victoria. *Journal of Zoology, London*, **160**, 405–36

Welcomme, R. L. (1972) An evaluation of the acadja method of fishing as practiced in the coastal lagoons of Dahomey (West Africa). *Journal of Fish Biology*, **4**, 39–55

Welcomme, R. L. (1976) Some general and theoretical considerations on the fish yield of African rivers. *Journal of Fish Biology*, **8**, 351–64

Welcomme, R. L. (1979) *Fisheries Ecology of Floodplain Rivers*. London: Longman

Welcomme, R. L. (1985) *River fisheries*. Rome, FAO Fisheries Technical Paper (262), 330 pp.

West-Eberhard, M. J. (1979) Sexual selection, social competition and speciation. *Proceedings of the American Philosophical Society*, **123**, 222–34

West-Eberhard, M. J. (1983) Polymorphism, speciation and phylogeny. *Quarterly Review of Biology*, **58**, 155–83

Whitehead, P. J. P. (1959) The anadromous fishes of L. Victoria. *Revue de Zoologie et de Botanique Africaines*, **59**, 329–63

Whitehead, P. J. P. (1985) *FAO Species Catalogue*. vol. 7. *Clupeoid Fishes of the World (Suborder Clupeoidei)*. Part 1. *Chirocentridae, Clupeidae and Pristigasteridae*. FAO Fisheries Synopses, no. 125. Rome: FAO

Whittaker, R. H. & Woodwell, G. M. (eds) (1972) Evolution of natural communities. In *Ecosystem Structure & Function*, ed. J. A. Wiens, pp. 137–56. Oregon State University Press

Wickler, W. (1962) 'Egg dummies' as natural releasers in mouth-breeding cichlids. *Nature, London*, **194**, 1092–3

Wickler, W. (1967) Specialization of organs having a signal function in some marine fish. *Studies in Tropical Oceanography, Miami*, **5**, 539–48

Wiens, J. A. (1977) On competition and variable environments. *American Scientist*, **65**, 590–7

Wilhelm, W. (1980) The disputed feeding behaviour of a paedophagous haplochromine cichlid (Pisces) observed and discussed. *Behaviour*, **74**, 310–23

Williams, D. McB. & Sale, P. F. (1981) Spatial and temporal recruitment patterns of juvenile coral reef fishes to coral habitats within 'One Tree Lagoon', Great Barrier reef. *Marine Biology (Berlin)*, **65**, 245–53

Willoughby, N. G. (1974) The ecology of the genus *Synodontis* (Siluroidei) in L. Kainji, Nigeria. PhD thesis, University of Southampton.

Winn, H. E. & Bardach, J. E. (1959) Differential food selection by moray eels and a possible role of the mucous envelope of parrot fishes in reduction of predation. *Ecology*, **40**, 296–8

Winn, H. E., Marshall, J. A. & Hazlett, B. (1964) Behavior, diel activities, and stimuli that elicit sound production and reactions in the longspine squirrelfish. *Copeia*, **2**, 413–25

Winn, H. E., Salmon, M. & Roberts, N. (1964) Sun compass orientation by parrot fishes. *Zeitschrift für Tierpsychologie*, **21**, 798–812

Witte, F. (1981) Initial results of the ecological survey of the haplochromine cichlid fishes from the Mwanza Gulf of L. Victoria (Tanzania): breeding patterns, trophic and species distribution, with recommendations for commercial trawl-fishery. *Netherlands Journal of Zoology*, **31**, 175–202

Witte, F. (1984*a*) Ecological differentiation in L. Victoria haplochromines: comparison of cichlid species flocks in African lakes. In *Evolution of Fish Species Flocks*, ed. A. A. Echelle & I. Kornfield, pp. 155–67. Orono: University of Maine at Orono Press

Witte, F. (1984*b*) Consistency and functional significance of morphological differences between wild-caught and domestic *Haplochromis squamipinnis* (Cichlidae). *Netherlands Journal of Zoology*, **34**, 596–612

Witte, F. & Witte-Maas, E. L. M. (1981) *Haplochromis* cleaner fishes: a taxonomic and eco-morphological description of two new species. *Netherlands Journal of Zoology*, **31**, 203–29

Woodhead, P. M. J. (1966) The behavior of fish in relation to light in the sea. *Oceanography & Marine Biology*, **4**, 337–403

Woodwell, G. M. & Smith, H. H. (ed.) (1969) *Diversity and stability in Ecological Systems*. Brookhaven Symposia in Biology, no. 22. Brookhaven: Associated Universities Inc. and US Atomic Energy Commission

Worthington, E. G. (1937) On the evolution of fish in the great lakes of Africa. *Revue der Gesamten Hydrobiologie und Hydrographie*, **35**, 304–17

Wourms, J. P. (1981) Viviparity: the maternal–fetal relationship in fishes. *American Zoologist*, **21**, 473–515

Wyman, R. L. & Ward, J. A. (1972) A cleaning symbiosis between the cichlid fishes *Etroplus maculatus* and *E. suratensis*. I. Description and possible evolution. *Copeia*, **4**, 834–8

Wynne-Edwards, V. C. (1962) *Animal Dispersion in Relation to Social Behaviour*. Edinburgh: Oliver & Boyd

Yamaoka, K. (1982) Morphology and feeding behaviour of five species of genus *Petrochromis* (Teleostei, Cichlidae). *Physiology & Ecology, Kyoto*, **19**, 57–75

Yanagisawa, Y. (1985) Parental strategy of the cichlid fish *Perissodus microlepis* with particular reference to intraspecific brood 'farming out'. *Environmental Biology of Fishes*, **12**, 241–9

Yanagisawa, Y. & Nshombo, M. (1983) Reproduction and parental care of the scale-eating cichlid fish *Perissodus microlepis* in L. Tanganyika. *Physiology & Ecology, Kyoto*, **20**, 23–31

Yant, P. M., Karr, J. R. & Angermeier, P. L. (1984) Stochasticity in stream fish communities: an alternative interpretation. *American Naturalist*, **124**, 573–82

Zaret, T. M. (1979) Predation in freshwater fish communities. In *Predator–prey Systems in Fisheries Management*, ed. R. H. Stroud & H. Clepper, pp. 135–43. Washington, D.C.: Sport Fishing Institute

Zaret, T. M. (1980) Life history and growth relationships of *Cichla ocellaris*, a predatory South American cichlid. *Biotropica*, **12**, 144–57

Zaret, T. M. (1984*a*) Central American limnology and Gatun Lake, Panama. In *Lakes & Reservoirs*, ed. F. B. Taub, pp. 447–65. Ecosystems of the World, no. 23. Amsterdam: Elsevier

Zaret, T. M. (1984*b*) Fish/zooplankton interactions in Amazon floodplain lakes. *Verhandlungen der internationalen Vereinigung für theoretische und angewandte Limnologie*, **22**, 1305–9

Zaret, T. M. (ed.) (1984*c*) *Evolutionary Ecology of Neotropical Freshwater Fishes*. The Hague: Dr W. Junk

Zaret, T. M. & Paine, R. T. (1973) Species introduction in a tropical lake. *Science, New York*, **182**, 449–55

Zaret, T. M. & Rand, A. S. (1971) Competition in stream fishes: support for the competitive exclusion principle. *Ecology*, **52**, 336–42

Zumpe, D. (1965) Laboratory observations on the aggressive behaviour of some butterfly fishes (Chaetodontidae). *Zeitschrift für Tierpsychologie*, **22**, 226–36

INDEX

FISH GROUPS
(Groups in main contexts; families listed pp. 329–36)

LAKES